AF356658

Geophysics for Mineral Exploration

Geophysics for Mineral Exploration

Editor

Michael S. Zhdanov

MDPI • Basel • Beijing • Wuhan • Barcelona • Belgrade • Manchester • Tokyo • Cluj • Tianjin

Editor
Michael S. Zhdanov
Geology and Geophysics
University of Utah
Salt Lake City
United States

Editorial Office
MDPI
St. Alban-Anlage 66
4052 Basel, Switzerland

This is a reprint of articles from the Special Issue published online in the open access journal *Minerals* (ISSN 2075-163X) (available at: www.mdpi.com/journal/minerals/special_issues/GME).

For citation purposes, cite each article independently as indicated on the article page online and as indicated below:

LastName, A.A.; LastName, B.B.; LastName, C.C. Article Title. *Journal Name* **Year**, *Volume Number*, Page Range.

ISBN 978-3-0365-1740-7 (Hbk)
ISBN 978-3-0365-1739-1 (PDF)

© 2021 by the authors. Articles in this book are Open Access and distributed under the Creative Commons Attribution (CC BY) license, which allows users to download, copy and build upon published articles, as long as the author and publisher are properly credited, which ensures maximum dissemination and a wider impact of our publications.

The book as a whole is distributed by MDPI under the terms and conditions of the Creative Commons license CC BY-NC-ND.

Contents

About the Editor vii

Preface to "Geophysics for Mineral Exploration" ix

Michael S. Zhdanov
Editorial for Special Issue "Geophysics for Mineral Exploration"
Reprinted from: *Minerals* **2021**, *11*, 692, doi:10.3390/min11070692 1

Jianfei Fu, Sanshi Jia and Ende Wang
Combined Magnetic, Transient Electromagnetic, and Magnetotelluric Methods to Detect a
BIF-Type Concealed Iron Ore Body: A Case Study in Gongchangling Iron Ore Concentration
Area, Southern Liaoning Province, China
Reprinted from: *Minerals* **2020**, *10*, 1044, doi:10.3390/min10121044 5

Fouzan A. Alfouzan, Abdulrahman M. Alotaibi, Leif H. Cox and Michael S. Zhdanov
Spectral Induced Polarization Survey with Distributed Array System for Mineral Exploration:
Case Study in Saudi Arabia
Reprinted from: *Minerals* **2020**, *10*, 769, doi:10.3390/min10090769 23

**Yunpeng Zhang, Baoshan Wang, Guoqing Lin, Yongpeng Ouyang, Tao Wang, Shanhui Xu,
Lili Song and Rucheng Wang**
Three-Dimensional P-wave Velocity Structure of the Zhuxi Ore Deposit, South China Revealed
by Control-Source First-Arrival Tomography
Reprinted from: *Minerals* **2020**, *10*, 148, doi:10.3390/min10020148 49

P.V. Sunder Raju and K. Satish Kumar
Magnetic Survey for Iron-Oxide-Copper-Gold (IOCG) and Alkali Calcic Alteration Signatures
in Gadarwara, M.P, India: Implications on Copper Metallogeny
Reprinted from: *Minerals* **2020**, *10*, 671, doi:10.3390/min10080671 67

**Fatim-Zahra Ihbach, Azzouz Kchikach, Mohammed Jaffal, Driss El Azzab, Oussama Khadiri
Yazami, Es-Said Jourani, José Antonio Peña Ruano, Oier Ardanaz Olaiz and Luis Vizcaíno
Dávila**
Geophysical Prospecting for Groundwater Resources in Phosphate Deposits (Morocco)
Reprinted from: *Minerals* **2020**, *10*, 842, doi:10.3390/min10100842 87

Nikolay Yavich, Mikhail Malovichko and Arseny Shlykov
Parallel Simulation of Audio- and Radio-Magnetotelluric Data
Reprinted from: *Minerals* **2019**, *10*, 42, doi:10.3390/min10010042 103

Nikolay Khokhlov and Polina Stognii
Novel Approach to Modeling the Seismic Waves in the Areas with Complex Fractured
Geological Structures
Reprinted from: *Minerals* **2020**, *10*, 122, doi:10.3390/min10020122 123

Yidan Ding, Guoqing Ma, Shengqing Xiong and Haoran Wang
Three-Dimensional Regularized Focusing Migration: A Case Study from the Yucheng Mining
Area, Shandong, China
Reprinted from: *Minerals* **2020**, *10*, 471, doi:10.3390/min10050471 141

Qingfa Meng, Guoqing Ma, Taihan Wang and Shengqing Xiong
The Efficient 3D Gravity Focusing Density Inversion Based on Preconditioned JFNK Method
under Undulating Terrain: A Case Study from Huayangchuan, Shaanxi Province, China
Reprinted from: *Minerals* **2020**, *10*, 741, doi:10.3390/min10090741 . . . . **159**

Michael Jorgensen and Michael S. Zhdanov
Recovering Magnetization of Rock Formations by Jointly Inverting Airborne Gravity
Gradiometry and Total Magnetic Intensity Data
Reprinted from: *Minerals* **2021**, *11*, 366, doi:10.3390/min11040366 . . . . **173**

About the Editor

Michael S. Zhdanov

Michael S. Zhdanov holds degrees in geophysics, physics, and mathematics from the Moscow State University. He is Distinguished Professor at The University of Utah, Director of the Consortium for Electromagnetic Modeling and Inversion (CEMI) sponsored by major international energy resources, petroleum and mining corporations and research organizations. His contributions are reflected in more than 300 articles, 16 books published in English, Russian, and Chinese, and dozens patents. He is an Honorary Gauss Professor of the Göttingen Academy of Sciences, Full Member of the Russian Academy of Natural Sciences, Fellow of the Electromagnetics Academy USA, and Honorary Professor of the China National Centre for Geological Exploration Technology. Professor Zhdanov received one of the highest awards of the international Society of Exploration Geophysicists, Honorary Membership Award, in recognition of his distinguished contributions to exploration geophysics.

Preface to "Geophysics for Mineral Exploration"

This Special Issue contains ten papers which focus on emerging geophysical techniques for mineral exploration, novel modeling, and interpretation methods, including joint inversions of multi physics data, and challenging case studies. The papers cover a wide range of mineral deposits, including banded iron formations, epithermal gold–silver–copper–iron–molybdenum deposits, iron-oxide–copper–gold deposits, and prospecting for groundwater resources.

Michael S. Zhdanov
Editor

Editorial

Editorial for Special Issue "Geophysics for Mineral Exploration"

Michael S. Zhdanov [1,2]

1 TechnoImaging, LLC, Salt Lake City, UT 84107, USA; mzhdanov@technoimaging.com
2 Consortium for Electromagnetic Modeling and Inversion (CEMI), University of Utah, Salt Lake City, UT 84112, USA

Citation: Zhdanov, M.S. Editorial for Special Issue "Geophysics for Mineral Exploration". *Minerals* **2021**, *11*, 692. https://doi.org/10.3390/min11070692

Received: 21 June 2021
Accepted: 23 June 2021
Published: 28 June 2021

Publisher's Note: MDPI stays neutral with regard to jurisdictional claims in published maps and institutional affiliations.

Copyright: © 2021 by the author. Licensee MDPI, Basel, Switzerland. This article is an open access article distributed under the terms and conditions of the Creative Commons Attribution (CC BY) license (https://creativecommons.org/licenses/by/4.0/).

Exploration geophysics plays a major role in unlocking mineral reserves. It is well recognized that many easily discovered large mineral deposits with a strong geophysical signature have already been identified. Future discoveries present significant challenges, being located undercover, in remote areas, and with less prominent geophysical signals. The modern-day challenges of exploration require developing novel geophysical techniques, which improve exploration success and lead to new discoveries.

This Special Issue contains ten papers which focus on emerging geophysical techniques for mineral exploration, novel modeling, and interpretation methods including joint inversions of multi physics data, and challenging case studies. The papers cover a wide range of mineral deposits, including banded iron formations, epithermal gold–silver–copper–iron–molybdenum deposits, iron-oxide–copper–gold deposits, and prospecting for groundwater resources.

Fu et al. [1] discuss an application of various deep-penetrating geophysical techniques to the exploration of ore deposits. In particular, they consider an important role of geophysical surveys in studying the banded iron formations (BIF). It is well known that the large-scale BIF-type iron mines represent one of the most important iron ore resources in the world. They constitute 70% of the world's high-grade iron ore reserves, and BIF-type iron mines produce over 90% of the world's iron ores. They are found all over the world, but mainly in Russia, Australia, Brazil, Canada, China, Africa, India, and the United States. They [1] present the results of integrated geophysical surveys in the Anshan-Benxi area of the North China Craton, where several major BIF-type iron deposits are located. The authors used deep-penetrating geophysical methods, including the high-precision ground magnetic survey (HPGMS), transient electromagnetic (TEM), and magnetotelluric (MT) methods. The results show that an optimal combination of these geophysical methods makes it possible to accurately determine the anomalous spatial locations and morphologies of the concealed iron ore bodies.

Alfouzan et al. [2] present the results of the Saudi Arabian Glass Earth Pilot Project. This project is a part of the geophysical exploration program to explore the upper crust of the Kingdom for minerals, groundwater, and geothermal resources. The project began with a large-scale airborne geophysical survey over approximately 8000 sq. km of green-field area, including electromagnetic (EM), magnetics, and gravity methods [3]. Based on the results of the airborne survey, several prospective mineralization targets were identified for follow-up exploration. A spectral induced polarization (SIP) survey was completed over one of the prospective targets. The field data were collected with a distributed array system, which had the potential for a strong inductive coupling (IC) effect. They [2] developed a method to fully include all 3D IC effects in the inversion of induced polarization (IP) data. The field SIP data were inverted using the generalized effective-medium theory of induced polarization (GEMTIP) in conjunction with integral equation-based modeling and inversion methods. The results of this inversion were interpreted and used to design a drill hole set up in the survey area, which intersected significant mineralization associated with gold, silver, and other base metals.

Zhang et al. [4] applied controlled-source first-arrival tomography to study the P wave velocity structure of the Zhuxi ore deposit, located in Jiangxi province, South China. Their velocity model identified the proven orebodies, mainly related to magmatic hydrothermal activities during the Yanshanian period. These were visible as high-velocity zones, corresponding to widespread copper–iron and a few tungsten–molybdenum orebodies. These results helped to further evaluation of the total reserves, suggesting that seismic tomography could be a useful tool for mineral exploration.

Raju and Kumar [5] demonstrated how airborne and ground magnetic survey data could be effectively used for studying the Iron–Oxide–Copper–Gold (IOCG) deposits in in Gadarwara, M.P, India. They [5] showed that such deposits could be inferred from the predictive magnetic exploration models combined with geological observations and petrophysical data.

Ihbach et al. [6] examined the water potential of aquifers within in the phosphatic series in Morocco, using a combination of several geophysical methods: magnetic resonance sounding (MRS), electrical resistivity tomography (ERT), time-domain electromagnetics (TDEM), and frequency-domain electromagnetics (FDEM). They [6] demonstrated the efficiency of the MRS method for prospecting groundwater resources, and evaluated the importance in the geological context of Youssoufia open-pit mining in Morocco. The ERT method was used to delineate the conductive horizons attributed to the groundwater aquifers. The TDEM and FDEM data were used for mapping and delimiting the aquifer potential recharge zones in the phosphate series. In summary, the authors confirmed the effectiveness of the developed approach to geophysical prospecting for groundwater resources in phosphate deposits.

Two papers of this Special Issue are dedicated to developing effective methods of computer simulation of geophysical data [7,8]. Yavich et al. [7] present a novel numerical method of simulating the controlled-source audio-magnetotellurics (CSAMT) and radio-magnetotellurics (CSRMT) data, which are widely used in mineral exploration. They [7] introduced an approach to 3D electromagnetic (EM) modeling based on the new type of preconditioned iterative solver for finite-difference (FD) EM simulation. This novel preconditioner combined Green's function preconditioner and a contraction operator transformation [9]. The effectiveness of the method was illustrated by the results of numerical simulation of the CSAMT and CSRMT responses in Moscow syneclise.

Khokhlov and Stognii [8] introduced a novel approach to modeling the propagation of seismic waves in a medium containing subvertical fractured inhomogeneities, typical for mineralization zones. It was based on the use of the grid-characteristic method [10], which could model the seismic responses from subvertical fractured inhomogeneities on a structural rectangular grid. This approach significantly simplified the construction of the numerical model and the use of the algorithms. The authors presented a numerical study to illustrate the effectiveness of the developed approach to modeling the seismic wave propagation in a medium with the fractured zones.

The final three papers of this Special Issue cover novel methods of focusing migration and inversion of potential field data. Ding et al. [11] applied 3D regularized focusing migration methods to image an entire gravity survey with a focusing stabilizer. They [11] used the concept of potential field migration and image focusing, introduced in [12], for producing high resolution migration images of the subsurface density distribution. The developed method was illustrated by interpretation of the gravity data collected from the skarn-type iron deposits in Yucheng, Shandong province of China.

Meng et al. [13] discussed the ways of improving the efficiency of 3D focusing inversion of the gravity data by using preconditioned the Jacobian-free Newton–Krylov (JFNK) method. They [13] also incorporated the unstructured meshes to improve modeling of the terrain and inhomogeneous density distribution. The practical effectiveness of this novel method was shown by the case study of processing gravity data collected in Huayangchuan, Shaanxi Province of China.

The paper by Jorgensen and Zhdanov [14] presents a method of joint inversion of airborne gravity gradiometry (AGG) and total magnetic intensity (TMI) data for a shared earth model. The authors discussed two novel techniques of joint inversion; one was based on the Gramian constraints [15], and the other used the joint focusing constraints. The paper also demonstrated how Gramian constraints can be effectively used for recovering the magnetization of rock formations from TMI data. The last problem is extremely important in the presence of remanent magnetization of the rocks produced by the ancient magnetic field. The paper demonstrated the power of the technique by constructing 3D density, magnetic susceptibility, and magnetization vector models of the Thunderbird V-Ti–Fe deposit in Ontario, Canada.

In summary, the Special Issue *"Geophysics for Mineral Exploration"* provides a snap shot of advanced modeling and inversion methods of geophysical data, and their applications to studying different types of mineral deposits.

Conflicts of Interest: The authors declare no conflict of interest.

References

1. Fu, J.; Jia, S.; Wang, E. Combined Magnetic, Transient Electromagnetic, and Magnetotelluric Methods to Detect a BIF-Type Concealed Iron Ore Body: A Case Study in Gongchangling Iron Ore Concentration Area, Southern Liaoning Province, China. *Minerals* **2020**, *10*, 1044. [CrossRef]
2. Alfouzan, F.; Alotaibi, A.; Cox, L.; Zhdanov, M. Spectral Induced Polarization Survey with Distributed Array System for Min-eral Exploration: Case Study in Saudi Arabia. *Minerals* **2020**, *10*, 769. [CrossRef]
3. Zhdanov, M.S.; Alfouzan, F.A.; Cox, L.; Alotaibi, A.; Alyousif, M.; Sunwall, D.; Endo, M. Large-Scale 3D Modeling and Inversion of Multiphysics Airborne Geophysical Data: A Case Study from the Arabian Shield, Saudi Arabia. *Minerals* **2018**, *8*, 271. [CrossRef]
4. Zhang, Y.; Wang, B.; Lin, G.; Ouyang, Y.; Wang, T.; Xu, S.; Song, L.; Wang, R. Three-Dimensional P-wave Velocity Structure of the Zhuxi Ore Deposit, South China Revealed by Control-Source First-Arrival Tomography. *Minerals* **2020**, *10*, 148. [CrossRef]
5. Raju, P.; Kumar, K. Magnetic Survey for Iron-Oxide-Copper-Gold (IOCG) and Alkali Calcic Alteration Signatures in Gadarwara, M.P, India: Implications on Copper Metallogeny. *Minerals* **2020**, *10*, 671. [CrossRef]
6. Ihbach, F.Z.; Kchikach, A.; Jaffal, M.; El Azzab, D.; Khadiri Yazami, O.; Jourani, E.S.; Peña Ruano, J.; Olaiz, O.; Dávila, L.V. Geophysica Prospecting for Groundwater Resources in Phosphate Deposits (Morocco). *Minerals* **2020**, *10*, 842. [CrossRef]
7. Yavich, N.; Malovichko, M.; Shlykov, A. Parallel Simulation of Audio- and Radio-Magnetotelluric Data. *Minerals* **2020**, *10*, 42. [CrossRef]
8. Khokhlov, N.; Stognii, P. Novel Approach to Modeling the Seismic Waves in the Areas with Complex Fractured Geological Structures. *Minerals* **2020**, *10*, 122. [CrossRef]
9. Yavich, N.; Zhdanov, M. Contraction pre-conditioner in finite-difference electromagnetic modelling. *Geophys. J. Int.* **2016**, *206*, 1718–1729. [CrossRef]
10. Favorskaya, A.V.; Zhdanov, M.S.; Khokhlov, N.I.; Petrov, I.B. Modelling theWave Phenomena in Acoustic and Elastic Media with Sharp Variations of Physical Properties Using the Grid-Characteristic Method. *Geophys. Prospect.* **2018**, *66*, 1485–1502. [CrossRef]
11. Ding, Y.; Ma, G.; Xiong, S.; Wang, H. Three-Dimensional Regularized Focusing Migration: A Case Study from the Yucheng Mining Area, Shandong, China. *Minerals* **2020**, *10*, 471. [CrossRef]
12. Zhdanov, M.S. *Geophysical Inverse Theory and Regularization Problems*, 1st ed.; Elsevier: Amsterdam, The Netherlands, 2002; Volume 36, pp. 42–186.
13. Meng, Q.; Ma, G.; Wang, T.; Xiong, S. The Efficient 3D Gravity Focusing Density Inversion Based on Preconditioned JFNK Method under Undulating Terrain: A Case Study from Huayangchuan, Shaanxi Province, China. *Minerals* **2020**, *10*, 741. [CrossRef]
14. Jorgensen, M.; Zhdanov, M.S. Recovering Magnetization of Rock Formations by Jointly Inverting Airborne Gravity Gradiom-etry and Total Magnetic Intensity Data. *Minerals* **2021**, *11*, 366. [CrossRef]
15. Zhdanov, M.S. *Inverse Theory and Applications in Geophysics*; Elsevier: Amsterdam, The Netherlands, 2015.

Article

Combined Magnetic, Transient Electromagnetic, and Magnetotelluric Methods to Detect a BIF-Type Concealed Iron Ore Body: A Case Study in Gongchangling Iron Ore Concentration Area, Southern Liaoning Province, China

Jianfei Fu [1,2], Sanshi Jia [1,3,]* and **Ende Wang [1,2]**

1   Key Laboratory of Ministry of Education on Safe Mining of Deep Metal Mines, Northeastern University, Shenyang 110819, China; fujianfei@mail.neu.edu.cn (J.F.); wnd@mail.neu.edu.cn (E.W.)
2   School of Resource and Civil Engineering, Northeastern University, Shenyang 110819, China
3   School of Resource and Materials, Northeastern University at Qinhuangdao, Qinhuangdao 066004, China
*   Correspondence: jiasanshi@neuq.edu.cn

Received: 27 September 2020; Accepted: 18 November 2020; Published: 24 November 2020

Abstract: The detection and evaluation of concealed mineral resources deep in metallic mines and in the surrounding areas remain technically difficult. In particular, due to the complex topographic and geomorphic conditions on the surface, the detection environments in these areas limit the choices of detection equipment and data collection devices. In this study, based on metallogenic theory and the metallogenic geological characteristics of banded iron formation (BIF)-type iron ores, equipment for surface geophysical surveys (i.e., the high-precision ground magnetic survey method, the transient electromagnetic method, and the magnetotelluric method) and data collection devices capable of taking single-point continuous measurements were employed to detect the concealed iron ore bodies in the transition zone CID-1 between the Hejia and Dumu iron deposits in the Gongchangling iron ore concentration area in the Anshan-Benxi area (Liaoyang, China), a representative area of BIF-type iron ores. The results showed that an optimal combination of these geophysical survey methods accurately determined the anomalous planar spatial locations and anomalous profile morphologies of the concealed iron ore bodies. On this basis, we determined their locations, burial depths, and scales. Two anomalous zones induced by concealed iron ore bodies, YC-1 and YC-2, were discovered in zone CID-1. Two concealed iron-bearing zones, one shallow (0–150 m) and one deep (300–450 m), were found in YC-1. A 100 m scale drilling test showed that the cumulative thickness of the shallow iron-bearing zone was over 23.6 m.

Keywords: iron deposit; mineral exploration; transient electromagnetic method; magnetotelluric method; single point continuous motion detection

---

1. Introduction

Over 5000 years ago, human societies gradually transitioned from the Stone Age to the Metal Age and began to develop and utilize metallic mineral resources (e.g., gold, copper, and iron). Particularly, the inhabitants of the Aegean region and China discovered and started to use copper and bronze in 3200 BC [1,2]. Large-scale development and the use of various metallic mineral resources ensued. Especially, outcropping ores were primarily developed and mined, and they remained dominant metallic mineral sources [3]. However, with continuous economic and social development, particularly since the beginning of the third industrial revolution, there has been an increasing demand for metallic mineral resources. This situation is met with the dilemma that all the outcropping ores have basically been exhausted.

Against this background, theoretical prediction methods and prospecting techniques for concealed ore deposits and ore bodies have become a focal but challenging topic of prospecting research. This research is focused primarily on the application of various deep-penetrating survey techniques, including fundamental geological, geochemical, and geophysical techniques, to the exploration of concealed ores [4–6]. In particular, geophysical survey techniques have played a tremendous role. On the basis of the research results and metallogenic theories, numerous concealed metallic ore deposits [7–9] have been discovered, such as the world-class pebble porphyry deposit in Alaska (USA), the Olympic Dam copper-gold-uranium-rare-earth element (Cu-Au-U-REE) deposit and the Ernest Henry copper-gold deposit in Australia, and the Nihe and Luohe iron deposits in Anhui (China) [10–12]. All the above research results concern the detection and discovery of concealed metallic ore deposits of a certain scale in large areas using geophysical methods. By contrast, the application of geophysical survey techniques to the class of concealed ore bodies in the existing metallic ore deposits has been rarely investigated. In fact, as a result of many years of continuous mining, many metal mines are becoming exhausted as their proven resources are depleting, whereas the concealed resources in the deep and surrounding areas of these mines have yet to be effectively detected and evaluated. This phenomenon is most common for iron, the metal with the longest mining history, particularly banded iron formation (BIF)-type iron deposits.

BIF-type iron mines, characterized by large-scale, as well as easy mining and beneficiation, are the most important iron ore resources in the world. High-grade iron ores in BIF-type iron mines constitute 70% of the world's high-grade iron ore reserves, and BIF-type iron mines produce over 90% of the world's iron ores [13]. BIF-type iron mines are found mainly in Russia, Australia, Brazil, Canada, China, Africa, India, and the United States. In China, BIF-type iron mines are mostly in the North China Craton (Figure 1) and are concentrated largely in the Anshan-Benxi area, eastern Hebei, the Huoqiu-Wuyang area, Wutai, western Shandong, and Guyang.

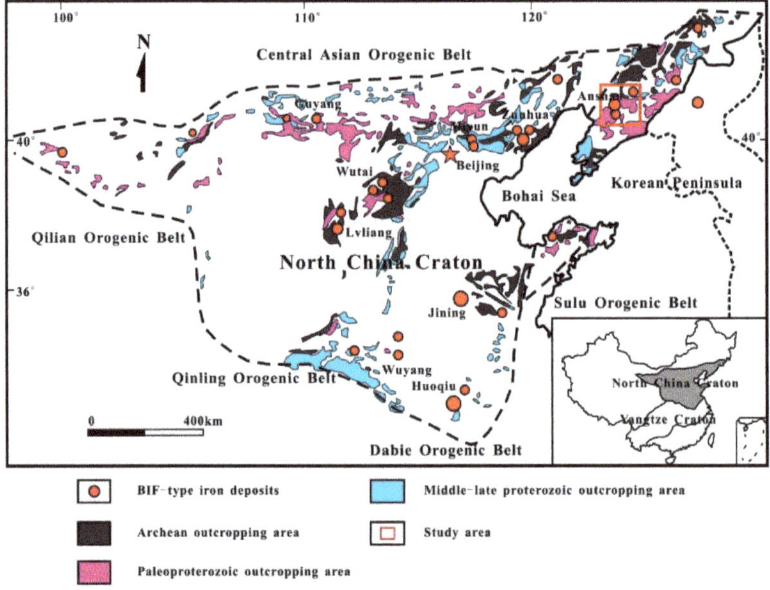

Figure 1. The distribution sketch of the Precambrian geology banded iron formation (BIF)-type iron deposits (modified from Zhang et al., 2012 [13]).

The Anshan-Benxi area is home to the most representative BIF-type iron mines, that is, the world-famous "Anshan-type" iron mines (Figure 2). BIF-type iron deposits, mostly discovered through engineering verification of the early outcrops, as well as airborne gravity and magnetic anomalies, have the longest mining histories and the largest mining scale. However, the concealed resources in the deep and surrounding areas of these deposits have never been effectively detected or evaluated. The available resource reserves are unable to meet the continuously increasing demand for resources. The use of gridding-type engineering drilling alone to evaluate the deep and surrounding areas of existing mines is often too costly, and thus impractical, whereas the use of older monotonous aeromagnetic and gravity test results is unable to meet the high-precision evaluation requirements for concealed ore bodies. Notably, recent results have demonstrated that based on the metallogenic geological characteristics of existing deposits, it is feasible to make full use of multiple geophysical survey methods (e.g., the high-precision ground magnetic survey method (HPGMS), the transient electromagnetic method (TEM), the magnetotelluric method (MT), and the controlled source audio-frequency magnetotelluric method (CSAMT)) to accurately locate and evaluate concealed ore bodies in existing metal mines [14–16]. While corroborating the research results for metallogenic theories, sizeable metallic ore bodies have been discovered. However, this type of research is rarely conducted on BIF-type iron mines. In particular, no such research has been performed on the BIF-type iron mines in the Anshan-Benxi area, which has complex cultural and geological settings. Hence, facing the crisis of the inability to ensure a continuous supply of iron ore resources in the future, how to efficiently, economically, and reasonably select multiple surface geophysical survey methods to evaluate concealed BIF-type iron bodies based on current metallogenic theories, as well as the current understanding of metallogenic geological characteristics, has become a technical difficulty that urgently needs to be addressed.

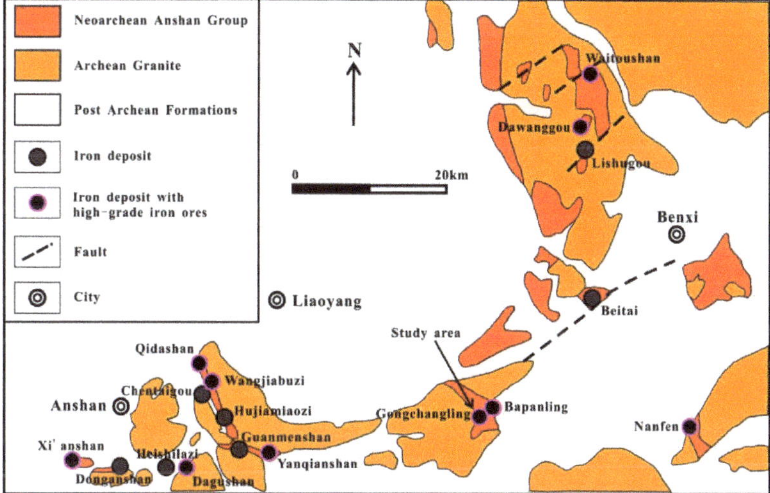

Figure 2. Geological map showing the distribution of the BIF-related high-grade iron deposits in the Anshan-Benxi area, Liaoning Province, China (modified from Li et al., 2015 [17]).

In view of this, this study investigated the detection and evaluation of the concealed iron ore bodies in the Gongchangling iron ore concentration area in the Anshan-Benxi area (Figure 3). Specifically, based on the current status of mining and the spatial distribution patterns of the ore bodies in the Hejia and Dumu iron mines, and considering the complex cultural, geological, topographic, and geomorphic settings that ground surveying must account for, this study employed an optimal combination of geophysical survey methods, namely, the HPGMS, TEM, and MT methods, and their respective data collection devices that can take single-point, continuous, mobile survey measurements in order to detect the geophysical anomalies caused by the presence of the concealed iron ore bodies. The location, burial depth, and scale of each concealed iron ore body were accurately determined. This study addressed the technical difficulty in detecting and evaluating concealed BIF-iron ore bodies in complex topographic and geomorphic settings on the surface. This detection application study can provide a technical reference and theoretical guidance for evaluating concealed ore bodies in similar metal mines.

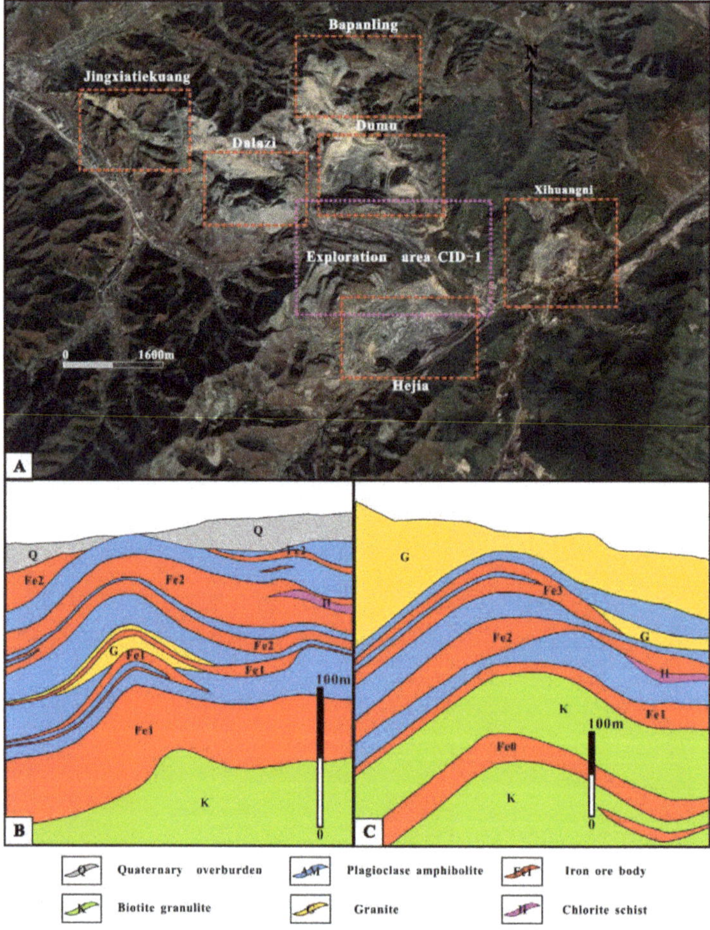

Figure 3. (**A**) Sketch map of the spatial distribution of iron deposits in the Gongchangling area; (**B**) Schematic diagram of a typical geological section showing the spatial distribution and occurrence characteristics of iron ore bodies in the Hejia iron deposit; (**C**) Schematic diagram of a typical geological section showing the spatial distribution and occurrence characteristics of iron ore bodies in the Dumu iron deposit.

2. Geological Setting of the Gongchangling Iron Ore Concentration Area

Located in the eastern section of the north margin of the North China Craton, the Anshan-Benxi area is home to dozens of large to super-large BIF-type iron deposits, including the Waitoushan, Nanfen, Dong'anshan, Qidashan, Gongchangling, Hujiamiaozi, and Dagushan deposits (Figure 2), as well as several hundred medium-sized and small iron deposits. The total proven reserves of iron ores in the

Anshan-Benxi area has surpassed 12.5 billion tons [18]. The iron ores in this area are located mainly in the strata of the Archean Anshan group. From bottom to top, the Anshan group can be divided into five formations, namely, the Shipengzi, Tongshicun, Cigou, Dayugou, and Yingtaoyuan formations, which differ in scale and outcropping degree in the Anshan-Benxi area. Each of these five formations contains BIF-type iron-bearing formations. The Cigou and Yingtaoyuan formations are important horizons that bear large and super-large iron ores and contain high-grade iron deposits of varying sizes, which are typical high-grade iron deposits among the deposits across the world and are represented by the Jingxiatiekuang iron deposits in the Gongchangling area. These iron deposits are also known as Anshan-type iron deposits.

The iron ores in the Gongchangling area are mainly located within the Cigou formation of the Archean Anshan group. Spatially, six large iron deposits are distributed in the Gongchangling area, namely, the Jingxiatiekuang, Dalazi, Bapanling, Dumu, Hejia, and Xihuangni iron deposits (Figure 3A). The Jingxiatiekuang iron deposit is home to world-renowned large Gongchangling-type high-grade iron ores (Anshan-type high-grade iron ores). Spatially, the iron ores in the Gongchangling area are densely distributed in a steady, continuous, layered manner and crop out on the surface on a large scale. As a result, they can be easily discovered and have been mined for a long time. With regard to prospecting, there is only one "blank" zone (CID-1), which is located between the Hejia and Dumu iron deposits. Zone CID-1 was selected as the survey area of this study. Zone CID-1 is covered with clastic sedimentary rocks formed by intense erosion as a result of tectonic development and is also the core of a large anticlinorium composed of the Cigou formation. The south and north limbs of the anticlinorium are the present-day Hejia and Dumu iron deposits, respectively. The lithology of the Cigou formation, the primary ore-hosting strata in the area, is composed predominantly of plagioclase amphibolite, biotite leptynite, biotite-amphibole plagiogneiss, hornblende schist, chlorite schist, tremolite schist, biotite albite leptynite, monzonite leptite, and magnetite quartzite. The ore-bearing lithological horizon of the Cigou formation is primarily a plagioclase amphibolite layer. Various types of schist, such as chlorite schist, quartz schist, hornblende schist, and mica schist, have developed in the areas where the plagioclase amphibolite layer comes in contact with the ore bodies. In addition, sizeable reserves of iron ore resources have been discovered in recent years in the biotite leptynite layer at the bottom of the ore-bearing horizon. This biotite leptynite layer is referred to as the iron-bearing zone in the Fe0 bed. For example, currently, the ore bodies in the Fe0 bed are primarily mined from the Dumu iron deposit.

Zone CID-1 was selected as the survey area of this study. The ore-bearing lithological horizon of the Hejia iron deposit, located south of zone CID-1, is consistent with that of the Dumu iron deposit, located north of zone CID-1. Owing only to the difference in the extent of later-stage uplifting and erosion, the Hejia iron deposit contains no granite layer and no iron-bearing zone in the Fe3 bed (Figure 3B,C). Currently, the iron bodies in the Fe1 and Fe2 beds are primarily mined from the Hejia iron deposit. Due to the difference in the stage of development and utilization, apart from the iron ore bodies mined in the Fe1, Fe2, and Fe3 beds from the Dumu iron deposit, a thick, large magnetite body layer in the Fe0 bed has been discovered in the biotite leptynite bed at the bottom of the Dumu iron deposit during the exploration and development process. In comparison, only a scattered magnetite body layer has been found in the biotite leptynite bed at the bottom of the Hejia iron deposit during the exploration process. Due to the conventional understanding that no mineralization occurs in biotite leptynite beds, no in-depth evaluation has been performed.

Survey zone CID-1 is the transition zone between the Hejia and Dumu iron deposits. In the center of zone CID-1, there exists a sizeable north-west trending Hejia fault, which has long been under erosion. Sporadic outcropping residual iron ores are visible only in the Quaternary coverage area on the surface. The extent of the erosion of zone CID-1 has yet to be systematically investigated. In addition, whether there exist concealed iron bodies beneath zone CID-1 has never been systematically evaluated. Due to the limitations of metallogenic theories and survey techniques, the evaluation of the resource potential of the deep and surrounding areas of many BIF-type iron mines has suffered the same shortcoming. For these reasons, zone CID-1 was selected, in this study, to investigate the detection of concealed iron bodies. First, concealed iron ore beds could be discovered and located in the Quaternary coverage area. Moreover, new geological discoveries and understandings could be verified to help break away from the conventional prospecting approach.

Table 1 summarizes the physical parameters of the main rocks and ores of CID-1. There are notable differences between their physical parameters. In particular, the iron ores have a notably lower resistivity and are more magnetic than the wall rocks. These traits provide a precondition for relevant geophysical surveys.

Table 1. Petrophysical properties in the Gongchangling iron ore concentration area.

Rocks and Ores	Susceptibility (k)/4Π × 10⁻⁶ SI	Resistivity (p)/Ω·m
	Regular Value	**Regular Value**
Biotite granulite	70.3	2262
Plagioclase amphibolite	31.8	5188
Granite	133.6	3500
Chlorite schist	15,093.9	13,362
Magnetic iron ore	209,000	1708
Haematite iron ore	22,900	3165

3. Methods

The concealed iron ore bodies in the transition zone CID-1 between the Hejia and Dumu iron deposits were detected primarily using deep-penetrating geophysical survey methods, that is, the HPGMS, TEM, and the MT methods, based on the observed differences between the physical parameters of the rocks and ores. First, the HPGMS method was employed to determine the spatial distribution pattern of the concealed iron ore bodies and to locate the geophysical anomalies caused by the presence of these concealed iron bodies. Second, portable TEM and MT data collection devices were used to survey the anomalous zones detected using the HPGMS method at fixed points and depths to determine the burial depths and sizes of the geophysical anomalies caused by the presence of the concealed iron bodies. Finally, engineering drilling verification was performed based on the determined spatial planar and burial-depth profile anomalies caused by the concealed iron bodies to obtain accurate location and scale information about the concealed iron bodies.

During the detection of the concealed iron bodies, the coordinates were set based on the coordinate parameters provided by the mine, except in HPGMS-based surveys. An S750 handheld geographic information system acquisition system (South Surveying & Mapping Instrument Co., Ltd., Guangzhou, China) with sub-meter-level positioning accuracy was used to collect coordinate data for spatial locations. The built-in global positioning system (GPS) with positioning accuracy better than 3 m was used to conduct HPGMS-based surveys.

Notably, to make full use of the advantages (i.e., high efficiency, low cost, and convenience, and speediness) of the geophysical survey methods, based on the in-situ survey environment (Figure 4A), portable equipment and data collection devices capable of taking single-point mobile surveys were used in this study. These devices were used in coordination to conduct flexible, efficient geophysical scans, and deep profile measurements on BIF-type concealed iron ore bodies.

3.1. High-Precision Ground Magnetic Method

A GSM-19T standard proton precession magnetometer (GEM, Canada) was employed to perform magnetic surveys. This magnetometer is equipped with an all-way probe with no north-pointing requirements and a high-precision GPS (Figure 4B) and has a resolution of 0.01 nT, an absolute precision of ±0.2 nT, and a dynamic measuring range of 20,000–120,000 nT.

Figure 4. (**A**) Close-up view of the detection area; (**B**) The high-precision ground magnetic method; (**C**) The transient electromagnetic method; (**D**) The magnetotelluric method.

Magnetic survey data were obtained from multiple south-north trending survey lines passing through the study area (i.e., zone CID-1) within the space where magnetic surveys could be conducted. The spacing between magnetic survey lines was 50 m. The spacing between magnetic survey points varied from 5 to 20 m. The magnetic survey lines and points covered the whole study area selected for surveying. The magnetic survey points were positioned using the built-in GPS (positioning precision: 3 m) of the magnetometer. Magnetic survey data were collected in mobile mode. The mobile observation wait time was set to 5 s. Magnetic survey data were collected from 8417 magnetic survey points on 64 magnetic survey lines.

The magnetic survey data were processed using the GeoExpl multivariate geospatial data management and analysis system developed by the China Geological Survey. The measured magnetic field data were corrected, subjected to a reduction-to-the-pole (RTP) treatment, and upward continuation, and their vertical derivatives were also calculated. These treatments eliminated various interfering factors and allowed the magnetic anomalies to accurately reflect anomalous magnetic bodies within a certain depth from the surface.

3.2. Transient Electromagnetic Method

The TEM method is a time-domain electromagnetic method. A multi-turn coincident minor loop device capable of being disassembled and assembled can be employed to facilitate fixed-point, continuous TEM-based geophysical surveys of narrow, finite spaces [19,20]. The TerraTEM system (Monash GeoScope, Australia) was used in this study as the TEM-based survey device (Figure 4C). This system can handle over 140 time-gate windows and collects high-resolution information within a short time. When equipped with a rapid, enhanced turn-off-time device, the TerraTEM system has a short turn-off time. In addition, this system is internally equipped with many combinations of detection devices, and therefore is suitable for the detection of concealed geological bodies at various depths.

The TEM survey lines were arranged such that they mainly passed through the long-axis direction of the anomalous planar bodies detected using the HPGMS method. The spacing between TEM survey lines was 100 m. The spacing between TEM survey points varied from 10 to 20 m. An S750 system with positioning precision better than 1 m was used to determine the locations of the TEM survey points. A multi-turn coincident minor loop device with a side length of 10 m, a 10-turn transmission coil, a seven-turn receiving coil, and a transmission voltage supply of 24 V, which could be disassembled and assembled on site, was used to collect TEM data. TEM data were collected and stacked 16–512 times. The single-point data sampling window density was 147, far exceeding the sampling density (interval) of conventional transient electromagnetic devices. As a result, this device could accurately identify anomalous changes in physical properties within a certain depth. Two TEM survey lines, TEM-1 and TEM-2, were set in the survey area. TEM data were collected from 21 (spacing 20 m) and 28 (spacing 10 m) points on survey lines TEM-1 and TEM-2, respectively.

The collected TEM data were processed using the built-in transient electromagnetic workstation. Specifically, the TEM data were subjected to treatments such as cleansing, preprocessing (shearing and filtering enhancement as well as distortion elimination), apparent resistivity conversion, and time-depth conversion. Finally, an apparent resistivity contour profile was generated and interpreted based on geological and geophysical data.

3.3. Magnetotelluric Method

An EH4 system, developed jointly by EMI (USA) and Geometrics (USA), was used in this study to conduct MT surveys. Capable of detecting concealed geological bodies at shallow and moderate depths [21], the EH4 system is a dual-source electromagnetic system with a working frequency range of 10 Hz–100 kHz. When in operation, the EH4 system collects the signals of a natural or artificial electric field (E, a group of orthogonal E_x and E_y components) with a corresponding frequency f, as well as the signals of a magnetic field (H, a group of orthogonal H_x and H_y components) with a frequency of f. The resistivity ρ corresponding to f was calculated using Equation (1):

$$\rho = (1/5f)|E/H|^2 \tag{1}$$

Thus, the resistivities of multiple underground layers were measured based on the relationship between frequency and depth. Notably, the EH4 system can take continuous, single-point measurements, and thus can be conveniently and flexibly applied to various adverse topographic and geomorphic settings (Figure 4D).

The survey lines of the EH4 system were arranged mainly in parallel to and close to the TEM survey lines. On the one hand, this arrangement facilitated comparison with the TEM survey results. On the other hand, it let us detect and discover deep anomalous geological bodies. The spacing between the survey lines of the EH4 system was 100 m. The spacing between the survey points of the EH4 system was 20 m. An S750 system with positioning precision better than 1 m was used to determine the locations of the survey points of the EH4 system. The following procedure was used to collect data: laying a survey grid, testing the equipment, placing electrodes, placing magnetic bars, placing an analogue front-end amplifier (AFE), placing the main frame, and collecting data. Two survey lines of the EH4 system, MT-1 and MT-2, were set in the survey area. Data were collected from 20 and 10 points along survey lines MT-1 and MT-2, respectively.

The time-series data collected by the EH4 system were first preprocessed, and then subjected to a fast Fourier transform (FFT). Thus, data for the imaginary and real components of the electric and magnetic fields, as well as the phases, were obtained. Finally, the built-in software of the EH4 system was employed to perform one-dimensional inversion and two-dimensional joint inversion on the data. A two-dimensional resistivity profile was produced based on the processed data.

4. Results

The collected HPGMS, TEM, and MT data were systematically processed and geophysically interpreted using relevant software. In particular, the cross-validation of the survey results obtained using two geophysical methods, that is, the TEM and MT methods for the shallow and deep areas, let us make full use of the technical advantages of geophysical surveying in single-point, continuous measurements within a certain depth and overcame the technical difficulty of surveying narrow spaces and complex topographic and geomorphic environments. In addition, the geophysical survey results were preliminarily verified by drilling.

4.1. High-Precision Ground Magnetic Method Data

A total of 8417 HPGMS data points were collected from the survey area (zone CID-1) between the Hejia and Dumu iron deposits. These raw data were first corrected and processed, and then used to produce a magnetic anomaly map (Figure 5A). When the geomagnetic field background of the survey area was 53,527 nT, the value of magnetic anomaly data varied from −6092 to 7872 nT. Evidently, the data satisfactorily reflected the scale of the concealed anomalous magnetic bodies.

To eliminate the effects of oblique magnetization and allow the location of a magnetic anomaly to reflect a hidden magnetic body directly beneath it, all the data were subjected to RTP treatment. Figure 5B shows the magnetic anomaly map produced, based on the RTP-treated HPGMS data. Clearly, the highly magnetic anomaly zones are distributed in a densely scattered pattern in the survey area. This distribution pattern, on the one hand, results from the scattered distribution of the residual magnetic ore bodies on the surface and, on the other hand, is a consequence of the poor continuity of ores caused by the later-stage tensile failure of the Hejia fault zone. These, to a certain extent, suppress and affect the reflection of the anomalies caused by the large concealed magnetic bodies.

To better show the location, scale, and approximate burial depth of each concealed anomalous magnetic body, the data were processed by upward continuation to various heights (20, 60, 100, 150, and 200 m) (Figure 6). This treatment eliminated the effects of the surface and shallow-surface residual anomalous magnetic bodies. The first-order vertical derivatives of the data were calculated to subdue the effects of the background field in the deep area and to highlight the local anomalies caused by the anomalous magnetic bodies in the survey area [22]. Calculating the derivatives of the first-order vertical derivatives, that is, the second-order vertical derivatives, eliminated or weakened the background field and facilitated the delimitation of the anomalous magnetic bodies [23,24]. In the research, we processed the magnetic data via upward continuation at different heights and rasterized the obtained data to get gridding data. Subsequently, we took the vertical derivative of the gridding data and obtained first-order and second-order vertical derivatives. Two notably planar positive magnetic anomaly zones, YC-1 and YC-2, were discovered through the above systematic treatment. Zone YC-1 is located in the center of the survey area, and its long axis is distributed along the north-west direction, which is consistent with the strike of the main body of the ores in the area. The anomalies in zone YC-1 steadily extend several hundred meters from the shallow area to the deep area. Zone YC-2 is in the northern survey area, and its long axis is distributed along the north-west direction. Zone YC-2 adjoins the Dumu iron deposit and is separated from zone YC-1 by the main Hejia fault zone. Moreover, based on the extension of the HPGMS data to various heights, as well as the determined boundaries of the anomalous bodies, the anomalous magnetic bodies within zones YC-1 and YC-2 are, overall, inclined towards the NE direction. These are the anomalous spatial characteristics of the concealed magnetic ore bodies detected in this study.

The planar spatial locations of the concealed iron ore bodies were accurately determined by processing the HPGMS data. The burial depths and profile variations of the concealed iron ore bodies were determined primarily from the measurements of the TEM and MT survey lines. Considering that zone YC-2 adjoins the Dumu iron deposit, geophysical profile measurements were taken primarily in zone YC-1. These results adequately show the effectiveness of the geophysical survey methods in detecting concealed metallic ore bodies.

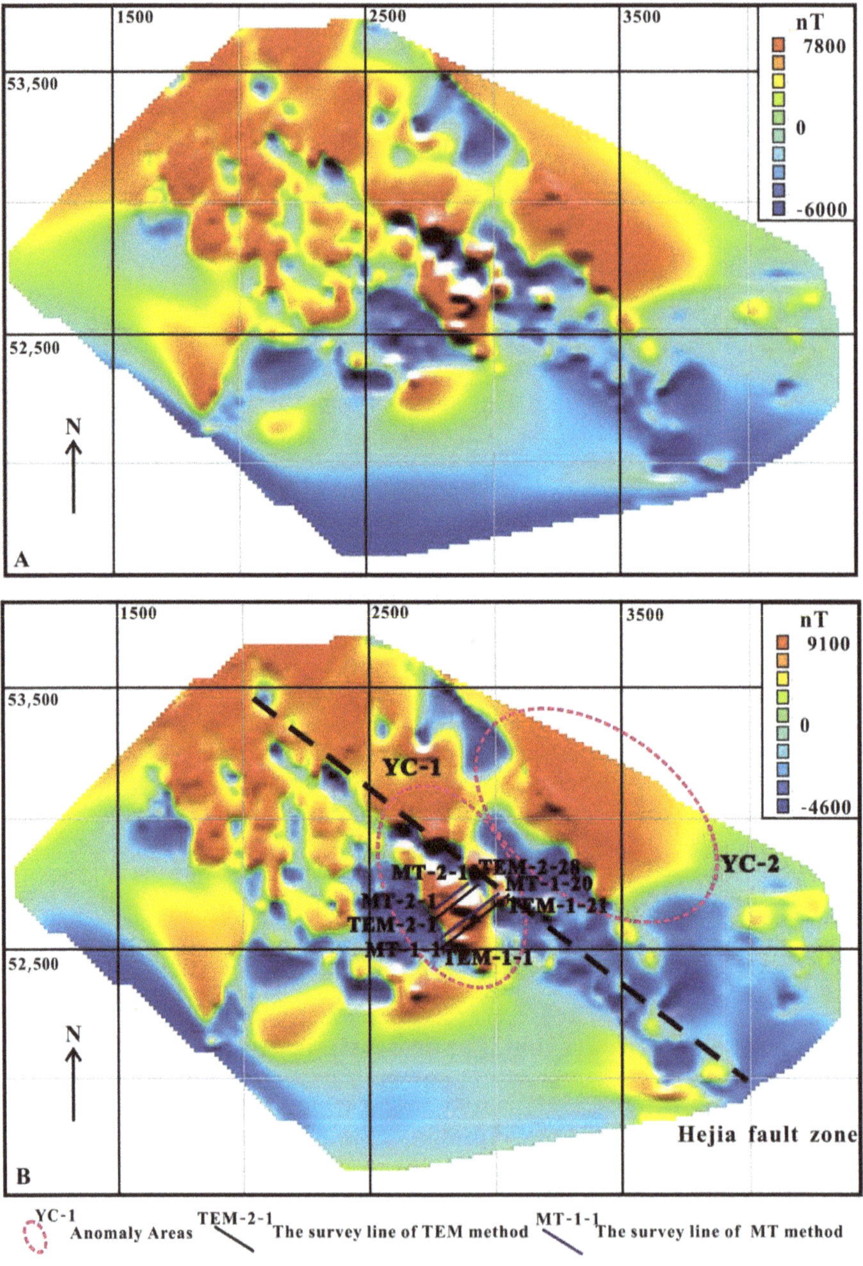

Figure 5. The magnetic anomaly map of the detection area in the Gongchangling iron ore concentration area. (**A**) Before reduction-to-the-pole treatment; (**B**) After reduction-to-the-pole treatment.

4.2. Transient Electromagnetic Method Data

Figure 5B shows the arrangement of the TEM survey lines. Figure 7 shows the processed TEM data. Due to the notable turn-off-time effect of the multi-turn coincident minor loop device [25], there is a blind detection zone at depths of 0–10 m in the shallow area. Notable low-resistivity anomaly zones (LRAZs) are present at burial depths of 0–150 m on both survey lines TEM-1 and TEM-2. The LRAZs are distributed in discontinuous, thick layers and are located within strong positive magnetic anomaly zones. According to the spatial distribution patterns of the iron ore bodies in the Hejia and Dumu iron deposits and, particularly, the sensitivity of the TEM to low-resistivity bodies [26,27], we can infer that the LRAZs are concealed magnetic ore bodies and that the high-resistivity anomalous bodies discontinuously distributed and intercalated within the LRAZs are wall rocks. This is primarily a result of the tectonic activity of the Hejia fault in the area.

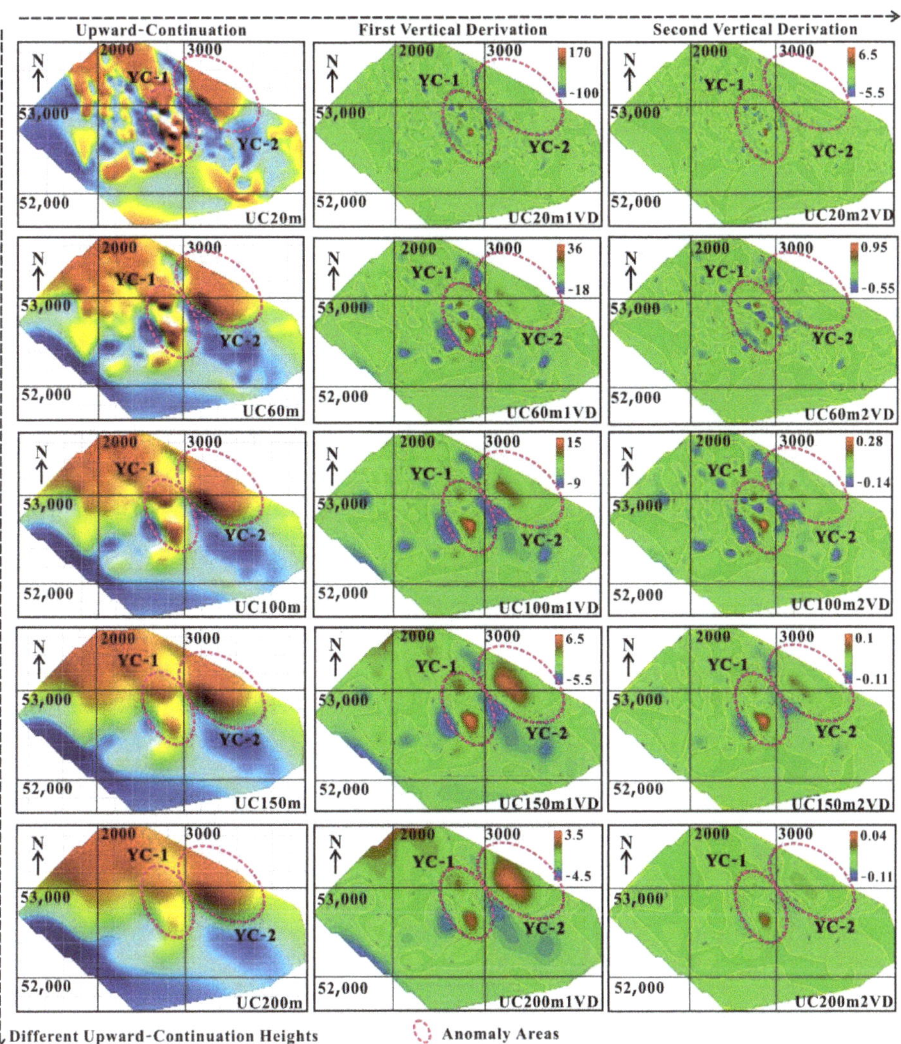

Figure 6. Diagram of the upward continuation and vertical derivation of magnetic measurement data in the Gongchangling iron ore concentration area.

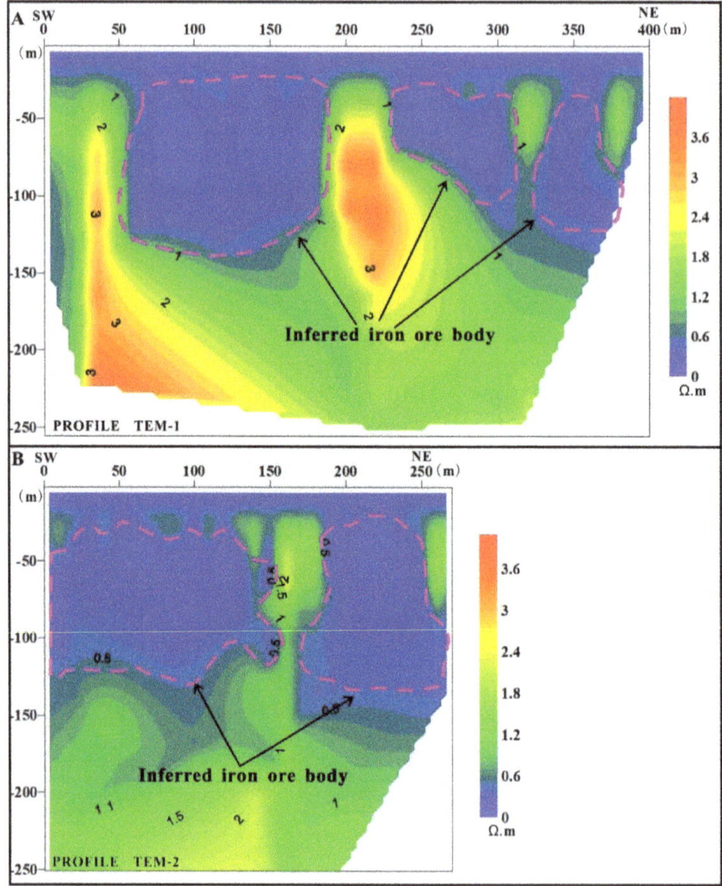

Figure 7. Transient electromagnetic method (TEM) data inversion results from the TerraTEM survey. (**A**) The resistivity section of the TEM profile TEM-1; (**B**) The resistivity section of the TEM profile TEM-2.

4.3. Magnetotelluric Data

Figure 5B shows the arrangement of the MT survey lines, which were adjacent and parallel to the TEM survey lines. Figure 8 shows the processed MT data. Compared with the TEM survey, the MT survey reached larger depths, up to 1000 m. The MT and TEM survey results both demonstrate that LRAZs are present at burial depths of 0–150 m, that they are distributed in a thick layer-like pattern, and that the deep areas of these LRAZs extend towards the north-east direction overall. Thus, we can infer that these LRAZs are concealed magnetite ore bodies. In addition, the MT survey results show that, as in the shallow area, LRAZs are present at burial depths of 300–450 m. We can infer that these LRAZs are also concealed magnetite ore bodies. However, compared with the TEM data, the MT data for the shallow area have relatively low resolution and are only able to show the outlines of the low-resistivity bodies. Nevertheless, the MT method can be used to perform geophysical surveys at greater depths. Moreover, the MT survey results show that the metamorphic basement in this area already appears at a burial depth greater than 500 m. This indirectly suggests that the thickness of the detected iron-bearing formation is 500 m.

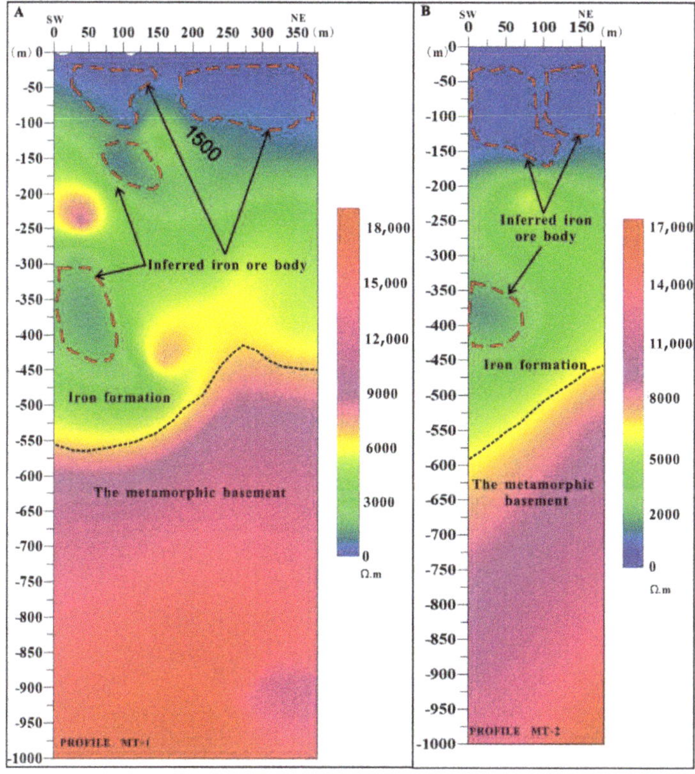

Figure 8. Magnetotelluric method (MT) data inversion results from the EH4 survey. (**A**) The resistivity section of the MT profile MT-1; (**B**) The resistivity section of the MT profile MT-2.

4.4. Drilling Verification

To verify the HPGMS, TEM, and MT survey results and to reduce the uncertainty in detecting concealed ore bodies, a verification borehole, ZKY1-1, was designed at 50 m on survey line TEM-2. Figure 9 shows the borehole verification results. Clearly, indeed, there exist concealed magnetite ore bodies in zone YC-1. The minimum burial depth of the ore bodies is 9.1 m. These ore bodies are overlain by Quaternary sediments. Compared with the TEM and MT survey results, which show the presence of concealed magnetite ore bodies of an anomalous scale distributed in a single, thick layer-like pattern at depths of 0–150 m, the drilling test results show that multiple layers of magnetite ore bodies of varying thicknesses (0.9–8.9 m) are present at depths of 0–100 m and that the total thickness of the magnetite ore bodies at 0–100 m is 23.6 m.

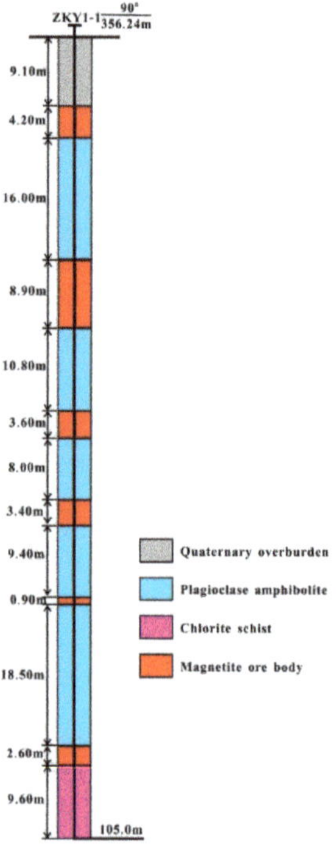

Figure 9. Diagrammatic cross-section from the validation by drilling.

5. Discussion

As an important geophysical survey method, the magnetic method can often be employed to scan large areas to directly find concealed deposits that contain ferromagnetic minerals, and it is extensively used to detect iron and iron polymetallic deposits and accurately determine the spatial locations of anomalous magnetic bodies [28–30]. In particular, the HPGMS method can be employed to accurately depict the spatial planar morphology of a concealed iron body [31]. In the magnetic method, a device close to the surface is used to take single-point, continuous scans, and therefore it is spatially closer to any anomalous magnetic bodies than devices used in other methods are. This can effectively avoid any shielding and interference from the strong magnetic fields of large deposits (e.g., Hejia and Dumu iron deposits) existing on the surface during the aeromagnetic scanning and measurement process, thereby highlighting the magnetic anomalies caused by the presence of the concealed iron ore bodies surrounding the existing iron deposits.

By taking full advantage of this quality, two sizeable, planar, concealed anomalous magnetic bodies were discovered, in this study within the transition zone CID-1 between the Hejia and Dumu iron deposits using the HPGMS method. On the basis of the HPGMS method survey results, as well as the sensitive response of the electromagnetic methods to metallic ore bodies [32,33], the electromagnetic methods (TEM and MT) were successfully employed, in this study, to survey the profiles of the concealed metallic ore bodies and determine their burial depths, morphologies, and scales [34–37]. The TEM and MT data collection devices, used in this study, can take single-point, continuous measurements, are adaptive to complex topographic and geomorphic environments, and can obtain high-quality geophysical survey data. The TerraTEM system was used to obtain data on shallow anomalies, while the EH4 system was employed to acquire data on deep anomalies. The data obtained on the shallow area using these two systems were cross validated. The TerraTEM system outperformed the EH4 system in survey resolution in shallow profile sections, whereas the EH4 system outperformed the

TerraTEM system in survey depth. When used in close coordination, the TerraTEM and EH4 system can satisfactorily locate anomalous bodies caused by mineralization.

Notably, the electromagnetic methods often overestimate the scale of anomalous bodies [21,38], particularly that of metallic ore bodies to which the electromagnetic methods are sensitive. In other words, the electromagnetic methods can effectively detect large-scale metallic ore body, but it has some limitations in accurately distinguishing specific anomalies from anomalous assemblage in current studies [39,40]. This quality was demonstrated by the electromagnetic survey results (TEM and MT) obtained in this study. Hence, when reasonably interpreting anomalously thick bodies detected by the electromagnetic methods based on metallogenic geological characteristics, it is necessary to set suitable verification boreholes to correct and improve the geophysical interpretations.

Importantly, the advantages of comprehensive geophysical survey methods for detecting concealed metallic ore bodies were fully availed in this study to detect concealed iron ore bodies [15,16]. In particular, owing to their convenience and sensitive response to anomalous bodies, the magnetic and electromagnetic survey-based devices can rapidly determine the anomalous spatial planar locations and anomalous profile distribution patterns of the concealed iron ore bodies. The MT method is advantageous for deep prospecting. For example, the MT method helped us to determine the thickness of the iron-bearing formation in the survey area, as well as the top boundary of the metamorphic basement, that is, it helped delimit the depth range of the prospecting space. The MT method can accurately evaluate the amounts of resources in an area and facilitate the verification of metallogenic theories, as well as resource planning and development. The MT method can sometimes even locate a metallogenic system [41,42], verify metallogenic theories, and expand the prospecting space.

6. Conclusions

An optimized combination of geophysical survey methods capable of taking single-point, continuous measurements (e.g., HPGMS, TEM, and MT methods) can be employed to accurately determine the anomalous planar spatial locations, anomalous profile morphologies, and burial depths of concealed iron ore bodies. In this study, two concealed iron ore body-induced anomalous zones, YC-1 and YC-2, were discovered using the HPGMS method within the transition zone CID-1 between the Hejia and Dumu iron deposits. Two concealed iron-bearing zones (a shallow zone and a deep zone) were found within the anomalous zone selected for investigation (YC-1) using the TEM and MT methods. The shallow and deep iron-bearing zones are buried at depths of 0–150 and 300–450 m, respectively. A 100 m scale drilling test confirmed the presence of multiple layers of iron ore bodies in the shallow iron-bearing zone. The cumulative thickness of these iron ore bodies is over 23.6 m.

Author Contributions: Conceptualization, J.F. and S.J.; methodology, J.F.; software, S.J.; validation, J.F., S.J. and E.W.; formal analysis, E.W.; investigation, J.F.; resources, E.W.; data curation, S.J.; writing—original draft preparation, J.F.; writing—review and editing, S.J.; visualization, E.W.; supervision, S.J.; project administration, J.F.; funding acquisition, J.F. and S.J. All authors have read and agreed to the published version of the manuscript.

Funding: This study was supported by the National Key Research and Development Program of China (grant no. 2016YFC0801603) and the Natural Science Foundation of Hebei Province (grant no. D2020501002).

Acknowledgments: We would like to thank Haitao Fu and the two anonymous reviewers for their constructive comments, which have helped us improve this paper.

Conflicts of Interest: The authors declare no conflict of interest.

References

1. Butt, C.R.M.; Hough, R.M. Why gold is valuable? *Elements* **2009**, *5*, 277–280. [CrossRef]
2. Gusentsova, T.M.; Kulkova, M.A. The subsistence strategy and paleoenvironment on the Stone Age site Podolye 1 in the southern Ladoga Lake region (Eastern Baltic). *Quat. Int.* **2020**, *541*, 41–51. [CrossRef]
3. Cheng, H.J.; Chen, C.; Zhan, X.Z.; Chang, J.Y.; Ding, Y.L.; Kuwanixibieke, M.; Yang, R.H.; Jia, N.E.; Fu, H.Z. New progress in the prediction theory and prospecting method for concealed deposits. *Geol. Explor.* **2017**, *53*, 456–463, (In Chinese with English abstract).

4. Hosseini-Dinani, H.; Aftabi, A. Vertical lithogeochemical halos and zoning vectors at Goushfil Zn-Pb deposit, Irankuh district, southwestern Isfahan, Iran: Implications for concealed ore exploration and genetic models. *Ore Geol. Rev.* **2016**, *72*, 1004–1021. [CrossRef]

5. Meng, G.X.; Lv, Q.T.; Yan, J.Y.; Deng, Z.; Qi, G.; Xue, R.H. The research and application of explorational technology of "Penetrating" to geology and mineral investigation in overburden area. *Acta Geosci. Sin.* **2019**, *40*, 637–650, (In Chinese with English abstract).

6. Kang, H.; Chen, Y.L.; Li, D.P.; Zhao, J.X.; Cui, F.R.; Xu, Y.L. Deep-penetrating geochemistry for concealed sandstone-type uranium deposits: A case study of Hadatu uranium deposit in the Erenhot Basin, North China. *J. Geochem. Explor.* **2020**, *211*, 106464. [CrossRef]

7. Anand, S.P.; Rajaram, M. Aeromagnetic data analysis for the identification of concealed uranium deposits: A case history from Singhbhum uranium province, India. *Earth Planets Space* **2006**, *58*, 1099–1103. [CrossRef]

8. Wang, Q.; Wang, X.B.; Yang, J.; Min, G.; Guo, J. Locating of the concealed skarn iron ore deposit based on multivariate information constraints: A case study of the Beiya gold mine in Yunnan province. *Chin. J. Geophys.* **2016**, *59*, 4771–4781.

9. Liu, J.; Zhang, J.Z.; Jiang, L.; Lin, Q.; Wan, L. Polynomial-based density inversion of gravity anomalies for concealed iron-deposit exploration in North China. *Geophysics* **2019**, *84*, 325–334. [CrossRef]

10. Wu, M.A.; Wang, Q.S.; Zheng, G.W.; Cai, X.B.; Yang, S.X.; Di, Q.S. Discovery of the Nihe iron deposit in Lujiang, Anhui, and its exploration significance. *Acta Geol. Sin.* **2011**, *85*, 802–809, (In Chinese with English abstract).

11. Shah, A.; Bedrosian, P.A.; Anderson, E.D.; Kelley, K.D.; Lang, J. Intergrated geophysical imaging of a concealed mineral deposit: A case study of the world-class Pebble porphyry deposit in southwestern Alaska. *Geophysics* **2013**, *78*, 317–328. [CrossRef]

12. Cave, B.W.; Lilly, R.; Glorie, S.; Gillespie, J. Geology, apatite geochronology, and geochemistry of the Ernest Heny Inter-lens: Implications for a re-examined deposit model. *Minerals* **2018**, *8*, 405. [CrossRef]

13. Zhang, L.C.; Zhai, M.G.; Wan, Y.S.; Guo, J.H.; Dai, Y.P.; Wang, C.L.; Liu, L. Study of the Precambrain BIF-iron deposits in the North China Craton: Progresses and questions. *Acta Petrol. Sin.* **2012**, *28*, 3431–3445.

14. Guo, Z.W.; Hu, X.P.; Liu, J.X.; Liu, C.M.; Xiao, J.P. Geophysical field data interpolation using stochastic partial differential equations for gold exploration in Dayaoshan, Guanxi, China. *Minerals* **2018**, *9*, 14. [CrossRef]

15. Guo, Z.W.; Hu, L.Y.; Liu, C.M.; Cao, C.H.; Liu, J.X.; Liu, R. Application of the CSAMT method to Pb-Zn mineral deposit: A case study in Jianshui, China. *Minerals* **2019**, *9*, 726. [CrossRef]

16. Zhang, J.M.; Zeng, Z.F.; Zhao, X.Y.; Li, J.; Zhou, Y.; Gong, M.X. Deep mineral exploration of the Jinchuan Cu-Ni sulfide deposit based on aeromagnetic, gravity, and CSAMT methods. *Minerals* **2020**, *10*, 168. [CrossRef]

17. Li, H.M.; Yang, X.Q.; Li, L.X.; Zhang, Z.C.; Liu, M.J.; Yao, T.; Chen, J. Desilicification and iron activation-reprecipitation in the high-grade magnetite ores in BIFs of the Anshan-Benxi area, China: Evidence from geology, geochemistry and stable isotopic characteristics. *J. Asian Earth Sci.* **2015**, *113*, 998–1016. [CrossRef]

18. Fu, H.T.; Wang, E.D.; Liu, L.S.; Fu, J.F. *The Ore-Control Condition and Ore-Prospecting Model of Sedimentary-Metamorphic Iron Deposits in Anshan-Benxi Area, China*; People's Publishing House: Shenyang, China, 2016; pp. 1–231, (In Chinese with English abstract).

19. Tang, H.Z.; Yang, H.Y.; Lu, G.Y.; Chen, S.E.; Yue, J.H.; Zhu, Z.Q. Small multi-turn coils based on transient electromagnetic method for coal mine detection. *J. Appl. Geophys.* **2019**, *169*, 165–173. [CrossRef]

20. Chang, J.H.; Su, B.Y.; Malekian, R.; IEEE, S.M.; Xing, X.J. Detection of water-filled mining goaf using mining transient electromagnetic method. *IEEE Trans. Ind. Inform.* **2020**, *16*, 2977–2984. [CrossRef]

21. Zeng, Q.D.; Di, Q.Y.; Liu, T.B.; Li, G.M.; Yu, C.M.; Shen, P.; Liu, H.T.; Ye, J. Exploration of gold and lead-zinc deposits using a magnetotelluric method: Case studies in the Tianshan-Xingmeng Orogenic Belt of Northern China. *Ore Geol. Rev.* **2020**, *117*, 103283. [CrossRef]

22. Hu, Z.W. *The Study on Magnetic Anomalies Characteristic and the Cause in the Philippine Sea and the South Sea*; China University of Geoscience: Wuhan, China, 2014; pp. 1–67, (In Chinese with English abstract).

23. Wang, W.Y.; Pan, Y.; Qiu, Z.Y. A new edge recognition technology based on the normalized vertical derivative of total horizontal derivative for potential field data. *Appl. Geophys.* **2009**, *6*, 226–233. [CrossRef]

24. Tan, J.Q.; Jiang, Y.L.; Li, S.; Li, C. Higher derivatives in the separation of gravity anomalies superimposed research and application. *Comput. Tech. Geophys. Geochem. Explor.* **2015**, *37*, 40–44, (In Chinese with English abstract).

25. Jia, S.S.; Shao, A.L.; Wang, H.L.; Wang, E.D. Detection and Evaluation of Underground Iron Ore Goaf Based on TEM. *J. Northeast. Univ. (Nat. Sci.)* **2011**, *32*, 1340–1343, (In Chinese with English abstract).

26. Suzuki, K.; Kusano, Y.; Ochi, R.; Nishiyama, N.; Tokunaga, T.; Tanaka, K. Electromagnetic exploration in high-salinity groundwater zones: Case studies from volcanic and soft sedimentary sites in coastal Japan. *Explor. Geophys.* **2017**, *48*, 95–109. [CrossRef]

27. El-kaliouby, H. Mapping sea water intrusion in coastal area using time-domain electromagnetic method with different loop dimensions. *J. Appl. Geophys.* **2020**, *175*, 103963. [CrossRef]

28. Christiansen, R.; Kostadinoff, J.; Bouhier, J.; Martinez, P. Exploration of iron ore deposits in Patagonia. Insights from gravity, magnetic and SP modelling. *Geophys. Prospect.* **2018**, *66*, 1751–1763. [CrossRef]

29. Ives, B.T.; Mickus, K. Using gravity and magnetic data for insights into the Mesoproterozoic St. Francois Terrane, Southeast Missouri: Implications for Iron Oxide deposits. *Pure Appl. Geophys.* **2018**, *176*, 297–314. [CrossRef]

30. Jackisch, R.; Madriz, Y.; Zimmermann, R.; Pirttijärvi, M.; Saartenoja, A.; Heincke, B.H.; Salmirinne, H.; Kujasalo, J.P.; Andreani, L.; Gloaguen, R. Drone-borne hyperspectral and magnetic data integration: Otanmäki Fe-Ti-V deposit in Finland. *Remote Sens.* **2019**, *11*, 2084. [CrossRef]

31. Liu, S.; Hu, X.Y.; Zhu, R.X. Joint inversion of surface and borehole magnetic data to prospect concealed ore bodies: A case study from the Mengku iron deposit, northwestern China. *J. Appl. Geophys.* **2018**, *154*, 150–158. [CrossRef]

32. Li, R.X.; Wang, H.; Xi, Z.Z.; Long, X.; Hou, H.T.; Liu, Y.Y.; Jiang, H. The 3D transient electromagnetic forward modeling of volcanogenic massive sulfide ore deposits. *Chin. J. Geophys.* **2016**, *59*, 725–733.

33. Kalscheuer, T.; Juhojuntti, N.; Vaittinen, K. Two-dimensional magnetotelluric modelling of ore deposits: Improvements in model contraints by inclusion of borehole measurements. *Surv. Geophys.* **2018**, *39*, 467–507. [CrossRef]

34. Bastani, M.; Malehmir, A.; Ismail, N.; Pedersen, L.B.; Hedjazi, F. Delineating hydrothermal stockwork copper deposits using controlled-source and radio-magnetotelluric methods: A case study from northeast Iran. *Geophysics* **2009**, *74*, 167–181. [CrossRef]

35. Pare, P.; Gribenko, A.V.; Cox, L.H.; Cuma, M.; Wilson, G.A.; Zhdanov, M.S.; Legault, J.; Smit, J.; Polome, L. 3D inversion of SPECTREM and ZTEM airbone electromagnetic data from the Pebble Cu-Au-Mo porphyry deposit, Alaska. *Explor. Geophys.* **2012**, *43*, 104–115. [CrossRef]

36. Malovichko, M.; Tarasov, A.V.; Yavich, N.; Zhdanov, M.S. Mineral exploration with 3D controlled-source electromagnetic method: A synthetic study of Sukhoi Log gold deposit. *Geophys. J. Int.* **2019**, *219*, 1698–1716. [CrossRef]

37. Shi, Y.; Xu, Y.X.; Yang, B.; Peng, Y.; Liu, S.Y. Three-dimensional audio-frequency magnetotelluric imaging of Zhuxi copper-tungsten polymetallic deposits, South China. *J. Appl. Geophys.* **2020**, *172*, 103910. [CrossRef]

38. Xu, S.; Xu, F.M.; Hu, X.Y.; Zhu, Q.; Zhao, Y.D.; Liu, S. Electromagnetic characterization of epithermal gold deposits: A case study from the Tuoniuhe gold deposit, Northeast China. *Geophysics* **2020**, *85*, 49–62. [CrossRef]

39. Yang, D.K.; Fournier, D.; Kang, S.; Oldenburg, D.W. Deep mineral exploration using multiscale eclectromagnetic geophysics: The Lalor massive sulphide deposit case study. *Can. J. Earth Sci.* **2019**, *56*, 544–555. [CrossRef]

40. Di, Q.Y.; Xue, G.Q.; Zeng, Q.D.; Wang, Z.X.; An, Z.G.; Lei, D. Magnetotelluric exploration of deep-seated gold deposits in the Qingchengzi orefield, Eastern Liaoning (China), using a SEP system. *Ore Geol. Rev.* **2020**, *122*, 103501. [CrossRef]

41. Heinson, G.S.; Direen, N.G.; Gill, R.M. Magnetotelluric evidence for a deep-crustal mineralizing system beneath the Olympic Dam iron oxide copper-gold deposit. *Geology* **2006**, *34*, 573–576. [CrossRef]

42. Jiang, W.P.; Korsch, R.J.; Doublier, M.P.; Duan, J.M.; Costelloe, R. Mapping deep electrical conductivity structure in the Mount Isa region, Northern Australia: Implication for mineral prospectivity. *J. Geophys. Res. Sold Earth* **2019**, *124*, 10655–10671. [CrossRef]

Publisher's Note: MDPI stays neutral with regard to jurisdictional claims in published maps and institutional affiliations.

© 2020 by the authors. Licensee MDPI, Basel, Switzerland. This article is an open access article distributed under the terms and conditions of the Creative Commons Attribution (CC BY) license (http://creativecommons.org/licenses/by/4.0/).

Article

Spectral Induced Polarization Survey with Distributed Array System for Mineral Exploration: Case Study in Saudi Arabia

Fouzan A. Alfouzan [1], Abdulrahman M. Alotaibi [1], Leif H. Cox [2] and Michael S. Zhdanov [2,3,*]

[1] King Abdulaziz City for Science and Technology, Riyadh 11442, Saudi Arabia; alfouzan@kacst.edu.sa (F.A.A.); aotaibi@kacst.edu.sa (A.M.A.)
[2] TechnoImaging, LLC, Salt Lake City, UT 84107, USA; leif@technoimaging.com
[3] Consortium for Electromagnetic Modeling and Inversion (CEMI), University of Utah, UT 84112, USA
* Correspondence: michael.s.zhdanov@gmail.com

Received: 19 July 2020; Accepted: 28 August 2020; Published: 30 August 2020

Abstract: The Saudi Arabian Glass Earth Pilot Project is a geophysical exploration program to explore the upper crust of the Kingdom for minerals, groundwater, and geothermal resources as well as strictly academic investigations. The project began with over 8000 km^2 of green-field area. Airborne geophysics including electromagnetic (EM), magnetics, and gravity were used to develop several high priority targets for ground follow-up. Based on the results of airborne survey, a spectral induced polarization (SIP) survey was completed over one of the prospective targets. The field data were collected with a distributed array system, which has the potential for strong inductive coupling. This was examined in a synthetic study, and it was determined that with the geometries and conductivities in the field survey, the inductive coupling effect may be visible in the data. In this study, we also confirmed that time domain is vastly superior to frequency domain for avoiding inductive coupling, that measuring decays from 50 ms to 2 s allow discrimination of time constants from 1 ms to 5 s, and the relaxation parameter C is strongly coupled to intrinsic chargeability. We developed a method to fully include all 3D EM effects in the inversion of induced polarization (IP) data. The field SIP data were inverted using the generalized effective-medium theory of induced polarization (GEMTIP) in conjunction with an integral equation-based modeling and inversion methods. These methods can replicate all inductive coupling and EM effects, which removes one significant barrier to inversion of large bandwidth spectral IP data. The results of this inversion were interpreted and compared with results of drill hole set up in the survey area. The drill hole intersected significant mineralization which is currently being further investigated. The project can be considered a technical success, validating the methods and effective-medium inversion technique used for the project.

Keywords: inversion; induced polarization; electromagnetics; three-dimensional; effective medium; discovery

---

1. Introduction

The Saudi Arabian Glass Earth Pilot Project is a multiyear Earth observation program. The goal is to explore for resources in the relatively unexplored regions of Saudi Arabia for minerals, groundwater, and geothermal resources as well as strictly academic investigations. The initial stage of the project involved the acquisition, processing, and interpretation of airborne electromagnetic, gravity, and magnetic geophysical data over an 8000 square kilometer area [1]. The project helped to identify a dozen potential mining targets, which can be associated with gold and other base metal deposits, including zinc and copper. A follow-up ground spectral induced polarization (SIP) survey was proposed to further investigate one of the identified anomalies. The SIP data were collected by Dias Geophysical with their distributed array system.

The spectral induced polarization (SIP) data collected in 3D present a very rich data set. Modern distributed array systems are instruments which allow nearly arbitrary survey geometry. They can collect 10,000 s of sounding points over several decades of bandwidth points (e.g., [2,3]). These data sets contain information about the 3D distribution of time dependent or frequency dependent complex conductivity of the medium. The advantage of this information is that it can be translated into knowledge about mineralization which goes well beyond that contained in conventional conductivity. In fact, it may be possible to discriminate economic and non-economic mineralization from the surface using this technique (e.g., [4,5]).

However, the major challenges to a widespread use of this technology were a lack of a model which describes and properly characterizes the complex conductivity response of different media (microscopic effects), and a lack of suitable software and methods to process and interpret large volumes of the SIP data with a technique which is rigorous enough to honor all the macroscopic physical effects. We address these two issues with our new inversion method and code which use the new induced polarization model: the generalized effective medium theory of induced polarization (GEMTIP) introduced by Zhdanov [5]. The inversion code also honors all the macroscopic physics of electromagnetic field which is required for modeling.

We first consider the physical model describing the induced polarization phenomenon. The induced polarization is related to a complex (amplitude and phase or real and imaginary) electrical resistivity which varies with frequency. Many different models have been proposed to replicate the complex resistivity spectra seen in real rocks. These include many models such as Debye dispersion [6], but most prevalent is the Cole–Cole model [7]. It was introduced to geophysics by [8,9], and [10]. A review of this method was done by [11]. Indeed, the Cole–Cole model is the most widely applied model for SIP data (e.g., [12–16]). However, it does not connect the mineralization and rock composition directly to the spectral induced polarization spectrum.

In 2008, Zhdanov introduced the GEMTIP model which does allow direct connection of the rock matrix to the complex resistivity spectrum [5]. In this paper, Zhdanov showed that the Cole–Cole model could be derived analytically from the general effective medium theory for electromagnetic field. Indeed, the Cole–Cole model is a specific solution of the GEMTIP model when conductive inclusions are spherical and there is only one type or phase of a mineral which is giving rise to the observed induced polarization. In the GEMTIP model, the petrophysical properties of the rock can be directly modeled, including mineralization, porosity and fluid content, anisotropy, and multiple mineral phases. This was confirmed by [17,18], who used QEMScan mineralogical testing along with laboratory measurements of the complex spectra of the samples.

There are also challenges with modeling all EM induction and IP effects accurately and then inverting the data on a large scale. To simplify the problem, most inversion methods are based on a direct current (DC) approximation to Maxwell's equations, and then a linear approximation of the IP effects with conductivity [19,20]. The problem with this approach is three-fold: (1) inductive coupling (IC) between the transmitter and receiver wires and also within the earth are ignored while these effects can be large [21]; (2) the assumption that IP effects are linearly related to conductivity is often invalid; and (3) other IP parameters (time constant, GEMTIP parameters, etc.) cannot be included using this method. Other researchers have included the full solution to Maxwell's equations to eliminate the traditional limitations to inductive coupling. For example, [22] uses a full solution but simplified the solver for a single in-phase and quadrature resistivity. The authors of [23] approached this problem with a full Cole–Cole relaxation model and accurate transmitter waveform and inductive coupling modeling, but the method was of limited applicability because the solution was restricted to 1D conductivity structures. [24] used a half-space approximation to fit the IC and then remove this part before inversion. Others have modeled the full electromagnetic (EM) and IP response for a

3D conductive body for ground-based methods (e.g., [25]) and airborne EM (e.g., [26]). But no one has previously solved the full EM problem in 3D during inversion to include inductive coupling in a ground-based IP survey.

The full solution to this problem is difficult. Computational times and memory requirements are problematic, especially when including full wire paths with large 3D data sets and 3D earth models. This is exacerbated because many forward modeling runs are required for each inversion run. Also, because of the ill-posedness of the problem [27–29], many inversion runs are required to properly explore the range of models which are acceptable to the observed data.

These problems have been overcome in part by using a very efficient integral equation solution [30] and a re-weighted regularized conjugate method to keep memory requirements at a minimum. Two papers have been written on this inversion method which includes both full IC and determination of all four Cole–Cole parameters: [15,31]. This work was extended by [32] to apply full 3D inversion including EM effects and complex transmitter wire paths, and then to solve for four parameters of the GEMTIP model.

In this paper, we applied these advanced modeling and inversion methods to the spectral IP data collected in Saudi Arabia during the Glass Earth Pilot Project. We used an efficient integral equation forward modeling scheme, including full EM effects and wire paths, for the forward modeling. The earth is represented as voxels with the complex conductivity of each voxel being parameterized with the GEMTIP model. We solved for the four dominate GEMTIP parameters for each voxel. The non-uniqueness of the problem is mitigated with regularized inversion. The re-weighted conjugate gradient method was used to optimize the search for the minimum of the objective functional.

Based on the results of the inversion, the potential target was identified in a green-field area of Saudi Arabia. The target was drilled, and mineralization was confirmed in a drill hole representing a significant technical success.

GEMTIP Resistivity Relaxation Model

For completeness, we begin our discussion with a brief review of the GEMTIP resistivity relaxation model. In a general case, the effective conductivity of rocks is not necessarily a constant and real number but is complex and may vary with frequency. A general approach to constructing the resistivity relaxation model is based on the rock physics and description of the medium as a composite heterogeneous multiphase formation. The complex resistivity is parameterized using the GEMTIP model [5], which is based on effective-medium representation of a composite heterogeneous multi-phase rock formation. The GEMTIP model was developed to characterize the complex resistivity of multiphase heterogeneous rocks and their petrophysical and structural properties, including grain size, grain shape, porosity, anisotropy, polarizability, volume fraction, and conductivity of the inclusions in the pore space.

In the papers [17,18], we have demonstrated that the GEMTIP model of the complex resistivity for a three-phase medium with elliptical inclusions can be described by the following formula:

$$\sigma(\omega) = \sigma_0 \left(1 + \sum_{l=1}^{2} \sum_{\alpha=x,y,z} \frac{f_l}{3\gamma_{l\alpha}} \left[1 - \frac{1}{1 + s_{l\alpha}(i\omega\tau_l)^{C_l}} \right] \right) \tag{1}$$

where σ_0 is the matrix conductivity.

Parameters τ_l and C_l are similar to the Cole–Cole model and represent the time constants and relaxation parameters, respectively. The coefficients $\gamma_{l\alpha}$ and $s_{l\alpha}$ are the structural parameters defined by the geometrical characteristics of the ellipsoidal inclusions, and f_l is their volume fraction. In a special case of only spherical grains with one mineral phase in a background rock matrix, Equation

(1) reduces to a Equation (2) which is very similar to the widely used Cole–Cole formula for complex resistivity [5]:

$$\sigma(\omega) = \sigma_0\left(1 + 3f\left[1 - \frac{1}{1 + (i\omega\tau)^C}\right]\right). \tag{2}$$

where σ_0 is the DC conductivity (S/m); f is volume fraction of mineral phase; ω is the angular frequency (rad/s), τ is the time parameter; and C is the relaxation parameter.

Equations (1) and (2) for complex and frequency dependent conductivity make it possible to consider simultaneously the electromagnetic induction and IP effects in observed EM data within both forward modeling and inversion stages of interpretation. Therefore, in the framework of this approach we do not need to apply any special technique of separating those effects, which is often done in interpretation of the IP data.

2. Spectral IP Survey Description

2.1. Geological Background

There are three distinct geologic regimes in the original study area (Figures 1 and 2). The western most part contains Arabian shield with mountains up to 1 km high. The central part is formed by Cenozoic volcanic flows that are several hundred meters thick and underlain by the Arabian shield. Eruptions were recorded as recently as 1256 AD. The eastern part of the survey area is the focus of this study. It exhibits Arabian shield with sediment filled basins and hills hundreds of meters high. The Quaternary sediments in the areas are predominately alluvium, Aeolian sand, and slightly consolidated alluvium of old wadi deposits. There are many historic mines in the areas, including the largest mine in Saudi Arabia, the Mahd Ad Dahab ("Cradle of Gold"). A geologic map of the survey area is shown in Figure 1. The legend for this figure is shown in Figure 2.

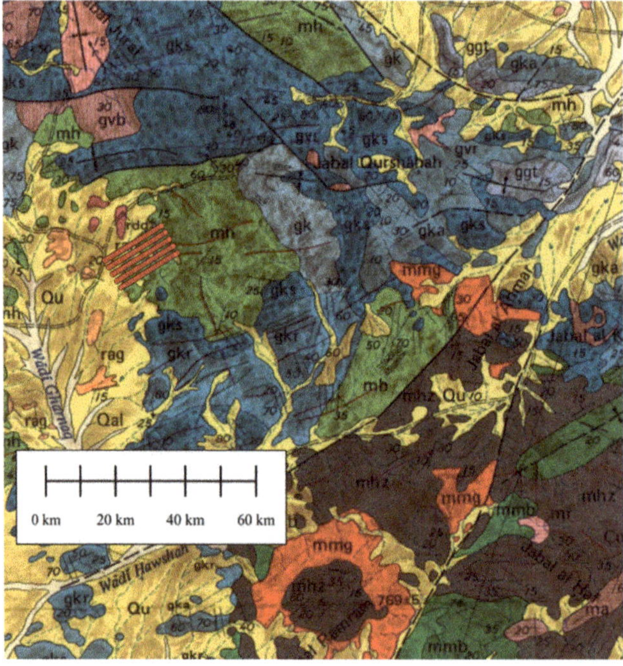

Figure 1. Geologic map of the survey area. The red lines show the 2D IP lines.

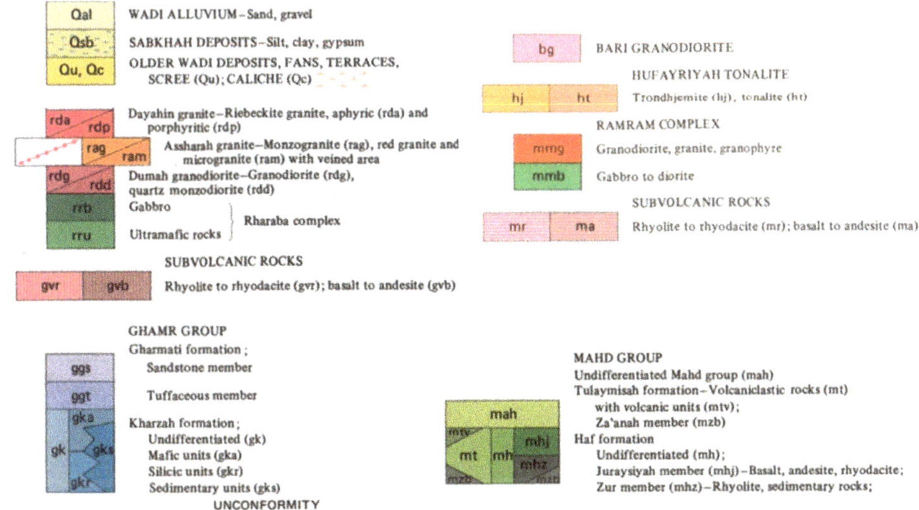

Figure 2. Geologic map legend for Figure 1.

2.2. Geophysical Background

The original regional study [1] area was covered by airborne EM, magnetics, and gravity surveys. From the airborne data, several large-scale regional features were discovered [1]. Many small-scale anomalies (hundreds of meters areal extent) were also seen based on the Glass Earth models of the physical properties of the subsurface. Note that, most of these anomalies are located on or very close to the large regional features that have been identified by geophysical data. Sulfides, such as pyrite, pyrrhotite, and chalcopyrite, are commonly associated with gold and copper mineralization. These minerals are electrically conductive and can often be detected by airborne surveys. We believe some of these anomalies may be sourced from sulfides. The fact that both copper and gold mineralization is known in the surrounding area is very encouraging.

An example of one of these anomalies is identified in the airborne EM data shown in Figure 3. The targets that we choose to follow-up with are all in exposed shield, all are conductive anomalies, and many have associated magnetic signatures. Each anomaly has been checked for geophysical significance and is not related to any cultural artifacts, neither present in the satellite images, nor marked by the aircrew.

We made one of the selected targets the highest priority target. The reasons for this were because the shape of the target suggested alteration or mineralization, and not saline groundwater or overburden, which are likely very common conductive features in this region. Also, there are magnetic anomalies associated with the conductive anomaly, which is another strong sign for a possible mineralization. The conductivity associated with this deposit, as inferred from the inversion of airborne data, is shown in Figure 3.

From these images, it appears the body exhibits both a magnetic response and an elevated electrical conductivity. The body is cigar shaped dipping steeply to the south–southeast. It is located near a large regional structure mapped with both the gravity and electromagnetic methods. A conductive overburden can be seen as a near surface target, but the prospective mineralization body is clearly hosted within the resistive bedrock.

The outstanding question is "Does this body contain mineralization?". This cannot be wholly established by geophysics alone but determining if the body has an induced polarization response increases the certainty that this conductivity anomaly is due to mineralization, and it is possibly an economic mineral deposit. The spectral induced polarization method is a great compliment to airborne

EM. Being a resistive method, it does an excellent job of imaging targets of moderate resistivity or conductivity. It also can differentiate between disseminated sulfides, massive sulfides, and clays or groundwater of similar conductivity.

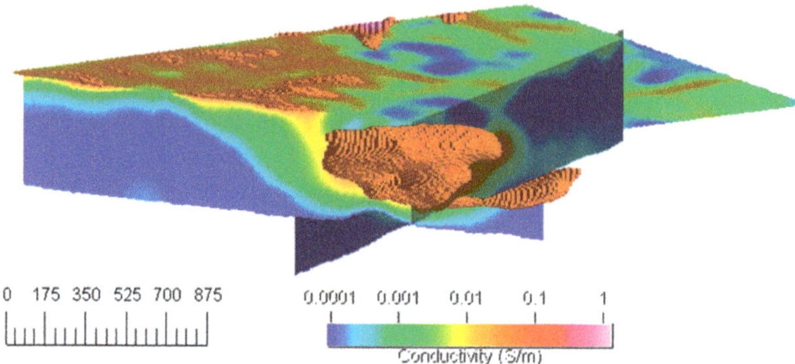

Figure 3. Conductivity volume of prospective target from the results of the airborne survey inversion. The cut-off shows conductivity greater than 0.05 S/m. The conductive overburden is shown in the near surface. The color map is conductivity in S/m, and the scale on the left shows distance in meters.

2.3. Field Data Collection

Two SIP surveys were collected over the same area. One was a conventional 2D dipole–dipole type setup with five lines spaced 350 m apart and 3 km long and 100 m dipoles, and another was a true 3D type survey. Dias Geophysical collected both spectral IP surveys with their 3D distributed array system [33] in February of 2018. The ground IP survey was designed with five three-kilometer-long lines oriented perpendicular to the dominant geologic strike in the area. The survey used 100 m grounded dipoles as sources and receivers to best collect spectral IP data. These lines are shown in Figure 4. The SIP data are collected as a time series of voltage with each current injection. The majority of the current injection dipoles used a 50% duty cycle 8 s square wave. Each current injection point undergoes several minutes of current injection with the waveform being recorded at the injection points. The voltage at the receivers is recorded continuously throughout their entire deployment.

The second survey was very similar to the first one, but had three lines of receivers, and a single line of transmitting dipoles along the center line of receivers. Voltages were measured at each receiver potential electrode for every transmitter position, allowing a large number of combinations of off-line and on-line data points to be measured for each transmitter position. This survey layout is also shown in Figure 4.

2.4. Field Data Processing

The data reduction stage converts the receiver voltages measured at a common ground to dipole voltages. The receiver voltage time series is then paired with the injection current, stacked to reduce noise, and then integrated through the time windows. A primary voltage measured during the current on-time is also extracted. The data were processed to 20 time windows from 40 ms to 1.92 s window centers, plus Vp (on-time). Before inversion, spikes, and outliers were removed. However, due the distributed array geometry, visual quality control of the data is difficult. For the wire paths which are nearly perpendicular to the transmitters at the endpoints, 3D features have a large effect on the response, and negative and positive apparent resistivities are possible. To deal with this, during inversion these responses were examined and the sign of apparent resistivities in these specific cases was corrected. Example maps and decays of the observed data are shown in the field data analysis section.

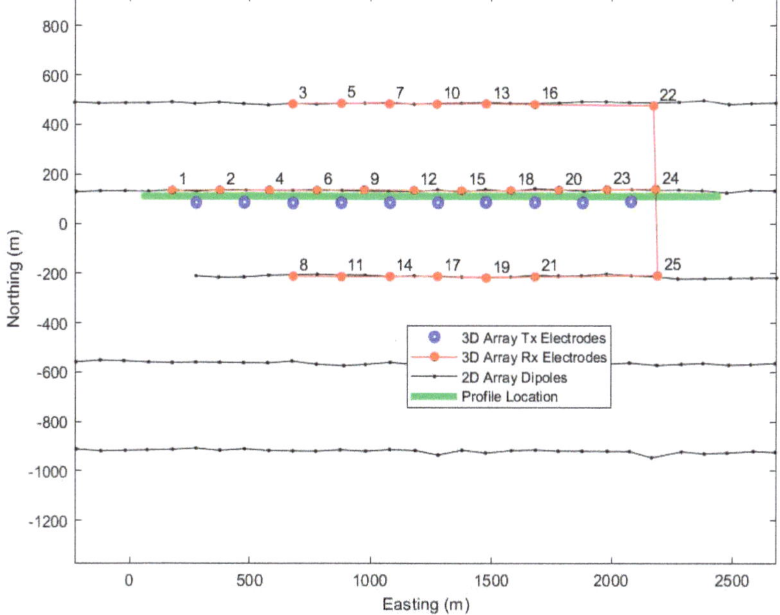

Figure 4. Location of spectral IP lines. The initial 2D survey is shown by the black lines and dots. The dots represent locations of transmitter and/or receiver dipoles for the dipoles/dipole survey. The red dots show receiver electrode positions for the 3D distributed array survey, and the red lines show the wire paths connecting these electrodes. The blue circles show transmitter electrode locations for the 3D distributed array survey. There were 9 transmitter dipoles. The green line indicates the location of the cross section shown in Figures 14–17. The receiver nodes correspond to the data plots in Figure 13.

3. Analysis of Inductive Coupling (IC) Effect

3.1. Inductive Coupling in IP Measurements

The collection of true broadband spectral IP data requires higher frequencies or early times, which increases the chances of incurring coupling. Additionally, as surveys become more complex and truly 3D, it is more difficult to eliminate the IC effect. The inductive coupling response can be larger than the IP response and it has been a major hindrance in the past to interpretation. The IC effect has been examined several times before, mostly in the context of standard array configuration (e.g., [21]). It has also been attempted to remove the IC effects and to correct the data to pure IP and resistivity responses (e.g., [33]).

In the ground SIP survey design implemented in this project, only one common receiver wire needs to be used for all receiver locations, greatly simplifying survey logistics and increasing the data collection speed. Each receiver can be referenced to every other receiver in the processing to create many receiver dipoles. However, this produces long receiver wire paths, some of which run parallel to transmitter current dipoles, and hence inductive coupling can become problematic if not modeled correctly.

In this section, we look at the effects of inductive coupling and advanced survey designs in the context of extracting spectral IP information. In order to illustrate our approach, we first consider the distributed field survey layout shown in Figure 5. There are nine transmitter dipoles operated in the case, and each neighboring current electrode pair operated as a current bipole. The voltage at each receiver point is measured relative to a common ground.

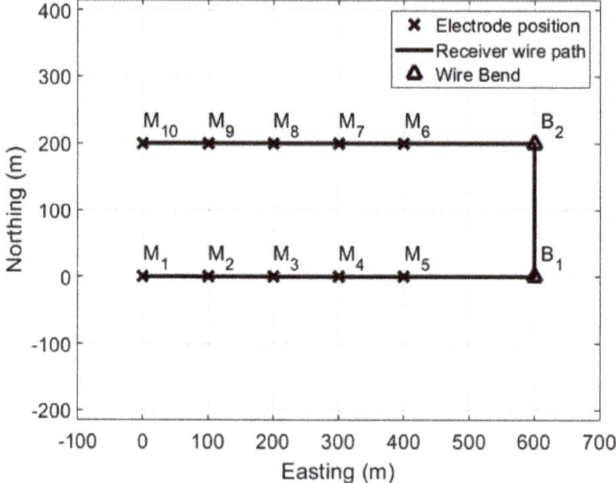

Figure 5. Hypothetical wire path. The electrode positions are labeled with an M, and the wire bends are given by a B. The wire path is shown as the black line.

A note on our nomenclature bipole and dipole: bipole is defined for our purposes as two poles connected by a length of wire. The separation between the poles is great enough that they cannot be considered at the same point. A current dipole is a theoretical construct in which the poles are proximal enough relative to the measurement distance that, for calculations, their separation distance approaches zero while the current strength approaches infinity in such a way to keep the moment constant.

We will demonstrate in this section that most of the IC effect is due to wire to wire, not EM effects in the earth. This means that simply using the full sets of Maxwell's equations but not including the true wire path is not sufficient for accurate modeling. In fact, in most cases using the straight wire path is nearly equivalent to using a DC approximation.

3.2. Three-Dimensional Forward Modeling Using Integral Equation Method

The forward modeling for electric field is based on the integral equation method. This has been covered extensively in the literature (e.g., [28,29]). For our purposes we start with the expression for the total electric field at point *j* at the surface of the domain of interest:

$$E(r\prime,\omega) = \iiint_D G_E(r\prime,r,\omega)\Delta\widetilde{\sigma}(r,\omega)E(r,\omega)dV + E^b(r\prime,\omega) \tag{3}$$

where r' is an arbitrary point along the receiver wire; E is the total electric field inside the domain of interest; E^b is the background electric field produced by a 1D earth; G_E is the domain-to-receiver electric Green's tensor; and $\Delta\widetilde{\sigma}$ is the anomalous conductivity, which is a departure from the 1D earth structure. The total electric field is a superposition of the anomalous field caused by the conductivity anomalies and the background field produced by the layered earth in the absence of any anomalous structure. Note the tilde over the anomalous conductivity indicating that this is a complex frequency dependent term to include the induced polarization effects, as described by Formula (1).

The electric field at the receiver's midpoint is commonly used as the predicted data for modeling and inversion (e.g., [32]). This is typically a good approximation for dipole data, which only may differ from the actual data for the very near offsets (when distance between the transmitter and receiver dipoles is less than two times their length). In a dipole–dipole survey, the voltage is measured as a difference between two adjacent electrodes. The voltage can be accurately approximated as the inline directed electric field at the midpoint multiplied by the distance between the two points.

However, with modern distributed array systems, the electric field at a receiver midpoint multiplied by the dipole length may be a poor approximation to the actual measured voltage. For example, the voltage may be measured between two distinct points on the different lines (e.g., M_2 and M_9 as shown in Figure 5). It will be much more accurate to integrate the electric field along the wire paths to obtain the voltage as given by Equation (3) than to use a midpoint electric field approximation. The voltage for the jth receiver combination is thus synthesized for a measurement from electrode M_k to electrode M_l as follows:

$$\Delta V_j = \Delta V_{M_k M_l}(\omega) = \int\limits_{M_k}^{M_l} \boldsymbol{E}(r', \omega) \cdot \boldsymbol{dl} = \begin{aligned} & \int\limits_{M_k}^{M_l}\left(\iiint\limits_{D_i} \boldsymbol{G}_E(r', r_i, \omega)\Delta\widetilde{\sigma}(r, \omega)\boldsymbol{E}(r_i, \omega)dV + \boldsymbol{E}^b(r', \omega)\right)\cdot \boldsymbol{dl} = \\ & \int\limits_{M_k}^{M_l}\left(\iiint\limits_{D_i} \boldsymbol{G}_E(r', r_i, \omega)\Delta\widetilde{\sigma}(r_i, \omega)\boldsymbol{E}(r_i, \omega)dV\right)\cdot \boldsymbol{dl} + \int\limits_{M_k}^{M_l} \boldsymbol{E}^b(r', \omega)\cdot \boldsymbol{dl} \end{aligned} \tag{4}$$

where $\boldsymbol{dl}$ is the wire path.

The first term in the last line of Equation (4) is the anomalous voltage:

$$\Delta V_j^a = \Delta V_{M_k M_l}^a(\omega) = \int\limits_{M_k}^{M_l}\left(\iiint\limits_{D_i} \overline{\overline{\boldsymbol{G}}}_E(r', r_i, \omega)\Delta\widetilde{\sigma}(r_i, \omega)\boldsymbol{E}(r_i, \omega)dV\right)\cdot \boldsymbol{dl} \tag{5}$$

and the second term is the background voltage:

$$\Delta V_j^b = \Delta V_{M_k M_l}^b(\omega) = \int\limits_{M_k}^{M_l} \boldsymbol{E}^b(r', \omega)\cdot \boldsymbol{dl} \tag{6}$$

The line integrals in Equations (5) and (6) are evaluated using Gaussian quadrature. This creates a computational problem, as the number of different combinations of receivers is rather large, as each receiver can be referenced to every other receiver to create a data point. For the array shown in Figure 5, this leads to $\frac{N(N-1)}{2} = 45$ unique voltage measurements where N is the number of receiver electrode locations. This is repeated for each transmitter position for this array. One can imagine that this number grows very rapidly for large, distributed arrays.

This computational difficulty can be overcome by breaking the line integrals in Equations (5) and (6) into individual wire segments. Each segment of wire is defined by a straight line that is broken by either receiver electrodes or bends in the wire:

$$\Delta V^b{}_{M_k M_l}(\omega) = \sum_{p=k}^{l-1} \int\limits_{M_p}^{M_{p+1}} \boldsymbol{E}^b(r', \omega)\cdot \boldsymbol{dl} \tag{7}$$

where p is the dummy index of the receiver wire segment end points (electrodes or bends). In Equation (7), each wire segment is evaluated with Gaussian quadrature for each transmitter position.

$$\Delta V^b{}_{M_p M_{p+1}}(\omega) = \sum_{q=1}^{N_G} w_q \boldsymbol{E}^b(r_q, \omega)\cdot \hat{\boldsymbol{l}} \tag{8}$$

where w_q are the weights for the Gaussian quadrature, r_q are the locations of the nodes, and N_G is the order of the quadrature.

The number of Gaussian quadrature points is adjusted from three for simple dipole–dipole type configurations to 50 when the receiver wire and transmitter wire are parallel and close (maximum

IC). For each unique combination of receiver positions, the voltage response is a summation of the appropriate wire segments. Every possible receiver pair can then be computed from a linear combination of the appropriate wire segments:

$$\Delta V^b{}_j(\omega) = \Delta V^b{}_{M_k M_l}(\omega) = \sum_{p=k}^{l-1} \Delta V^b{}_{M_p M_{p+1}}(\omega) \tag{9}$$

In this way, excellent accuracy is maintained, yet the maximum number of segments is the number of receiver positions plus the number of bends in the wire. This is much less than the number of unique receiver pairs. This greatly reduces the computational load for the background field computation.

3.3. Numerical Modeling of Inductive Coupling in the Frequency and Time Domain

We begin with the simple case of a polarizable half-space for simplicity in developing some intuition about the IP and IC problems across a variety of source-receiver geometries and time and frequencies. The transmitter wire is modeled as a series of electric dipoles on the surface of the homogeneous earth and integrated with Gaussian quadrature. The receiver wires are modeled by integrating the electric field along the wire path from receiver electrode M to electrode N to synthesize the voltage difference which would be measured in the field:

$$\Delta V_{MN} = \int_M^N E \cdot dl \tag{10}$$

where electric field, E, is computed as discussed in Section 3.2 above.

We consider first a simple case of a half-space characterized by the Cole–Cole model parameters listed in Table 1.

Table 1. Base Cole–Cole model parameters for half-space IC testing.

Parameter	Value
DC Resistivity (ρDC)	100 [Ohm·m]
Chargeability (η)	100 [mV/V]
Time Constant (τ)	0.1 [s]
Relaxation Factor (C)	0.5 [ ]

The current is 1 A. The time domain response is computed by using a cosine transform of the imaginary part of the frequency domain spectrum, which gives the step-off response. Note that, an infinite step-off is used for these calculations for simplicity; but using a 50% duty cycle repeating waveform of limited bandwidth allows us to draw the same conclusions, and the true waveform is used for the field data inversion. Also note that, the time range and frequency range shown here is greater than available commercial systems, but it is instructive to look at the expanded spectrum.

To isolate the effects of the inductive coupling and wire path, we have computed four different responses shown. One is the true response which includes all effects, as would be measured in the field. This includes the electromagnetic inductive response (IC), the true wire path, and the complex conductivity of the earth. The other three responses are a straight wire path from the M to N receiver electrode positions (ignoring the true wire path), setting $\frac{\partial B}{\partial t} = \nabla \times E = 0$ (ignoring inductive coupling), and ignoring induced polarization (chargeability is 0; earth has a constant conductivity over frequency). Note that, in the case of the DC approximation, the wire path does not matter, as the fields are conservative, and the line integral in Equation (10) is path independent.

We show an example of the severe inductive coupling effects in Figure 6. This shows the frequency domain apparent resistivity and phase as functions of frequency, and time domain voltage and apparent chargeability. Time domain apparent chargeability is the voltage at some time after turnoff normalized by the on-time voltage. There are several features to point out in this figure. Panel (a) shows the geometry of the setup. Panel (b) shows that the DC approximation is excellent to around 1 Hz, at which IC begins to dominate. Panel (c) shows the frequency domain quadrature response. As one should expect, the phase reflects both the polarization and IC much stronger than the in-phase response. With this configuration, the IC dominates the phase for nearly the entire frequency range. Only at the very lowest frequencies ($<3 \times 10^{-4}$ Hz) does the IP affect the response. Above this frequency, IC effects completely dominate the spectrum. This frequency is well below the range of commercial systems. The same IC domination of the response is also seen in the space of frequency domain apparent resistivity (panel (d)) and phase (panel (e)). In the time domain (panels (f) and (g)) the IC part completely dominates from the earliest times to 10 ms. By 20 ms, the response is free of inductive coupling and only the IP decay remains.

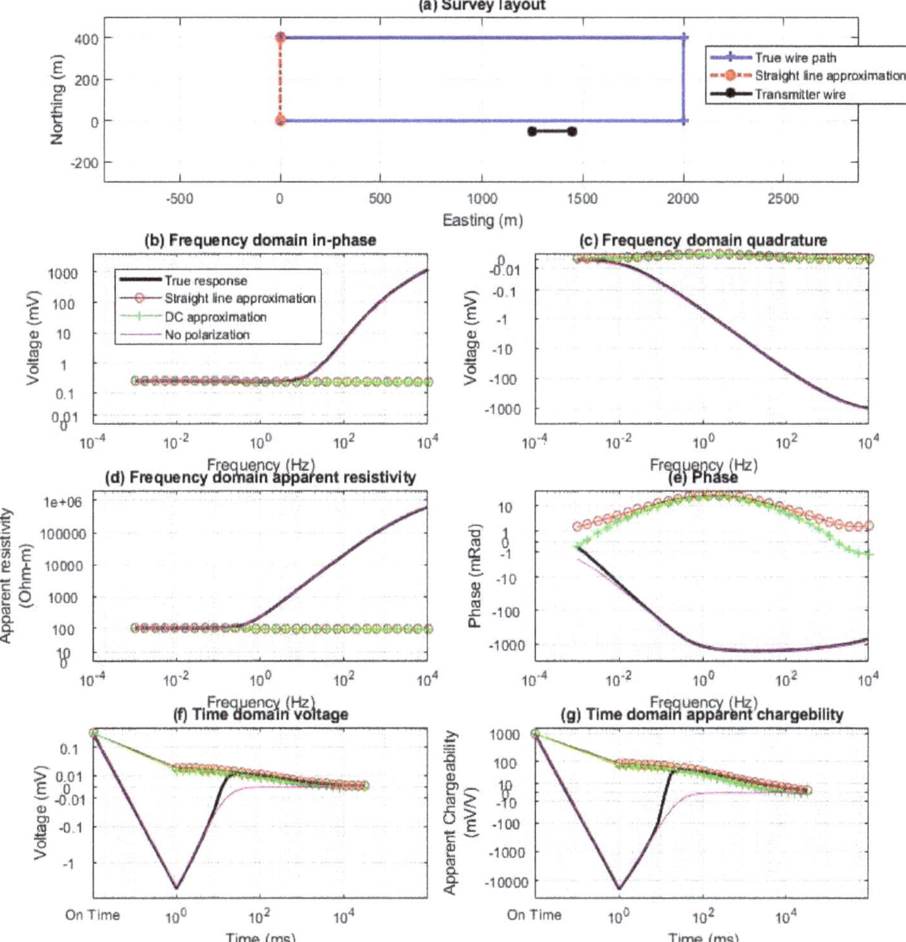

Figure 6. Frequency and time domain response with a non-conventional survey layout. Panel (**a**) shows the source–receiver layout. Note that, only active wires paths are shown and modeled. Panel (**b**) presents the frequency domain apparent resistivity. Panel (**c**) shows the frequency domain phase response. Panel (**d**) shows the frequency domain apparent resistivity and panel (**e**) shows the frequency domain phase. Panel (**f**) displays the time domain voltage decay. Panel (**g**) shows the apparent chargeability. The first tic on the time domain curve is the on-time channel.

Modeling shows that, as expected, longer offsets, receiver wire paths which traverse close to the transmitter dipoles, and more conductive media all increase the effects of the IC, both by increasing the amplitude and pushing the domination to later times or lower frequencies.

Figure 7 shows a similar configuration to Figure 6, but with the transmitter wire not running directly along the receiver wire. The responses share many characteristics of the response shown in Figure 6. In the frequency domain case (panels b–e), we can see that the induced polarization is important below 1 Hz, while above 3 Hz the IC dominates with little or no sensitivity to the IP. There is a small transition range where both IP and IC share contributions to the response spectrum. It is very interesting that, in the time domain response, the separation between the IP and no IP cases is complete and spans the entire time range of the field data collected (40 ms to 1.92 s). Beyond 20 ms, the true response, straight line approximation, and DC approximation all overlie each other. The only curve deviating from this behavior is the response which does not include IP effect. This means inductive coupling does not contaminate the IP data (for this configuration) after 20 ms. There is a small IC effect in the early times of the response, but it is relatively minor and disappears after a few milliseconds.

3.4. Discrimination of the IP Time Constant and Relaxation Coefficient When Considering the IC Effects

The goal of collecting the spectral IP data is in measuring the details about the polarization spectrum of the medium. Real rocks exhibit time constants that range for many orders of magnitude [17,18], while many time-domain surveys measure less than two decades of decay (e.g., 50 ms to 2 s). Additionally, inductive coupling can dominate the early times and higher frequencies, so this leads to the question if a reasonable range of time constants of the media can be determined from the real data.

Figure 8 shows the response of the half space with the parameters given in Table 1 with the same survey configuration as shown in Figure 6. Only the true wire path with full EM effects have been computed. Each curve shows a different time constant of the medium. These vary from 1 ms to 100 s, quite a wide and realistic range. Also shown is a case with no chargeability. Note that, for the frequency domain, there is no separation of the curves at frequencies above 20 mHz, which indicates that the time constant for this configuration would not be resolved in the frequency domain using conventional frequency ranges. In the time domain, there is separation of the curves for time constants from around 1 ms to 10 s. This occurs over the time window from 40 ms to 5 s. Outside of this range the time constant cannot be resolved. Longer time constants would require lower base frequencies and resolving shorter time constants would require accounting for the inductive coupling.

Figure 9 shows the response of the half space with the parameters given in Table 1 but with a 10 Ohm·m half-space. Again, the same survey configuration as shown in Figure 6 is used. For the frequency domain, there is no separation of the curves at frequencies above 2 mHz, which is a decade lower than when the earth has a resistivity of 100 Ohm·m. In the time domain, there is separation of the curves for time constants from around 10 ms to 10 s. This occurs over the time window from 300 ms to 5 s. This is much later than when the earth has a resistivity of 100 Ohm·m, and also is nearly a decade later in than the first time sample. In this case, inductive coupling would need to be accounted for to resolve short time constants and to make use of the earlier time channels.

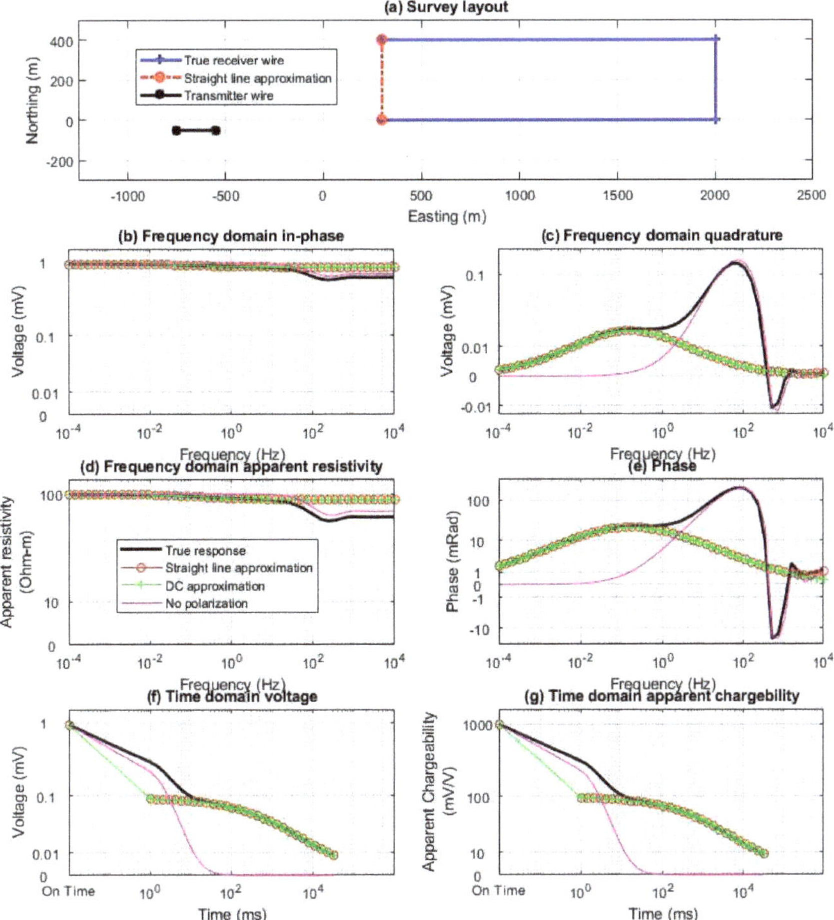

Figure 7. Frequency and time domain response with a non-conventional survey layout for the case of transmitter locating at some distance from the wire loop. Panel (**a**) shows the source-receiver layout. Panel (**b**) presents the frequency domain apparent resistivity (in-phase data). Panel (**c**) shows the frequency domain phase response (quadrature data). Panel (**d**) show the frequency domain apparent resistivity and panel (**e**) show the frequency domain phase. Panel (**f**) displays the time domain voltage decay. Panel (**g**) shows the time-domain apparent chargeability. The first tic on the time domain curve is the on-time channel.

Figure 10 examines the resolution of the relaxation parameter. The colored lines show the responses of half space with the properties shown in Table 1, but with C varying as shown in the legend. In the frequency domain, the effect of varying C is largely pushed outside of typical survey bandwidth, and the base frequencies would need to approach 1000 s to detect the variations. This is outside the patience of most field crews! The effect of increasing C from 0.1 to 0.5 in this figure is large enough to increase the magnitude of the response. Thus, over the limited time range available with most instruments, variations of C below 0.5 will be nearly indistinguishable from variations in chargeability. For large values of C (0.75 to 1.0), the inflection in the response becomes apparent (the dotted pink line), and this could be resolvable.

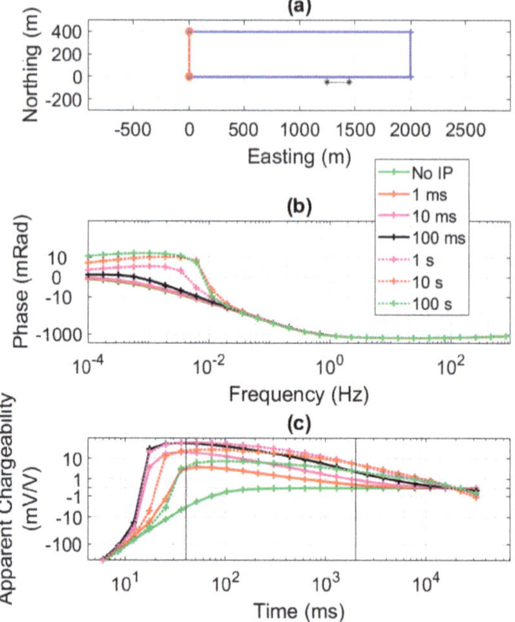

Figure 8. Time constant distinguishability for 3D configuration with the transmitter wire near the receiver wire and a 100 Ohm·m half space. Panel (**a**) shows the source-receiver layout. Panel (**b**) shows the frequency domain phase. Panel (**c**) shows the time-domain apparent chargeability. The color of the plots corresponds to a different time constant of the medium—from 1 ms (bold green line) to 100 s (dotted green line). The two vertical lines in panel (**c**) show the time range of the current system.

Thus, our study of the inductive coupling (IC) effect on the IP data shows that, the frequency-domain response has a much larger contribution from IC than the time-domain response. The time domain shows better sensitivity to IP and to the time constants. The wire path is less important in time domain modeling but can still have a large effect when transmitter receiver separations are large, and the medium is conductive (<100 Ohm·m).

In both frequency domain and time domain, there is about one decade of time or frequency response that contains information from both IC and IP. This means that by including wires paths and the full EM solution in forward modeling and inversion, we can extract IP information from the response one decade earlier than if the DC approximation is used. This is especially important when short time constants are expected.

Additionally, all the IP parameters are coupled, and this must be kept in mind when interpreting inversion results. The chargeability must be relatively large for the time constant and relaxation parameters to be resolved. The time constant is poorly resolved if the relaxation parameter is small (0.5 or less), and the relaxation parameter and chargeability can also be traded for an equivalent response when the relaxation parameter is small.

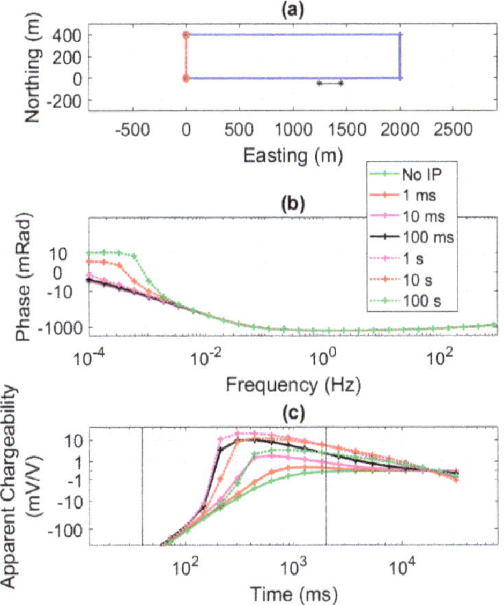

Figure 9. Time constant distinguishability for 3D configuration with the transmitter wire near the receiver wire and a 10 Ohm·m half space. Panel (**a**) shows the source-receiver layout. Panel (**b**) shows the frequency domain phase. Panel (**c**) shows the time-domain apparent chargeability. The color of the plots corresponds to a different time constant of the medium—from 1 ms (bold green line) to 100 s (dotted green line). The two vertical lines in panel (**c**) show the time range of the current system.

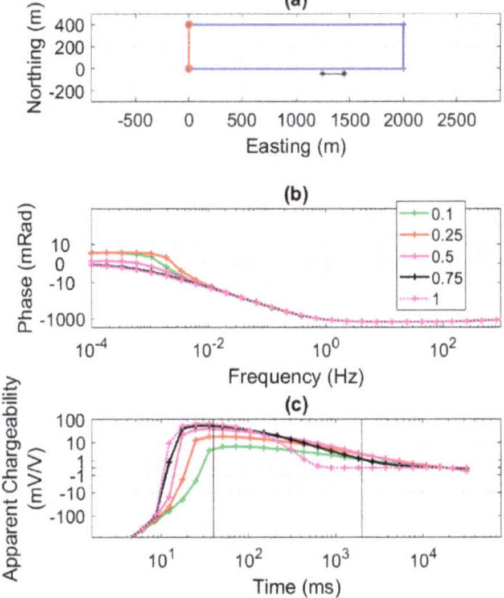

Figure 10. Relaxation constant distinguishability for 3D configuration with the transmitter wire near the receiver wire and a 100 Ohm·m half space. Panel (**a**) shows the source-receiver layout. Panel (**b**) shows the frequency domain phase. Panel (**c**) shows the time-domain apparent chargeability. The color of the plots corresponds to a relaxation constant of the medium-from 0.1 (bold green line) to 1 (dotted pink line). The two vertical lines in panel (**c**) show the time range of the current system.

4. Inversion of the IP Data

4.1. Modeling of the Inductive Coupling

The modeling method to obtain the domain fields uses the integral equation method [30], and our formulation includes the inductive coupling naturally. The modeling is based on the quasi-stationary (not static or DC) Maxwell's equations. The IP effect is completely represented by modeling the conductivity as a frequency dependent and complex term. The modeling is done in the frequency domain and the results are transformed to the time domain with a cosine transform. The sources are modeled as an integration of current dipoles along the length of the transmitter wires. The receivers are modeled by integrating the electric fields along the path of the receiver wire. This modeling therefore includes inductive and capacitive coupling from the transmitting wires to the ground, the ground to the active receiving wires, and the transmitter wires directly to the receiving wires. Because of the very high impedance, we assume there is no feedback from the active receiving wires to the earth and back to the receiving wires, or from the passive wires to the active receiving wires. All filters in the instruments are included by applying them to the frequency spectrum before transforming the result to the time domain.

4.2. Inversion

As we discussed above, in the case of studying the IP effect, the conductivity has to be treated as a complex and frequency dependent function, $\sigma = \sigma(\omega)$, which increases significantly the number of unknown parameters of the inversion [34,35]. We can reduce this number by approximating the conductivity distribution using a GEMTIP model (1), or its simplified version (2) described in Section 2 above. For this field data set, we invert for the GEMTIP parameters (σ_0, f, τ, C). These are all updated and included in the forward modeling during the inversion. At this point we are solving for the simplified GEMTIP model parameters, which is very similar to the Cole–Cole model. However, in future we plan on including elliptical parameters into the inversion, which provide a better representation of the complex resistivity of the rocks [16].

Our problem can now be written in the classic form of the operator equation:

$$\mathbf{d} = \mathbf{A}(\sigma f(\eta, \tau, C) - \sigma_b) = \mathbf{A}_G(\mathbf{m}) \tag{11}$$

where $\mathbf{A}_G$ is GEMTIP forward modeling operator, and $\mathbf{m}$ is a vector of the GEMTIP model parameters $[\sigma, \eta, \tau, C]$.

The inversion is based on minimization of the Tikhonov parametric functional, $P^\alpha(\mathbf{m})$, with the corresponding stabilizer $S(\mathbf{m})$ [32]:

$$P^\alpha(\mathbf{m}) = \|\mathbf{W}_d(\mathbf{A}_G(\mathbf{m}) - \mathbf{d})\|_{L_2}^2 + \alpha S(\mathbf{m}), \tag{12}$$

where $\mathbf{W}_d$ is the data weighting matrix, and α is a regularization parameter.

There are several choices for the stabilizer [28,29]. In this paper, for simplicity, we use the minimum norm stabilizer (S_{MN}), which is equal to the square L_2 norm of the difference between the current model $\mathbf{m}$ and an appropriate a priori model $\mathbf{m}_{apr}$:

$$S_{MN}(\mathbf{m}) = \|\mathbf{W}_m(\mathbf{m} - \mathbf{m}_{apr})\|_{L_2}^2 \tag{13}$$

where $\mathbf{W}_m$ is the weighting matrix of the model parameters.

We used the re-weighted regularized conjugate gradient (RRCG) method to find the minimum of the parametric functional P^α. The details of the inversion algorithm are described by [32,34]. This method is based on integral equation (IE) forward modeling method outlined above in Section 3.2, which requires the discretization of the volume containing the anomalous complex conductivity only. The background medium, surrounding the volume with the anomalous conductivity, is characterized by some known horizontally layered conductivity, σ_b. Another advantage of the IE method is that the sensitivities of the observed IP data to the changes of the anomalous conductivity can be expressed in close form using the electric and magnetic Green's tensors defined for the background conductivity model σ_b.

Indeed, following [34] the sensitivity of the electric field at point j due to perturbations in the conductivity of the ith cell is given by the following formula:

$$\frac{\partial E(r_j, \omega)}{\partial \sigma_i} = \iiint_{D_i} G_E(r', r_i, \omega) E(r_i, \omega) dV. \tag{14}$$

The right hand-side terms contain the domain-to-receiver Green's tensors (G_E, 2nd rank) and the total electric field $E(r)$. This equation is the extended Born sensitivity. Do not confuse this with the Born sensitivity or Born approximation, which would use a 1D background field approximation instead of the total, rigorously calculated electric field used here.

The sensitivities for the voltage along this wire path are described, according to (6), by the following expression:

$$\frac{\partial \Delta V_{M_k M_l}(\omega)}{\partial \sigma_i} = \frac{\partial}{\partial \sigma_i} \int_{M_k}^{M_l} E(r', \omega) \cdot dl = \int_{M_k}^{M_l} \left(\iiint_{D_i} \overline{\overline{G}}_E(r', r_i, \omega) E(r, \omega) dV \right) \cdot dl. \tag{15}$$

The line integral in Equation (14) is evaluated using Gaussian quadrature.

$$\frac{\partial \Delta V_{M_k M_l}(\omega)}{\partial \sigma_i} = \sum_{p=k}^{l-1} \int_{M_p}^{M_{p+1}} \left(\iiint_{D_i} \overline{\overline{G}}_E(r', r_i, \omega) E(r_i, \omega) dV \right) \cdot dl \tag{16}$$

where p is the dummy index of the receiver wire segment end points (electrodes or bends).

For the sensitivities, dl remains constant over each integration because the receiver wires are assumed to be straight (or broken into straight segments), so the order of the tensor-vector products can be changed. This gives the following expression:

$$\frac{\partial \Delta V_{M_k M_l}(\omega)}{\partial \sigma_i} = \sum_{p=k}^{l-1} \int_{M_p}^{M_{p+1}} \iiint_{D_i} \overline{\overline{G}}_E(r', r_i, \omega) dl \cdot E(r_i, \omega) dV. \tag{17}$$

Because the domain electric field is assumed to be constant over a cell, it can be removed from the domain integral. The domain electric field is also constant with respect to the line integral along the wire path and can be removed from integral as well:

$$\frac{\partial \Delta V_{M_k M_l}(\omega)}{\partial \sigma_i} = \sum_{p=k}^{l-1} \left(\int_{M_p}^{M_{p+1}} \iiint_{D_i} \overline{\overline{G}}_E(r', r_i, \omega) dV dl \right) \cdot E(r_i, \omega). \tag{18}$$

The term in the parenthesis is the Green's linear operator for the ith cell to the pth wire segment:

$$G_i^p(\omega) = \int_{M_p}^{M_{p+1}} \iiint_{D_i} \overline{\overline{G}}_E(r'', r_i, \omega) dV dl. \tag{19}$$

Note that, this is independent of both the total electric field in the domain, $E(r, \omega)$, and the source position. This means that this term only needs to be precomputed before the inversion and never updated, and that it only needs to be computed for each wire segment and frequency. The Green's linear operator for each unique set of receiver positions can be computed by linear combinations of the wire segments, as is done for the background fields. The line integral in Equation (19) is computed through a Gaussian quadrature using 15 points per cell the wire passes over. Note also that this term is a first rank tensor (vector) because of the evaluation of the line integral in Equation (19).

The sensitivities are then composed as follows:

$$J_{ij}(\omega) = \frac{\partial \Delta V_{M_k M_l}(\omega)}{\partial \sigma_i} = \sum_{p=k}^{l-1} G_i^p(\omega) \cdot E(r_i, \omega) \tag{20}$$

and the anomalous voltage response can be found by combining Equation (19) with Equation (5):

$$\Delta V_j^a(\omega) = \sum_{i=1}^{N_c} J_{ij}(\omega) \Delta \widetilde{\sigma}_i \tag{21}$$

Equations (9), (20), and (21) provide the required quantities for modeling and inversion of the 3D distributed SIP survey data using voltages and exact wire paths in a computationally efficient manner.

The sensitivities for the GEMTIP parameters that we invert for (e.g., σ_0, f, τ, C), are found by the chain rule with Equations (1) or (2), as shown by [32,34].

The examples of synthetic IP model study and a case study of application of this method for the Copper Deposit in Mongolia were discussed in [32], where the reader can find additional details about the developed method of the IP data inversion.

5. Results of Interpretation of SIP Survey Data

We have applied the developed inversion code to these field SIP data. The inversion was run according to the description above with horizontal cell sizes of 50 m × 50 m. There were eight cells in the z direction varying from 26 to 223 m in thickness giving a total depth range of 0–600 m. There were nine transmitters and 23 receiver electrodes for total of 482 combinations and these each had 20 time channels from 40 ms to 1.92 s plus the on-time channel. The center line had the nine transmitters (see Figure 4) which were 200 m dipoles. The receivers were combinations of electrodes from all three receiver lines. The receiver lines were 350 m apart and approximately 3 km long.

The error model used for the observed data is given by the following equation:

$$\varepsilon_i = |d_i| * 10\% + 4E - 6 \; V/m \tag{22}$$

where *i* indicates the *i*th data point or the error estimate of that point and $|d_i|$ is the absolute value of the *i*th data point. The data weights are the inverse values of these errors. The normalized χ^2 at each iteration during the inversion is shown in Figure 11. The inversion converged from a normalized χ^2 of around 3 to 1.1 at the 60th iteration, which was chosen to be optimal based on the noise level in the observed data, the flattening of the convergence curve, and examination of the resulting models. The optimal iteration is guided by theory but chosen by experience.

The field data collection was described in the previous sections. There is no standard plotting conversion for this type of distributed array data. We are plotting the physical fields to develop some intuition about the processes and to visually compare the observed and predicted data to analyze the data fit. Each image shows the voltage corresponding to one transmitter position. The plotting point is the average position of the two potential measurement points, as was shown in the synthetic example.

Figure 12 shows examples of the data fit for different transmitters and time channels. The observed and predicted data visually match very well. This is also demonstrated by the final χ^2 of 1.1. A perfect fit would theoretically have a value of 1.0. Figure 13 presents examples of observed (solid lines) and predicted (dotted lines) voltage decays from the on-time measurement to the latest time at 1.8 s.

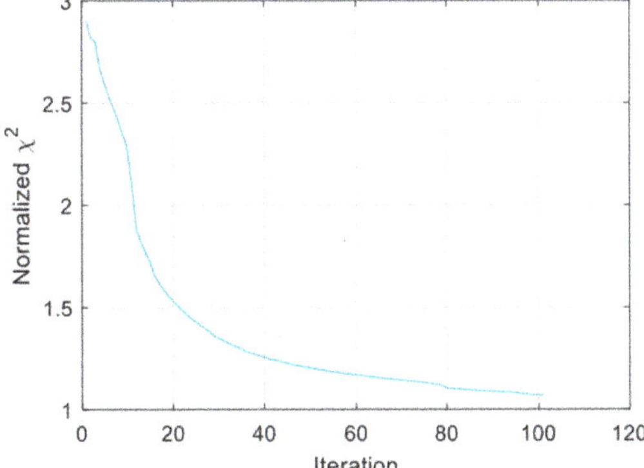

Figure 11. Normalized χ^2 convergence curve of the final inversion. Iteration 60 was chosen as the optimal inversion iteration based on the noise level in the observed data.

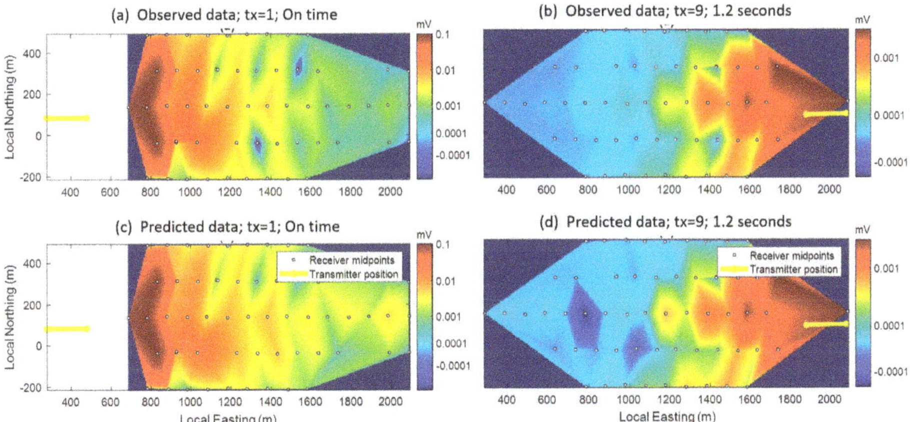

Figure 12. Maps of predicted (top panels) and observed (bottom panels) data. The columns of images show the results for different transmitter positions. The left column is transmitter one (indicated by the yellow line) during the on-time and the right column shows data for transmitter #9 at 1.2 s.

The inversion results under the transmitter line are shown in Figure 14. The inversion results of 3D distributed array data are shown in the top panel, the 3D inversion results from the 2D lines are given in the middle panel. The bottom panel shows the results of 3D from the Spectrum time domain airborne electromagnetic (TD AEM) data (inverted using the method of [36]). There were two boreholes drilled in the survey area. The position of one of these drill holes is shown in Figure 14 overlaid on the resistivity images. We do not have the coordinates of the other borehole.

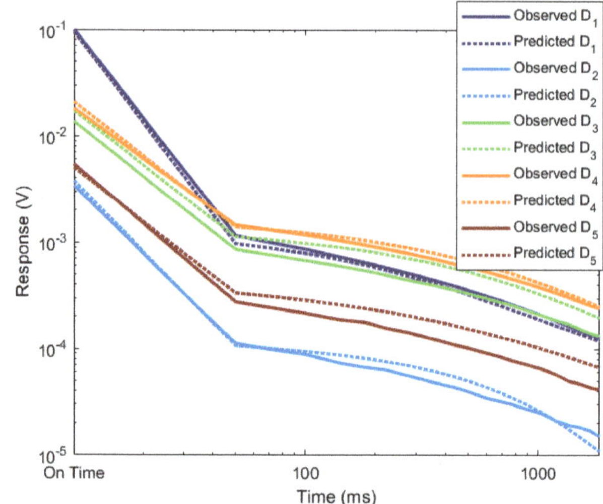

Figure 13. Examples of voltage decay from the on-time measurement to the latest time at 1.8 s. The solid line is the observed data, and the dotted line is the predicted data. D1 corresponds to the data from transmitter #1 and the receiver wire path from receiver #4 to #8; D2 is from transmitter #1 and the receiver path from receiver #5 to #7; D3 is from transmitter #9 and the receiver path from receiver #15 to #17; D4 is from transmitter #9 and the receiver path from receiver #15 to #10; and D5 is from transmitter #6 and the receiver path from receiver #5 to #7.

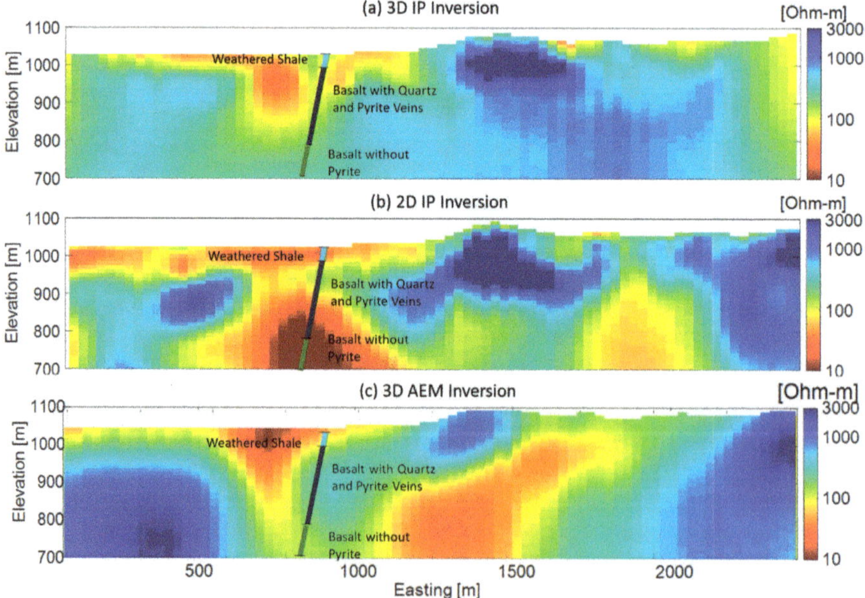

Figure 14. Inversion results under the center line of the 3D array, along the profile shown in Figure 4. The top panel (**a**) shows the resistivity recovered from the inversion of the distributed 3D array, the middle panel (**b**) shows the resistivity derived from the standard 2D dipole–dipole lines, and the bottom panel (**c**) shows the results from the Spectrum time domain airborne electromagnetic (TD AEM) data. The position of the borehole and preliminary qualitative estimates of the types of the rocks from the drilling are overlaid on the images.

All three inversion results show generally the same features. First, there is pervasive conductive overburden in this area. This is to be expected in this highly weathered environment. Most areas below this conductive overburden are resistive, but there are notable exceptions. The most interesting feature is a strong conductor near 700 mE. The depth of this conductor is different in the images produced by different datasets; however, the overall locations and sizes of the conductor are similar, which is remarkable considering that we compare the ground and airborne data. The top and bottom panels are the most similar in the top 150 m. A conductor is shown at depth in the middle panel, but note that the intersection of the borehole shows basalt with no pyrite at this location, which means this should probably be resistive. This is likely an off-line (3D) effect manifesting in the 2D results. All three methods image the conductive overburden. All three also show a strong resistor under the hill at 1400 mE. We expect the AEM to have a higher depth of investigation, and we have truncated the AEM results for this figure to 300 m for comparison. This greater depth of exploration explains the conductor at depth between 1200 and 1500 m in elevation which is only weakly imaged in both resistivity inversion results.

Figure 15 shows the chargeability results of the SIP inversions from the distributed array (top panel), and from the 2D dipole–dipole line data (bottom panel). The results have similarities, especially in the near surface from 1400 to 2000 mE, but they show differences in the strength of the chargeability and the 2D line shows the chargeable body much more broken. In the drilling location with pyrite confirmed by visual inspection of the core, the 2D dipole–dipole survey shows a small increase in chargeability while the distributed array clearly shows a significant increase of chargeability which also corresponds better to the drilling result. The large chargeability anomaly at 1500 mE has not yet been drilled due to logistical challenges for the location.

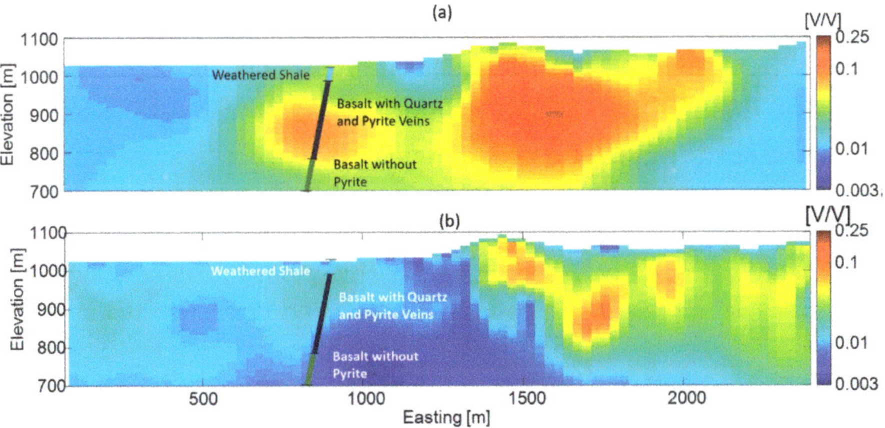

Figure 15. Inversion results of the chargeability under the center line of the 3D array, along the profile shown in Figure 4. The top panel (**a**) shows the results from the distributed array, and the bottom panel (**b**) shows the results from the 2D dipole–dipole lines. The position of the borehole and preliminary qualitative estimates of the types of the rocks from the drilling are overlaid on the images.

The fact that both the resistivity and the chargeability from the 3D IP inversion matches the drilling better than the 2D IP inversion demonstrates an importance of using the data collected by a distributed array for interpretation of the IP survey data and inverted in full 3D.

Regarding the time constant and relaxation parameter, these images contain very interesting features. The 3D inversion of the 2D line data (bottom panels) shows quite similar features and characteristics across the three parameters of chargeability, time constant, and relaxation parameter. This can be explained by the results in the synthetic section—C and chargeability are largely coupled and required low noise and accurate modeling to separate. Time constant is also difficult to separate from chargeability, especially when the relaxation parameter is 0.5 or less.

However, the 3D inversion of the distributed array data tells a different story. The four recovered parameters of the inversion all paint different pictures of the subsurface. We believe this is because they are all imaging different properties and provide independent information. The resolutions of the IP parameters are all linked and must be interpreted with this in mind. For example, neither the time constant nor the relaxation parameter has resolution when the chargeability is week. The time constant is also poorly resolved if the relaxation parameter is small (0.5 or less), and the relaxation parameter and chargeability are also coupled when the relaxation parameter is small.

Figures 14–17 shows the general schematic of the geology encountered while drilling (borehole is shown in these figures overlying the geophysical results). The uppermost lithology is a weathered shale, which is moderately conductive and not chargeable. The basalt with quartz has lower resistivity (due to the pyrite) and is chargeable (due to the pyrite). Below this is basalt with little presence of pyrite. This account for its lack of chargeability. The basalt is moderately weathered which could explain an elevated conductivity as shown by the 2D lines, but more likely there is alteration or sulfides which are being imaged and were not intersected (off-line effects).

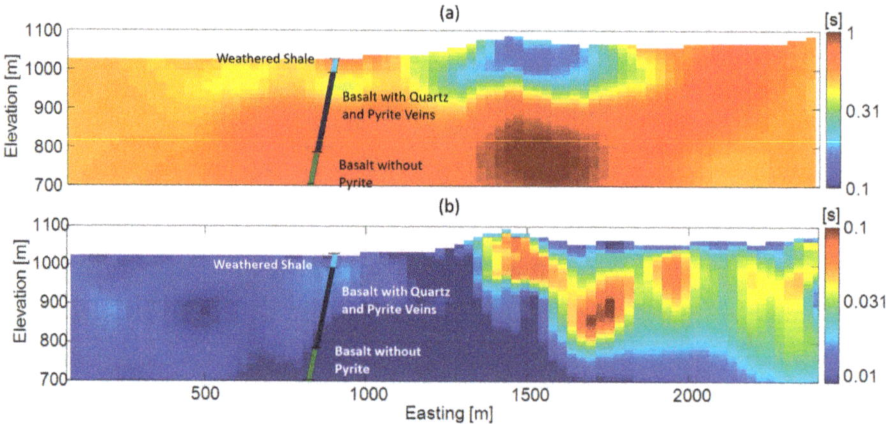

Figure 16. Inversion results of the times constant under the center line of the 3D array, along the profile shown in Figure 4. The top panel (**a**) shows the results from the distributed array, and the bottom panel (**b**) shows the results from the 2D dipole–dipole lines.

Figure 18 shows a picture of the quartz veining and associated sulfides minerals (pyrrhotite, pyrite, and chalcopyrite). The pyrite and chalcopyrite mineralization is responsible for the observed IP effect. The preliminary analysis of drilling results also shows quartz veins filled with the minerals, like pyrrhotite, pyrite, and chalcopyrite, with the presence of gold and silver. This points to a possible epithermal gold and silver vein type deposit. This is further encouraged by the proximity to the Mahd Ahd Dahab and Jabal Sayid deposits, aka the "Cradle of the Gold". In the Mahd Ahd Dahab deposit, the polymetallic mineralization of chalcopyrite, sphalerite, galena, and gold and silver tellurides, occurs in the quartz veins as well. More drilling and rock sample analysis, however, will be required for this prospect area to make an estimate of the economical potential of the discovery.

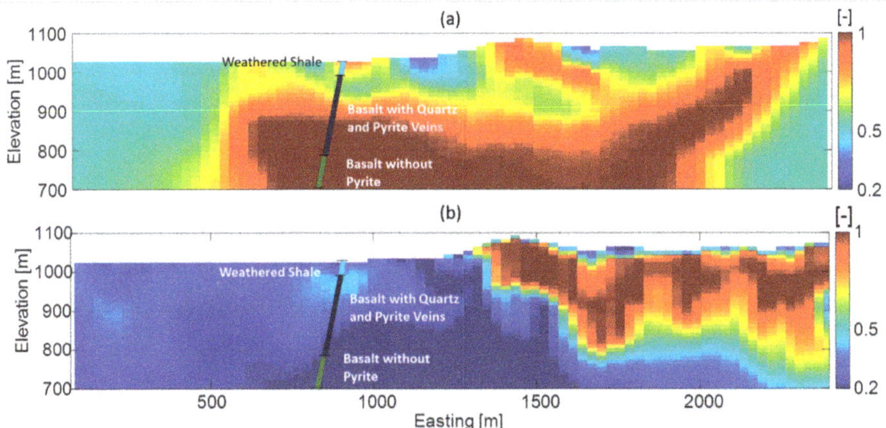

Figure 17. Inversion results of the relaxation coefficient under the center line of the 3D array, along the profile shown in Figure 4. The top panel (**a**) shows the results from the distributed array, and the bottom panel (**b**) shows the results from the 2D dipole–dipole lines.

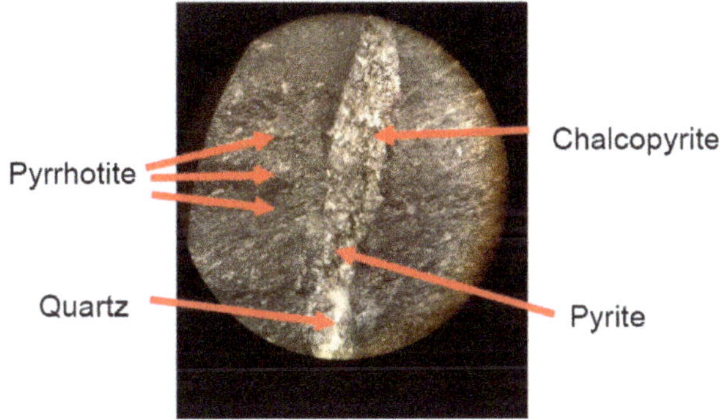

Figure 18. Example of the core sample from a depth of 158 m.

6. Conclusions

In this paper, we have presented the results of the SIP survey conducted over one of the prospective targets located by the Glass Earth Pilot Project in Saudi Arabia. We have developed a novel 3D SIP inversion method and code which consider all inductive coupling, the wires paths, and includes IP effects through the use of the GEMTIP model. We have also demonstrated an importance of taking into account the inductive coupling (IC) effect in interpretation of the data acquired by the SIP survey with distributed array system with the synthetic modeling study.

Since the resistivities in this survey are generally greater than 100 Ohm·m and with the time constants on the order of 30 to 100 ms, the IP and IC responses did not overlap with the time windows and geometry of this survey. This means that in this specific case history, inductive coupling could have been small. This is not the case in general and this must be determined based on each specific survey's geometry and expected physical properties. The important result of our study is that, it is impossible to know if the IC coupling affects the data until the survey has been completed and the data analyzed. Thus, it's best to take the IC effect into account in analyzing the data. The inversion algorithm presented in this paper makes it possible to take into account the IC effect automatically, if it exists in the observed data.

The distributed systems have become widely used in industry, which makes this result of special practical significance. The 3D GEMTIP inversion with automatic accounting for the inductive coupling recovered a resistivity model which closely matched independently recovered resistivity models obtained previously by 3D inversion of the airborne EM data and matched more closely to drilling results than a 3D inversion algorithm of the 2D line data which ignoring wire paths and IC.

The Glass Earth Pilot Project has followed an exploration workflow from country wide selection through ground SIP survey, borehole drilling and discovery of blind mineralization by using the advanced 3D inversion methods detailed in this paper.

While the discovery is fortuitous, it would not have happened without the techniques of modern geophysics. There is no surface expression of this mineralization. More geologic analysis will need to be done in this area before we have a solid understanding of how exactly the time constant and relaxation parameter are related to geology, but the initial images are promising. The 300 m drill hole were drilled in an original area of nearly 8000 km^2. The drilling results identified quartz veins filled with the minerals, like pyrrhotite, pyrite, and chalcopyrite, with the presence of gold and silver type of mineralization in our survey area which is similar to an epithermal, low-sulfidation, polymetallic type deposit of Mahd Ahd Dahab Mine.

Author Contributions: Conceptualization, M.S.Z. and F.A.A.; support of acquisition, M.S.Z. and F.A.A.; software, M.S.Z. and L.H.C., methodology, M.S.Z. and L.H.C.; validation, F.A.A. and L.H.C., investigation, M.S.Z., F.A.A., L.H.C., and A.M.A.; writing, M.S.Z. and L.H.C.; visualization, L.H.C.; supervision, M.S.Z. All authors have read and agreed to the published version of the manuscript.

Funding: This research was supported by King Abdulaziz City for Science & Technology (KACST) and TechnoImaging and received no external funding.

Acknowledgments: We acknowledge the support of King Abdulaziz City for Science & Technology (KACST), TechnoImaging, and the Consortium for Electromagnetic Modeling and Inversion (CEMI) at the University of Utah. We thank Dias Geophysical for collecting good quality geophysical data.

Conflicts of Interest: The authors declare no conflict of interest.

References

1. Zhdanov, M.; Alfouzan, F.; Cox, L.; Alotaibi, A.; Alyousif, M.; Sunwall, D.; Endo, M. Large-Scale 3D Modeling and Inversion of Multiphysics Airborne Geophysical Data: A Case Study from the Arabian Shield, Saudi Arabia. *Minerals* **2018**, *8*, 271. [CrossRef]

2. Sheard, N.; Ritchie, T.; Rowston, P. MIMDAS-A quantum change in surface electrical geophysics. In Proceedings of the PDAC Conference, Toronto, ON, Canada, 10–13 March 2002.

3. Eaton, P.; Anderson, B.; Queen, S.; MacKenzie, I.; Wynn, D. NEWDAS—The Newmont Distributed IP Data Acquisition System. In *SEG Technical Program Expanded Abstracts 2010*; Society of Exploration Geophysicists: Tulsa, OK, USA, 2010; pp. 1768–1772.

4. Zonge, K.L.; Wynn, J.C. Recent advances and applications in complex resistivity measurements. *Geophysics* **1975**, *40*, 851–864. [CrossRef]

5. Zhdanov, M. Generalized effective-medium theory of induced polarization. *Geophysics* **2008**, *73*, F197–F211. [CrossRef]

6. Nordsiek, S.; Weller, A. A new approach to fitting induced-polarization spectra. *Geophysics* **2008**, *73*, F235–F245. [CrossRef]

7. Cole, K.S.; Cole, R.H. Dispersion and Absorption in Dielectrics I. Alternating Current Characteristics. *J. Chem. Phys.* **1941**, *9*, 341–351. [CrossRef]

8. Pelton, W.H. Interpretation of Induced Polarization and Resistivity Data. Ph.D. Thesis, University of Utah, Salt Lake City, UT, USA, 1977.

9. Pelton, W.H.; Rijo, L.; Swift, C.M. Inversion of two-dimensional resistivity and induced-polarization data. *Geophysics* **1978**, *43*, 788–803. [CrossRef]

10. Pelton, W.H.; Ward, S.H.; Hallof, P.G.; Sill, W.R.; Nelson, P.H. Mineral discrimination and removal of inductive coupling with multifrequency IP. *Geophysics* **1978**, *43*, 588–609. [CrossRef]

11. Seigel, H.O.; Vanhala, H.; Sheard, S.N. Some case histories of source discrimination using time-domain spectral IP. *Geophysics* **1997**, *62*, 1394–1408. [CrossRef]
12. Yuval, W.; Oldenburg, D. Computation of Cole-Cole parameters from IP data. *Geophysics* **1997**, *62*, 436–448. [CrossRef]
13. Routh, P.S.; Oldenburg, D.W. Electromagnetic coupling in frequency-domain induced polarization data: A method for removal. *Geophys. J. Int.* **2001**, *145*, 59–76. [CrossRef]
14. Rowston, P.; Busuttil, S.; McNeill, G. Cole–Cole Inversion of Telluric Cancelled IP Data. *ASEG Ext. Abstr.* **2003**, *2003*, 1–4. [CrossRef]
15. Yoshioka, K.; Zhdanov, M.S. Three-dimensional nonlinear regularized inversion of the induced polarization data based on the Cole–Cole model. *Phys. Earth Planet. Inter.* **2005**, *150*, 29–43. [CrossRef]
16. Zhang, W.; Liu, J.-X.; Guo, Z.-W.; Tong, X.-Z. Cole-Cole model based on the frequency-domain IP method of forward modeling. In Proceedings of the Progress in Electromagnetics Research Symposium Proceedings, Xi'an, China, 5–8 July 2010; pp. 383–386.
17. Burtman, V.; Zhdanov, M.; Lin, W.; Endo, M. Complex resistivity of mineral rocks in the context of the generalized effective-medium theory of the IP effect. In *SEG Technical Program Expanded Abstracts 2016*; Society of Exploration Geophysicists: Dallas, TX, USA, 2016; pp. 2238–2242.
18. Zhdanov, M.S.; Burtman, V.; Endo, M.; Lin, W. Complex resistivity of mineral rocks in the context of the generalised effective-medium theory of the induced polarisation effect. *Geophys. Prospect.* **2018**, *66*, 798–817. [CrossRef]
19. Li, Y.; Oldenburg, D.W. Inversion of 3-D DC resistivity data using an approximate inverse mapping. *Geophys. J. Int.* **1994**, *116*, 527–537. [CrossRef]
20. Li, Y.; Oldenburg, D.W. 3-D inversion of induced polarization data. *Geophysics* **2000**, *65*, 1931–1945. [CrossRef]
21. Dey, A.; Morrison, H.F. Electromagnetic coupling in frequency and time-domain induced-polarization surveys over a multilayered earth. *Geophysics* **1973**, *38*, 380–405. [CrossRef]
22. Commer, M.; Newman, G.A.; Williams, K.H.; Hubbard, S.S. 3D induced-polarization data inversion for complex resistivity. *Geophysics* **2011**, *76*, F157–F171. [CrossRef]
23. Fiandaca, G.; Auken, E.; Christiansen, A.V.; Gazoty, A. Time-domain-induced polarization: Full-decay forward modeling and 1D laterally constrained inversion of Cole–Cole parameters. *Geophysics* **2012**, *77*, E213–E225. [CrossRef]
24. Fullagar, P.K.; Zhou, B.; Bourne, B. EM-coupling removal from time-domain IP data. *Explor. Geophys.* **2000**, *31*, 134–139. [CrossRef]
25. Smith, R.S.; Walker, P.W.; Polzer, B.D.; West, G.F. The time-domain electromagnetic response of polarizable bodies: An approximate convolution algorithm1. *Geophys. Prospect.* **1988**, *36*, 772–785. [CrossRef]
26. Marchant, D.; Kang, S.; McMillian, M.; Haber, E. Modelling IP effects in airborne time domain electromagnetics. *ASEG Ext. Abstr.* **2018**, *2018*, 1–6. [CrossRef]
27. Tikhonov, A.N.; Arsenin, V.I. *Solutions of Ill-Posed Problems*; Halsted Press: Washington, DC, USA; New York, NY, USA, 1977.
28. Zhdanov, M.S. *Inverse Theory and Applications in Geophysics*; Elsevier Science: Amsterdam, The Netherlands; Boston, MA, USA, 2015.
29. Zhdanov, M.S. *Geophysical Inverse Theory and Regularization Problems*; Elsevier Science: Amsterdam, The Netherlands; Boston, MA, USA, 2002.
30. Zhdanov, M.S. *Geophysical Electromagnetic Theory and Methods*, 1st ed.; Elsevier: Amsterdam, The Netherlands; Boston, MA, USA, 2009.
31. Xu, Z.; Zhdanov, M.S. Three-Dimensional Cole-Cole Model Inversion of Induced Polarization Data Based on Regularized Conjugate Gradient Method. *IEEE Geosci. Remote Sens. Lett.* **2015**, *12*, 1180–1184.
32. Zhdanov, M.; Endo, M.; Cox, L.; Sunwall, D. Effective-Medium Inversion of Induced Polarization Data for Mineral Exploration and Mineral Discrimination: Case Study for the Copper Deposit in Mongolia. *Minerals* **2018**, *8*, 68. [CrossRef]
33. Rudd, J.; Chubak, G. The Facility of a Fully-Distributed DCIP System with CVR. In Proceedings of the 15th SAGA Biennial Conference and Exhibition, Cape Town, South Africa, 31 July 2017.
34. Zhdanov, M.S. *Foundations of Geophysical Electromagnetic Theory and Methods*; Elsevier Science: Amsterdam, The Netherlands; Boston, MA, USA, 2018.

35. Zhdanov, M. New Geophysical Technique for Mineral Exploration and Mineral Discrimination Based on Electromagnetic Methods. US Patent No. 7,324,899, 29 January 2008.
36. Cox, L.H.; Wilson, G.A.; Zhdanov, M.S. 3D inversion of airborne electromagnetic data. *Geophysics* **2012**, *77*, WB59–WB69. [CrossRef]

 © 2020 by the authors. Licensee MDPI, Basel, Switzerland. This article is an open access article distributed under the terms and conditions of the Creative Commons Attribution (CC BY) license (http://creativecommons.org/licenses/by/4.0/).

Article

Three-Dimensional P-wave Velocity Structure of the Zhuxi Ore Deposit, South China Revealed by Control-Source First-Arrival Tomography

Yunpeng Zhang [1,2], Baoshan Wang [1,3,4,*], Guoqing Lin [2], Yongpeng Ouyang [5], Tao Wang [6], Shanhui Xu [1], Lili Song [1] and Rucheng Wang [6]

1   Institute of Geophysics, China Earthquake Administration, Beijing 100081, China; zhangyp@cea-igp.ac.cn (Y.Z.); xushanhuia@126.com (S.X.); songll@cea-igp.ac.cn (L.S.)
2   Department of Marine Geosciences, Rosenstiel School of Marine and Atmospheric Science, University of Miami, Miami, FL 33149, USA; glin@rsmas.miami.edu
3   School of Earth and Space Sciences, University of Science and Technology of China, Hefei 230026, China
4   Mengcheng National Geophysical Observatory, University of Science and Technology of China, Hefei 230026, China
5   912 Geological Team, Bureau of Geology and Mineral Resources of Jiangxi Province, Yingtan 335001, China; yongpeng0524@163.com
6   State Key Laboratory for Mineral Deposits Research, School of Earth Sciences and Engineering, Nanjing University, Nanjing 210046, China; twang0630@nju.edu.cn (T.W.); rcwang@nju.edu.cn (R.W.)
*   Correspondence: bwgeo@ustc.edu.cn

Received: 11 January 2020; Accepted: 6 February 2020; Published: 9 February 2020

Abstract: The Zhuxi ore deposit, located in Jiangxi province, South China, is the largest tungsten reserve in the world. To better understand the geological structure and distribution of orebodies, we conducted a high resolution three-dimensional P-wave velocity tomography of the uppermost 0.5 km beneath the Zhuxi ore deposit and adjacent area. Our velocity model was derived from 761,653 P-wave first arrivals from 998 control-source shots, recorded by a dense array. As the first 3D P-wave velocity structure of the Zhuxi ore deposit, our model agrees with local topographic and tectonic structures and shows depth-dependent velocity similar to laboratory measurements. The Carboniferous formations hosting the proven orebodies are imaged as high velocities. The high-velocity anomalies extend to a larger area beyond the proven orebodies, and the locations of high–low velocity boundaries are in accordance with the boundaries between the Neoproterozoic formation and the Carboniferous–Triassic formation. Seismic tomography reveals that high-velocity anomalies are closely related to the mineralized areas. Our results are helpful for further evaluating the total reserves and suggest that seismic tomography can be a useful tool for mineral exploration.

Keywords: Zhuxi ore deposit; control source; dense array; body wave tomography

---

1. Introduction

The Zhuxi ore deposit, located in the eastern part of the Jiangnan orogenic belt in Jiangxi Province, China, is a skarn-type mineral deposit [1]. The geological survey and borehole inspection in 2016 [2] estimated the tungsten reserves of this deposit to be more than 3.44 million tons, making it the largest proven tungsten reserve in the world [1].

The Zhuxi area experienced superimposed transformation from multi-stage tectonic–magmatic events from the Neoproterozoic to the late Mesozoic era [3–5]. The activity during the Yanshanian period was the most violent, which resulted in regional diagenesis and metallogenesis and was characterized by stretching, shearing, crustal thinning, granite intrusion, crustal rupture, and fluid activity [6,7]. The thrust nappe structure can cause interlaminar nappe slip of the ore-bearing stratum and form weak zones that are conducive to magma emplacement and orebody storage [8]. The Zhuxi ore deposit was formed after the multi-stage tectonic evolution of oblique intrusion of granitic magmas, skarn mineralization, hydrothermal cooling and alteration, and precipitation of metal sulfides [7]. The major orebodies beneath the Zhuxi ore deposit are located in the contacting zones between the Huanglong Formation and the Neoproterozoic slate (Figure 1), bounded by several NE-striking faults [1]. The ores are mostly hydrothermal-type copper–iron ores at shallow depths, whereas there are rich stratoid skarn scheelite (copper) ores at greater depths [7].

Until now, the reserves and distribution of the Zhuxi ore deposit have been mainly evaluated from the geological survey [2] and borehole explorations [1,7]. Two major ore belts—the Main Ore Belt (MOB) and the North Ore Belt (NOB) (Figure 1)—have been identified. As revealed by borehole inspections, most orebodies are NE-trending and dominated by a veinlet-disseminated structure [7]. Due to the limited number of inspecting boreholes, however, some geological problems are not fully understood. Such as the stratigraphic boundaries between different formations, the spatial distribution of the orebodies outside the prospecting survey area, and the shape of the concealed orebodies at depth reconstructed by nappe structure. These limited understandings restrict the study and prospecting of metallogenic regularity in the Zhuxi ore deposit and adjacent area. Imaging the subsurface structure with geophysical methods is helpful in overcoming these plights.

Recently, several geophysical explorations (e.g., gravity–magnetic, electromagnetic, and aeromagnetic survey) have been carried out in Zhuxi [9,10]. The ore zones are shown to have low Bouguer gravity anomalies, positive aeromagnetic anomalies, and high magnetic derivative norms [9,10]. However, gravity–magnetic, electrical, and electromagnetic explorations were mainly used to delineate the orebodies targeting horizontally, and it still need borehole inspection to find the depth and strata [11]. Three-dimensional fine images are still necessary to locate the orebodies and precisely estimate the reserves.

Seismic reflection and refraction are effective tools for exploring the subsurface structure and detecting reservoir related reflective interfaces [12,13]. However, seismic reflection is rarely used in orebody explorations because most metal mines have complex structures and seismic records in metal mine areas are subject to low signal-to-noise ratios, weak reflections, and strong scattering [11]. Fortunately, laboratory measurements indicate that strong seismic velocity differences exist between the underlying and overlying formations around the Zhuxi ore veins. Therefore, the fine-scale 3D velocity structure images around the Zhuxi ore deposit will be helpful in revealing the distribution of the metamorphic belt and assessing mineral reserves.

To image the fine-scale structure around the Zhuxi ore deposit, we conducted a geophysical experiment in Zhuxi and the adjacent region in November and December of 2017 (hereafter, referred to as the Zhuxi experiment). In this study, we present the 3D velocity structure around the Zhuxi ore deposit obtained from the control source seismic experiment data.

2. Zhuxi Experiment

From November to December 2017, a multi-disciplinary experiment was conducted around the Zhuxi ore deposit, including seismic, gravity, and electric surveys [14]. During this experiment, we deployed one 2D array (black triangles in Figure 1) and four dense seismic profiles (blue triangles in Figure 1) for the seismic survey. The 2D array covered an area of 30 km × 40 km with 178 portable three-component short period (5 s–150 Hz) seismometers (manufactured by the Chongqing Geological Instrument Co., Ltd, Chongqing, China) with station spacing of 1–5 km. The four profiles were 26.5 km long in total, equipped with 2311 vertical component 10 Hz geophones (manufactured by the Sercel, Nantes, France) with an average station spacing of 11.5 m.

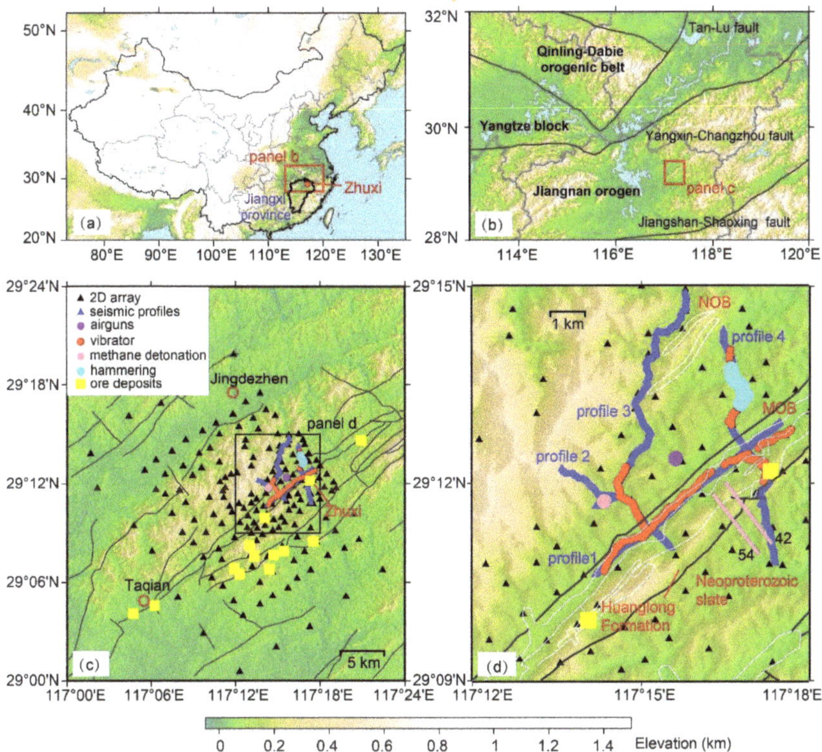

Figure 1. (**a**,**b**) Location and regional geologic map of our study area. (**c**,**d**) Distribution of control sources and seismic array for the Zhuxi experiment. Black and blue triangles represent 2D array and dense seismic profiles, respectively. The purple, red, pink, and cyan dots represent the airgun (215 shots at the same location), vibrator (738 shots), methane detonation (1 shot), and hammer (44 shots) excitation sites, respectively. The yellow squares denote the most recently discovered ore deposits [10]. The black lines indicate the geologic faults [1,7,15–17]. The white lines in (**d**) are the stratigraphic boundaries from Chen et al. [7]. The pink lines in (**d**) are the prospecting lines 54 and 42. MOB stands for the main ore belt and NOB for the north ore belt (envelope areas of the thin white contours). These abbreviations hold throughout this paper. The background is the 90 m topography from the Shuttle Radar Topography Mission [18].

Four types of control sources, namely, vibrator (28 tons force; supplied by the BGP Inc., China National Petroleum Corporation, Zhuozhou, Hebei, China), electric hammer (800 kg force; supplied by the Institute of Geophysics, China Earthquake Administration, Beijing, China), airgun (2000 in^3; supplied by the Institute of Geophysics, China Earthquake Administration, Beijing, China), and methane detonation source (supplied by the China Academy of Engineering Physics, Mianyang, Sichuan, China) [19], were used to generate seismic waves. The vibrator is an electronically controlled, hydro mechanical system driven by a servo-valve assembly that exerts ground force through its baseplate against the Earth's surface [20]. During the Zhuxi experiment, the vibrator was fired along four profiles at 738 points, with an average spatial interval of 40 m. At each point, we lunched one 12 s long sweep with frequency 1.5–96 Hz at 28 tons force. The hammer generates seismic waves by hitting a square iron plate fixed to the ground [21]. The airgun source can rapidly release compressed air to generate a pressure pulse and relatively long period air bubble oscillation that is then transmitted as seismic waves [22,23]. After being fired in a sealed container, methane detonation produces seismic waves when the high pressure air is quickly released to impact the surroundings. The methane detonation source was used for the first time in this experiment [19].

The vibrator and hammer sources were fired along profiles 1, 2, and 3 (Figure 1b) and were synchronized with the receivers (Sercel 428XL) with a wire connection. The seismometers on profile 4 and the 2D array (three-component, short-period seismographs) were continuously recorded and synchronized with the Global Positioning System (GPS; manufactured by the Trimble, Sunnyvale, CA, USA) timing. In addition, the origin time of the source was recorded using the GPS timing system with millisecond accuracy. The database using in the paper comes from the observations supported by the Institute of Geophysics, China Earthquake Administration, which can be requested from the authors.

3. First Arrival Picking

To remove contamination from the source time functions and improve the signal-to-noise (SNR), we first preprocessed the waveform data. For the vibrator, we reconstructed the record from the continuous signals using a cross-correlation method [24]. The airgun signals were linearly stacked to enhance the SNR [25,26], exploiting the high repeatability of the airgun source [23,27]. We then manually picked first P-wave arrivals from raw waveform for different sources and the P phases were clearly identified for all sources (Figure 2).

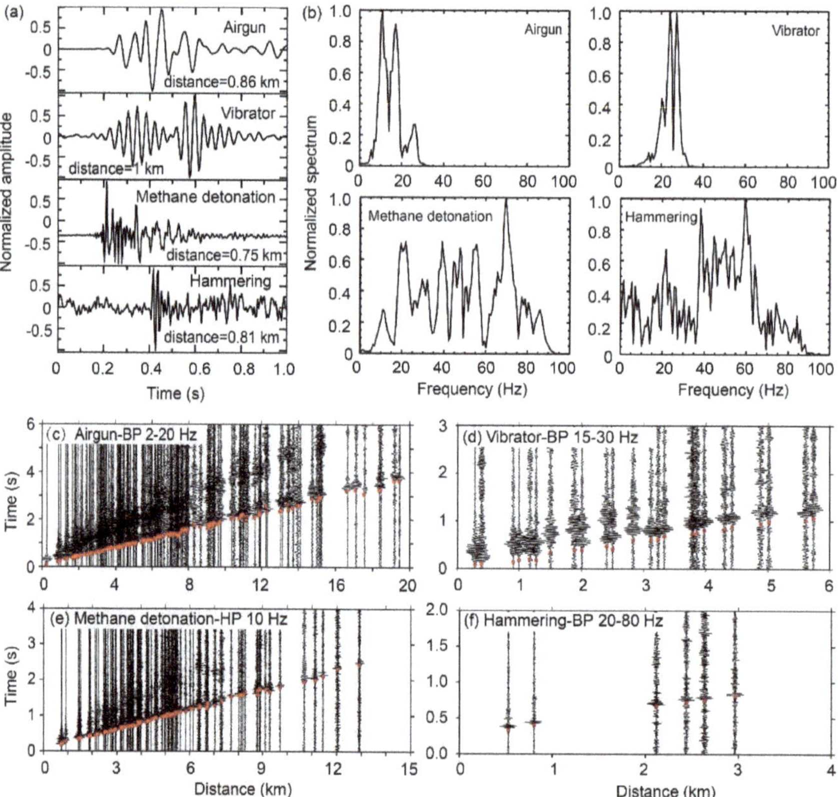

Figure 2. Examples of the raw waveforms (**a**) and spectrum characteristics (**b**) for four control sources with similar distances, and the filtered vertical component waveforms and phase picking (the red dots) of airgun (**c**), vibrator (**d**), methane detonation (**e**), and hammer (**f**), respectively. BP stands for band-pass filtering and HP stands for high-pass filtering.

We also filtered the seismic records to respective predominate frequency bands (determined from the spectrum characteristics analysis) for each seismic source to compare their waveform characteristics and propagation capabilities (Figure 2). The longest distance the airgun can propagate 20 km with relatively lower frequency. The vibrator source has clear P-wave first arrival phases with propagation distances up to 6 km. The methane detonation source can realize a detection of 10–5 km with higher frequency than the airgun source, which is suitable for small-scale fine-structure exploration [19]. The hammer signal has the highest frequency with the shortest propagation distance, and the SNR of the waveform is relatively lower compared with the signals from other sources (Figure 2). We did not use the low SNR data (Figure 2d) from the hammer sources in this study.

The manually picked travel times are distributed around a linear trend (black line in Figure 3a) with an average velocity of 5.62 km/s (obtained from the least-square fitting). There are some outliers in the phase picking, mainly due to poor SNRs (Figure 3a). We only kept the picks with deviations less than 0.3 s (discarding 1397 phases). In total, 761,653 reliable P-wave arrivals (Figure 3a) were used for further tomographic inversion.

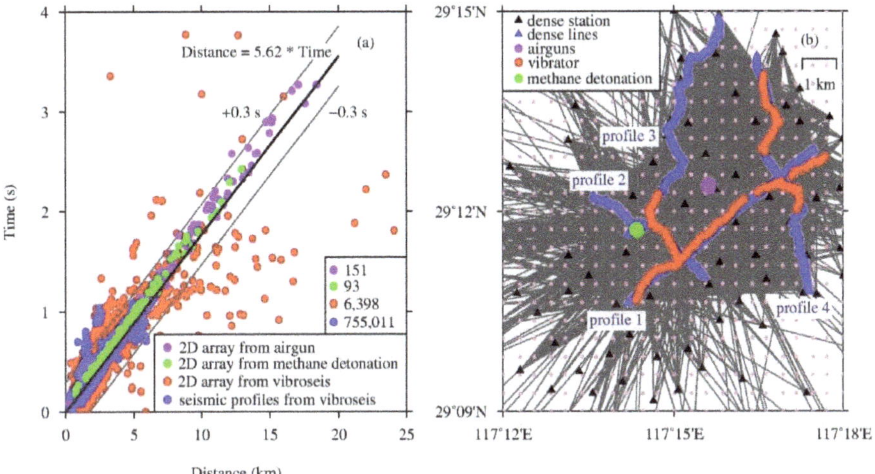

Figure 3. Travel time (**a**) and ray path coverage (**b**) of the P-wave first arrivals for different control sources. The straight black lines in (**a**) are obtained from the least-square fitting. Gray lines in (**b**) correspond to picks bounded by two gray lines (±0.3 s) in (**a**). The pink dots in (**b**) are grid nodes used in the 3D tomographic inversions.

4. Method and Input Velocity Model

4.1. Inversion Method

In this study, the 3D shallow P-wave velocity structures beneath the Zhuxi ore deposit and the surrounding areas were obtained by using the simul2000 program [28–31], which has been widely used in local seismic imaging [32–34].

In the simul2000 inversion, the arrival time residuals are represented as functions of seismic velocities:

$$r_{ij} = t_{ij}^{obs} - t_{ij}^{calc} \tag{1}$$

$$r_{ij} = \frac{\partial T_{ij}}{\partial x}\Delta x + \frac{\partial T_{ij}}{\partial y}\Delta y + \frac{\partial T_{ij}}{\partial z}\Delta z + \Delta\tau_i + \sum_{k=1}^{k=total}\frac{\partial T_{ij}}{\partial m_k}\Delta m_k \tag{2}$$

where t_{ij}^{obs}, t_{ij}^{calc}, r_{ij}, and T_{ij} are the P-wave observed arrival time, calculated arrival time, arrival time residual, and travel time from event i to station j, respectively. The earthquake location coordinate is

(x, y, z), and m_k is the seismic velocity of the k th grid node. The perturbation of a parameter from its initial value is denoted by Δ. Since we used control sources in this paper, the first four items in Equation (2) vanished. The simul2000 algorithm uses damped least-squares inversion to solve for the model perturbations and can calculate the covariance matrix to estimate the resolution of the final model.

4.2. Input Velocity Model

Seismic tomography is generally a linearized approximation of a nonlinear problem [31]; therefore, the initial velocity model is of key importance in the inversion process. The optimal one-dimensional (1D) P-velocity model can be obtained using the VELEST package [35,36]. This algorithm seeks the optimal 1D (layered) velocity model that minimizes the difference between the calculated and observed travel times [36] and is widely used for initial reference models in 3D local earthquake tomography (e.g., Matrullo et al. [37]).

We selected 19 starting 1D P-wave velocity models with different surface (0 km) velocities and velocity gradients (Figure 4) for the VELEST inversion procedure. Starting from these different initial models, all inversions converged to similar structures (red lines in Figure 4) after 20 iterations. The 19 inverted velocity models were further averaged and smoothed as the final 1D velocity model (green line in Figure 4) for 3D tomography.

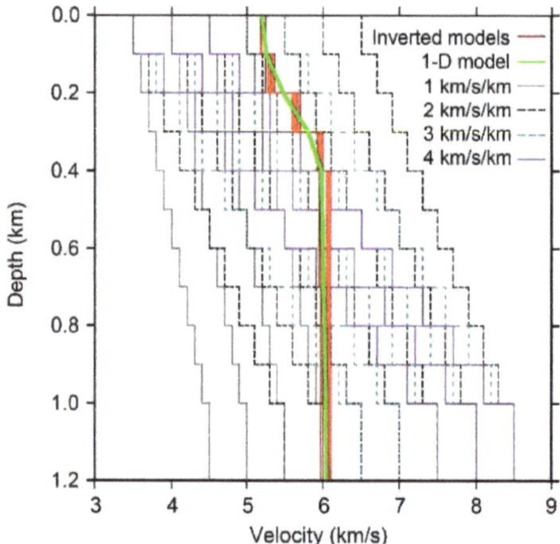

Figure 4. 1D initial velocity models obtained from the VELEST program. The models (red lines) were inverted from different starting models with four gradients (gray solid, black dashed, cyan dashed, and purple solid lines represent 1, 2, 3, and 4 km/s/km, respectively). The green line denotes the final 1D initial velocity model for our inversion.

5. Three-Dimensional P-Wave Velocity Structure

5.1. Model Setup and Parameter Selection

In inverting Equation (2), we set horizontal grid spacing of 0.5 km and 0.1 km spacing in depth (pink dots in Figure 3b). Considering that the highest elevation station in our study was 0.45 km above sea level, we placed an unmodeled layer at 0.5 km above sea level.

To stabilize the inversion of Equation (2), a damping factor is generally used, which plays a role in the inversion process. Damping factors are often selected by applying a trade-off analysis between data misfit and model variance [33,38]. We explored a wide range of damping factors to determine the optimal damping parameter (Figure 5). According to the trade-off curve, we chose 300 as the damping value for our tomography, producing a good compromise between data misfit and model variance.

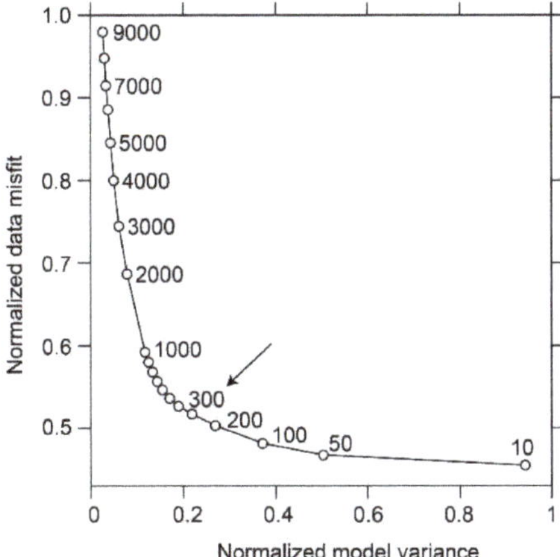

Figure 5. Trade-off curve between data misfit and model variance. A damping parameter of 300 was selected for our inversion, in both the checkerboard test and real data inversion.

5.2. Model Quality and Resolution Test

After 12 iterations, the tomographic inversion converged with the root–mean–square of arrival time residual reduced from 0.03 s to 0.01 s (Figure 6). Before the inversion, the residual varied from −0.06 s to 0.09 s with an average value of approximately 0.03 s, suggesting a systematic deviation in the initial velocity model. After the inversion, the residual followed a Gaussian distribution centered at 0 s.

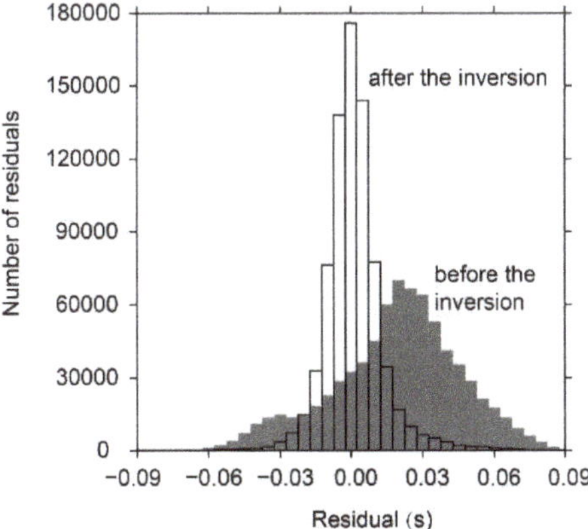

Figure 6. Histograms of the arrival time residuals before (gray bars) and after (hollow bars) the inversion. The root–mean–square of the residual decreased from 0.03 to 0.01 s.

To assess resolving power of the data and parametrization, we performed a checkerboard resolution test with the same inversion parameters as for the real data. We computed synthetic travel times through the 1D velocity model (green line in Figure 4) with alternative ±5% velocity perturbations across two adjacent grid nodes. The input and inverted-P-wave velocity structures are shown in Figure 7. The resolution matrix which is a stricter criterion than ray density (Lin et al.), [32–34] was calculated to estimate the resolution of the model and the uncertainties in the model parameters. The green contours enclose the well-resolved area with the diagonal element of the resolution matrix greater than 0.1 (1.0 represents the best resolution and 0.0 is not resolved at all [31,33]). As shown in Figure 7, the velocity structure of the uppermost 0.5 km is well resolved at the central part of the survey area.

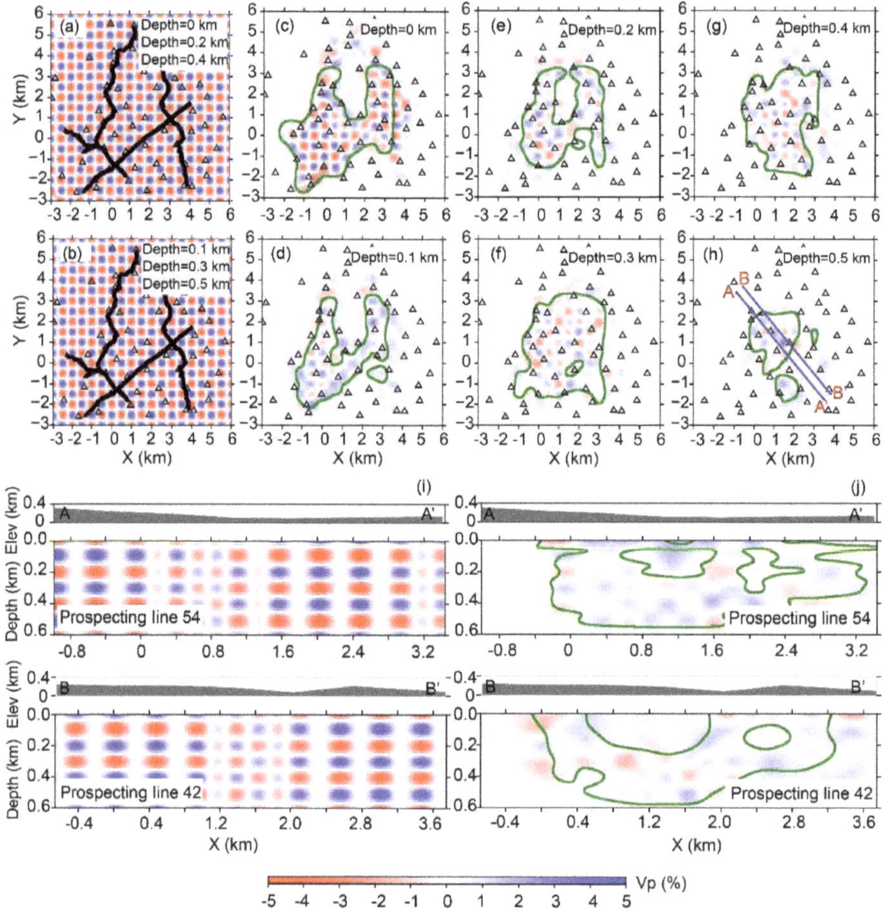

Figure 7. Checkerboard resolution test in which the synthetic travel times are computed from the 1D initial velocity model with ±5% velocity anomalies across two grid nodes. (**a**,**b**) Map views of the true model. (**c–h**) Map views of the inverted model at different depth layers. (**i**) Cross section of the true model along the prospecting lines 54 (**AA′**) and 42 (**BB′**) shown in (**h**). (**j**) Cross section of the inverted model along the same profile. The green contours enclose the well-resolved area with the diagonal element of the resolution matrix greater than 0.1.

5.3. Three-Dimensional P-Wave Velocity Structure

At the surface (0 km), the velocity structure shows a clear banded distribution, consistent with the geologic faults (NE-trending faults) and topographic distribution, as shown in Figure 1 (i.e., the high-velocity anomalies correlate with the high topography). From 0 to 0.5 km, the most notable features in our model are the high-velocity anomalies beneath the MOB and the NOB (Figure 8) (including Zhuxi, Tanling, and Hongmeiling). The MOB for the Zhuxi ore deposit corresponds to the high magnetic gradients, low Bouguer gravity anomalies [9,10], and low resistivity anomalies [14]. In our model, the MOB is imaged as a high-velocity anomaly from the surface to a depth of 0.5 km, and the majority of the high-velocity anomalies exist in Permian and Carboniferous formations (Figure 8).

The NOB is a small copper–molybdenum deposit in the cretaceous granitic porphyry [7,39,40]. Our velocity structure presents some sporadic high-velocity anomalies (black rectangles in Figure 8) beneath the NOB.

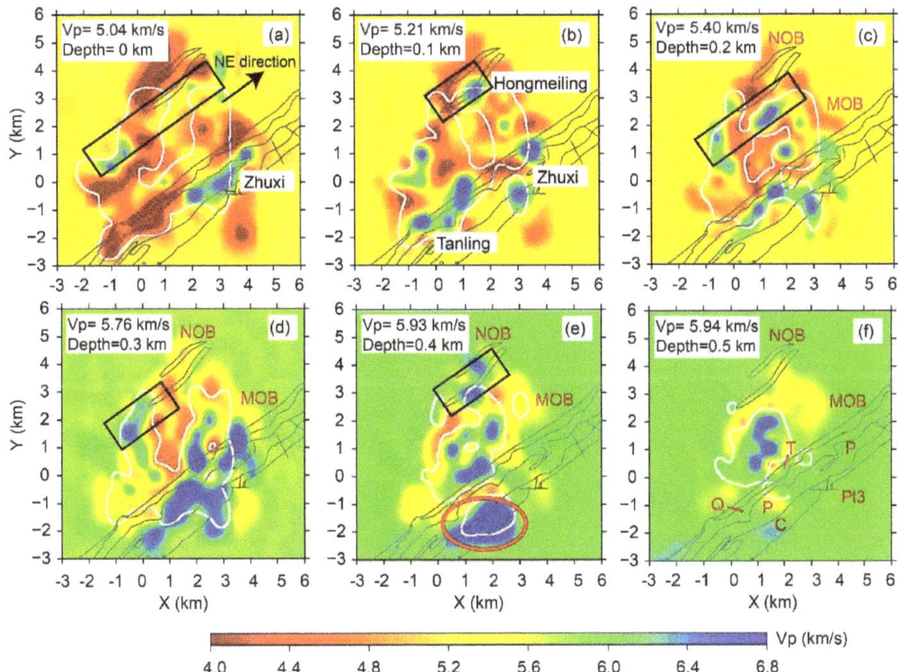

Figure 8. (**a–f**) P-wave velocity structures at different depths. Gray lines indicate geologic faults and boundaries. The white contours enclose the well-resolved area with the diagonal element of the resolution matrix greater than 0.1. MOB stands for the main ore belt and NOB for the north ore belt (envelope areas of the thin black contours). The average velocity value is also shown for each depth layer. Abbreviations in (**f**) are C: Carboniferous, P: Permian, Pt$_3$: Neoproterozoic, Q: Quaternary, and T: Triassic.

6. Discussion

6.1. Absolute Velocities of the Orebodies

As previously mentioned, many boreholes were drilled down to 2.2 km [41] to inspect the distribution of orebodies. These inspections provided an unprecedented opportunity to calibrate our velocity model. We collected 78 rock samples recovered from different strata along five prospecting lines (green lines in Figure 9a). The depth of rock samples ranged from the surface to 2.2 km (with 11 samples shallower than 0.6 km). P-wave velocities (black dash line in Figure 9b) were measured under atmospheric pressure and room temperature using the method proposed by Xu et al. [42].

We then compared our velocity profiles with the laboratory measurements. Three velocity profiles (initial 1D, inverted average of whole study area, and inverted average around the prospecting area) were used for comparison. The average velocity around the prospecting area (the magenta line in Figure 9b) is higher than the initial 1D model and the average velocity from 3D tomography, which reveals that the orebodies have higher velocities than other areas. The velocity profiles from the laboratory measurements (black dash line in Figure 9b) and our inverted model (magenta line in Figure 9b) have similar depth dependencies. The velocities measured in the laboratory are generally lower than the tomographic results (especially at 0.4 km), probably because the laboratory measurements were affected by pores and micro-cracks introduced during the core recovery and sample preparation. Furthermore, the laboratory measurements were performed under atmospheric pressure much lower than the in situ pressure. At 0 km depth, the tomographic velocity is lower than the laboratory measurements partly because of the wide spreading of surface sediments. At 0.5 km depth, P-wave velocity are an estimated 6 km/s from both tomography and laboratory.

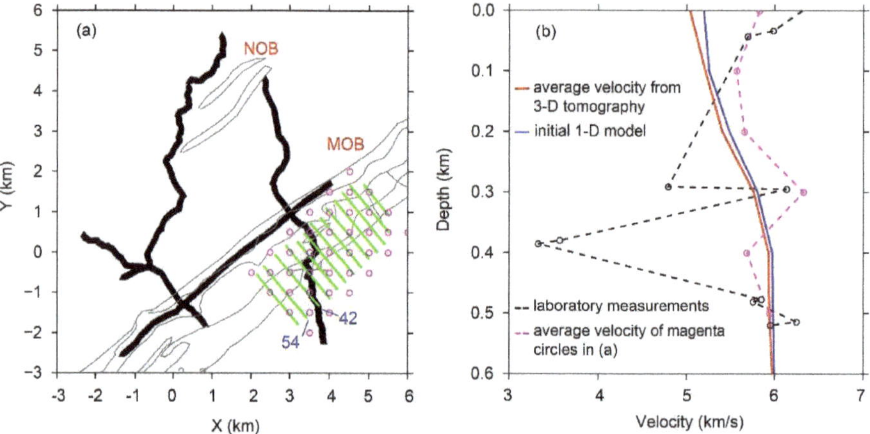

Figure 9. Comparison of velocity models and laboratory measurements. (**a**) The green lines indicate the prospecting lines and the blue letters represent the line codes, after Chen et al. [7]. Magenta circles represent some grid nodes located around the prospecting lines to compare our inversion result with drilling data. (**b**) The black dash line represents the result of laboratory measurements. The magenta line is the average velocity of the magenta circles in (**a**). The blue line is the 1D initial velocity model, and the red line is the average velocity from 3D tomography.

6.2. P-Wave Velocity Beneath Prospecting Lines

The borehole inspections along several lines also reveal the spatial variation of the lithology, and lines 54 and 42 (Figure 9) are the most well-studied [7]. In this section, we further compare our velocity model with these two sections across the Zhuxi ore deposit (Figure 10).

High-velocity anomalies are observed beneath the MOB for both prospecting lines 54 and 42 (Figure 10b,d). According to the borehole explorations, the Carboniferous Huanglong Formation and the Permian Qixia Formation beneath the MOB are the dominant reserve layers of the copper–tungsten ore from surface to deeper (~2 km), containing abundant Cu, Zn, Fe, W, Mo, and other elements [1,10,39,43]. These high-velocity anomalies (with red arrows, most in the Carboniferous formation) correspond to the hydrothermal vein-type copper and iron ores beneath the MOB in the shallow part (<0.2 km) and a small part of tungsten orebody beneath 0.2 km [1,7]. Our results also reveal that the orebodies of prospecting line 42 are shallower than those of prospecting line 54, which is in accordance with the results from borehole inspection [7,8].

Our velocity results show that there are obvious high–low velocity boundaries between the Neoproterozoic (Pt3 in Figure 10b,d, low velocity) and the Carboniferous–Triassic (C–T in Figure 10b,d, high-velocity), and the locations of velocity boundaries are consistent with stratigraphic classification (black lines in Figure 10) from geological survey and borehole inspection [8]. These velocity contrasts are in accordance with lithologic settings, where the Neoproterozoic is mostly phyllite with argilloarenaceous rock and lightly metamorphosed rock, and the Carboniferous–Triassic formation is mainly composed of carbonates, limestone, and dolomite [1,7]. Shi et al. [14] also revealed that the average resistivities for the Neoproterozoic are higher than those of Carboniferous–Permian.

The observed velocity contrasts extend NW and deep beyond the MOB (Figure 10b,d), and this trend may depict the depth extension of stratigraphic boundaries (thick dashed solid lines in Figure 10).

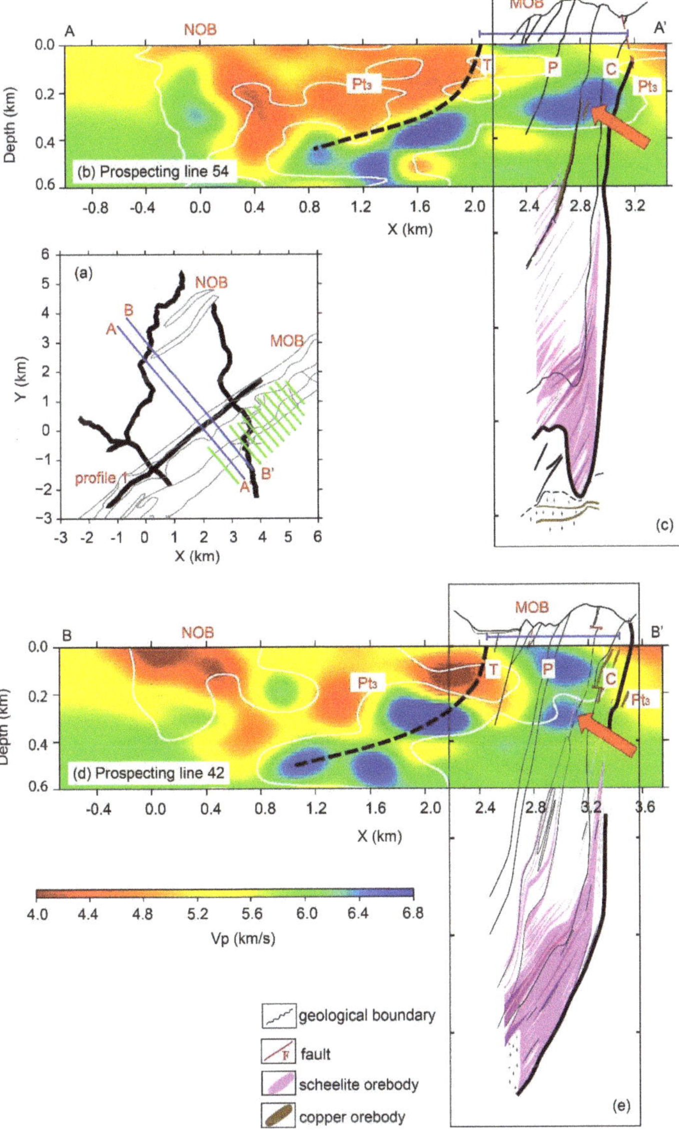

Figure 10. P-wave absolute velocity of sections across the Zhuxi ore deposit along the prospecting lines 54 (**AA′**) and 42 (**BB′**). The white contours enclose the well-resolved area with the diagonal element of the resolution matrix greater than 0.1. (**c**) and (**e**) are depth extensions of orebodies along the prospecting lines modified from Ouyang et al. [8]. The black solid lines are stratigraphic boundaries from Ouyang et al. [8]. The black dotted lines are predicted stratigraphic boundaries. Surface topography is shown at the top of each panel. The vertical exaggeration is 2. Abbreviations are the same as those in Figures 1 and 8.

6.3. Cross Sections of the Whole Zhuxi Ore Deposit

The correlation between the high-velocity structure and mineralized zone shown in the last section suggests that velocity structure is a good indicator for stratigraphic classification. To determine the spatial distribution of the high-velocity zone, we present seven more cross sections (CC′ to II′ parallel to AA′ and BB′).

Similar to the prospecting lines 54 and 42, high-velocity anomalies are observed beneath the MOB and NOB for all seven cross sections (Figure 11). The high-velocity anomalies along the CC′ to FF′ beneath the MOB (Figure 11b–e) are considerably higher than those along other cross sections. There is a high-velocity zone (red circle in Figure 8e) on the south edge of the MOB at 0.4 km (Figure 8e), where a low electrical resistivity structure is observed to be extending deeper than 2 km [14]. This is attributed to the Tanling deposit and may be a large concealed magmatic rock [39]. The range of the high-velocity zone beneath the NOB is significantly smaller than the MOB, which is consistent with the orebody reserve estimations. Other geophysical results also indicate that the MOB has lower Bouguer gravity anomalies and a higher magnetic field gradient than the NOB [9,10]. In addition, there are some high-velocity anomalies in the southwest of the NOB (black rectangles in Figure 8), which may correspond to the sporadic distribution of the cretaceous granitic porphyry. The low-velocity zones (0–0.2 km) beneath the MOB (Figure 11b,e,f) are likely the Quaternary sediment [1,7].

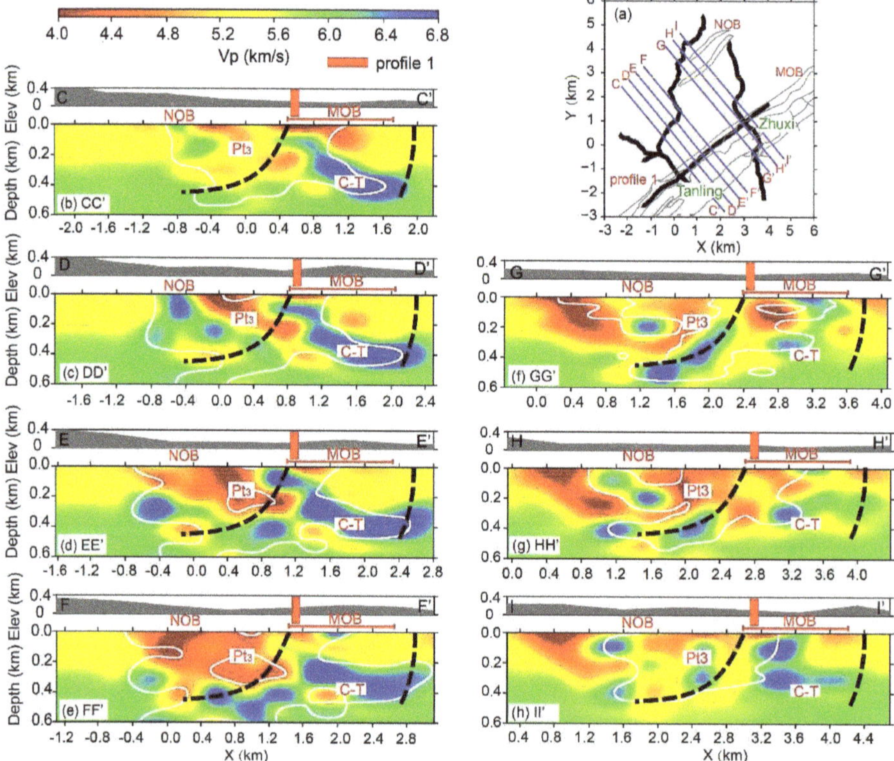

Figure 11. P-wave absolute velocity of sections outside of the prospecting lines. The white contours enclose the well-resolved area with the diagonal element of the resolution matrix greater than 0.1. The black dotted lines are predicted stratigraphic boundaries. Surface topography is shown at the top of each panel. The vertical exaggeration is 2. Abbreviations are the same as those in Figures 1 and 8.

The high-low velocity boundaries are visible and can be tracked in the cross sections (Figure 11), and these velocity boundaries may indicate the stratigraphic boundaries (black dotted lines in Figure 11) between the Neoproterozoic and the Carboniferous–Triassic. From all cross sections (including prospecting lines 54 and 42), we suggest that the MOB can span 2 km in accordance with the results of borehole explorations [7]. Even though the high-velocity anomalies between the MOB and the NOB become weaker from northeast to southwest, all of these belts present a north-dipping structure. Our results also indicate that the velocity beneath the Tanling is higher than Zhuxi (Figure 11). Shi et al. [14] also revealed that Tanling has lower resistivity than Zhuxi and suggested that the deep part of Tanling is probably of copper mineralization rather than tungsten mineralization.

6.4. Formation of the Zhuxi Ore Deposit

The borehole explorations show that the high-velocity bodies in the shallow part of the Zhuxi ore deposit (Figures 8, 10 and 11) are mainly chalcopyrite and pyrite [40]. There is a large amount of pyrite distribution (especially in the metamorphic rocks). These copper and iron orebodies in the shallow part are mainly lenticular and vein-type. The lenticular-type ore is caused by the difference in the ore-bearing space formed by tectonism and the vein-type ore is mainly originated from hydrothermal activity. Mineralization mainly occurred in the Carboniferous formation and the shallow part of the Permian Qixia Formation, whereas little metallogenic activity is observed beneath the Triassic formation. Whether the other high-velocity anomaly zones in the Permian have been mineralized needs further analysis from geological, geochemical, and geophysical aspects. Two mineralization mechanisms exist for the copper–tungsten and iron deposits; the hydrothermal activity is the dominant mechanism, and the sedimentation is the secondary one. The hydrothermal activity is described in detail below (Figure 12).

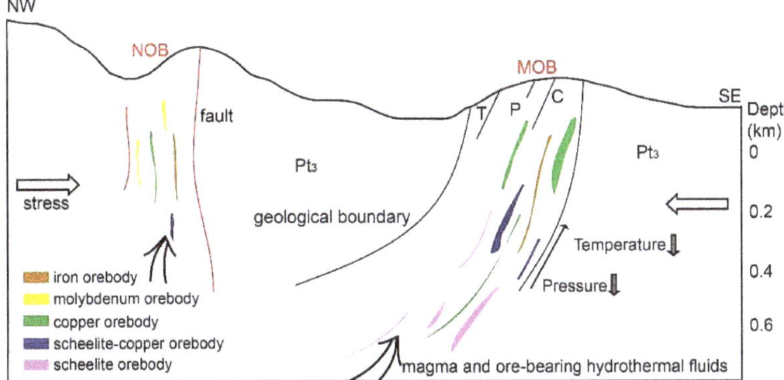

Figure 12. Simplified shallow metallogenic mechanism for the Zhuxi ore deposit. Abbreviations are the same as those in Figures 1 and 8. As the hydrothermal fluid flows upward, granitic magma will continuously extract tungsten elements and alter into scheelite. As the temperature and pressure gradually decrease, copper–zinc minerals are separated as sulfides, and then copper–tungsten (zinc) orebody and iron and copper (zinc) orebody are formed by metasomatism.

The granitic magma activities during the Yanshanian period [43–47] bring ore-bearing hydrothermal fluid for mineralization [1], and there are a series of faults (especially the NE-trending thrust faults) forming possible upwelling channels. In the early stage of mineralization, magma and ore-bearing hydrothermal fluids originated from the high-temperature and high-pressure environment. During the upward migration, hydrothermal fluids continuously extract metal elements (mainly tungsten) from carbonate rocks and then alter into scheelite during the latest stage of magma crystallization [48]. As the ore-bearing hydrothermal fluids continue to rise, the temperature and pressure will gradually decrease. A large portion of copper–iron–zinc minerals will be separated as sulfides, and then copper–tungsten (zinc) orebody and iron and copper (zinc) orebody are formed by metasomatism [1,49].

In short, the MOB and NOB were formed under the superimposed effects of tensile structure, magma upwelling, hydrothermal metasomatism, and mineralization [1,7,8,50]. Our 3D body wave tomography results suggest the extension of stratigraphic boundaries outside the prospecting survey area and reveal that these potential mineralized areas may be gradually connected in the deep area beneath the MOB and NOB. Therefore, the MOB and NOB may have originated from the same hydrothermal system.

Subject to the low power of seismic sources, our first-arrival P-wave tomography only presents the velocity structure of the uppermost 0.5 km. In a future study, we will use the continuous records from the 2D array to perform an ambient seismic noise tomography. The ambient seismic noise tomography is expected to provide S-wave velocity structure coarser but deeper than the current P-wave velocity model presented here. In addition, the S-wave velocity structure will provide more constraints on the deep distribution of tungsten orebodies.

7. Conclusions

In this study, we imaged the high-resolution 3D velocity structure beneath the Zhuxi ore deposit using the data of the Zhuxi experiment conducted in 2017. Our velocity model shows a clear banded distribution, consistent with the local topographic and tectonic structures, and a similar depth-dependent velocity trend with the laboratory measurements of rock samples. The proven orebodies, mainly related to magmatic hydrothermal activities during the Yanshanian period, are shown as high-velocity anomalies beneath the MOB and NOB, corresponding to widespread copper–iron and a few tungsten–molybdenum orebodies. According to the distribution of high-velocity anomalies, we suggest that the stratigraphic boundaries extending outside the prospecting survey area (presenting north-dipping structure) and these potential mineralized areas are probably connected at depth. Our results can help in further evaluating the total reserves, suggesting that seismic tomography can be a useful tool for mineral exploration.

Author Contributions: Conceptualization, B.W. and G.L.; methodology, Y.Z., B.W. and G.L.; software, G.L. and Y.Z.; resources, L.S.; data curation, S.X.; writing—original draft preparation, Y.Z., B.W. and G.L.; writing—review and editing, Y.O., T.W. and R.W.; supervision and project administration, B.W. All authors have read and agreed to the published version of the manuscript.

Funding: This work was supported by the National Natural Science Foundation of China (Grant Nos. 41804063, 41790462, and 41842042).

Acknowledgments: We thank the BGP Inc., China National Petroleum Corporation, Geophysical Exploration Center of China Earthquake Administration, Jiangxi Bureau of Geology and Mineral Exploration and Development for their collaboration on data collection. We are grateful to Q. Wang, H.J. Yao and two reviewers for their constructive comments. The figures are produced by GMT [51].

Conflicts of Interest: The authors declare no conflict of interest.

References

1. Chen, G.H.; Wan, H.Z.; Shu, L.S.; Zhang, C.; Kang, C. An analysis on ore-controlling conditions and geological features of the Cu-W polymetallic ore deposit in the Zhuxi area of Jingdezhen, Jiangxi Province. *Acta Petrol. Sin.* **2012**, *28*, 3901–3914, (in Chinese with English abstract).

2. No. 912 Geological Team. *Final report to detailed survey of Zhuxi copper metallogenic deposit in Fuliang County, Jiangxi province*; Bureau of Geology and Mineral Resources of Jiangxi Province (No. 912 Geological Team): Jiangxi, China, 2016; pp. 1–72.

3. Shu, L.S.; Faure, M.; Wang, B.; Zhou, X.M.; Song, B. Late palaeozoic–early mesozoic geological features of South China: Response to the Indosinian collision events in Southeast Asia. *C. R. Geosci.* **2008**, *340*, 151–165. [CrossRef]

4.   Shu, L.S.; Wang, Y.; Sha, J.G.; Jiang, S.Y.; Yu, J.H.; Wang, Y.B. Jurassic sedimentary features and tectonic settings of Southeastern China. *Sci. China Series D. Earth Sci.* **2009**, *52*, 1969–1978. [CrossRef]

5.   Shu, L.S.; Zhou, X.M.; Deng, P.; Wang, B.; Jiang, S.Y.; Yu, J.H.; Zhao, X.X. Mesozoic tectonic evolution of the Southeast China block: New insights from basin analysis. *J. Asian Earth Sci.* **2009**, *34*, 376–391. [CrossRef]

6.   Xia, Z.Z.; Wang, X.G. Metallogenic and diagenic mechanism of the detachment and nappe orogenesis at the Zhuxi super-large tungsten-copper deposit in Jiangxi Province. *J. Geol.* **2016**, *40*, 552–559, (in Chinese with English abstract).

7.   Chen, G.H.; Shu, L.S.; Shu, L.M.; Zhang, C.; Ouyang, Y.P. Geological characteristics and mineralization setting of the Zhuxi tungsten (copper) polymetallic deposit in the Eastern Jiangnan Orogen. *Sci. China Earth Sci.* **2016**, *59*, 803–823. [CrossRef]

8.   Ouyang, Y.P.; Rao, J.F.; Yao, Z.Y.; Zhou, X.R.; Chen, G.H. Mineralization and prospecting direction of the "Zhuxi type" skarn deposit. *Geol. Sci. Technol. Inform.* **2018**, *37*, 148–158, (in Chinese with English abstract).

9.   Yan, T.J. Comprehensive study of geophysical methods to the deep propecting of tungsten-copper polymetallic ore in ZhuXi area of Jiangxi. Ph.D. Thesis, Jilin University, Changchun, China, 2017. (in Chinese with English abstract).

10.   Wang, C.B.; Rao, J.F.; Chen, J.G.; Ouyang, Y.P.; Qi, S.J.; Li, Q. Prospectivity mapping for "Zhuxi-type" copper-tungsten polymetallic deposits in the Jingdezhen region of Jiangxi Province, South China. *Ore Geol. Rev.* **2017**, *89*, 1–14. [CrossRef]

11.   Lü, Q.T.; Zhang, X.P.; Tang, J.T.; Jin, S.; Liang, L.Z.; Niu, J.J.; Wang, X.B.; Lin, P.R.; Yao, C.L.; Gao, W.L.; et al. Review on advancement in technology and equipment of geophysical exploration for metallic deposits in China. *Chin. J. Geophys.* **2019**, *62*, 3629–3664, (in Chinese with English abstract).

12.   Xu, X.; Gao, R.; Guo, X.Y.; Keller, G.R. The crustal structure of the Longmen Shan and adjacent regions: An integrated analysis of seismic profiling and gravity anomaly. *Chin. J. Geophys.* **2016**, *51*, 26–40, (in Chinese with English abstract).

13.   Malehmir, A.; Durrheim, R.; Bellefleur, G.; Urosevic, M.; Juhlin, C.; White, D.; Milkereit, B.; Campbell, G. Seismic methods in mineral exploration and mine planning: A general overview of past and present case histories and a look into the future. *Geophysics* **2012**, *77*, WC173–WC190. [CrossRef]

14.   Shi, Y.; Xu, Y.X.; Yang, B.; Zhou, P.; Liu, S.Y. Three-dimensional audio-frequency magnetotelluric imaging of Zhuxi copper-tungsten polymetallic deposits, South China. *J. Appl. Geophys.* **2020**, *172*, 103910. [CrossRef]

15.   Pan, X.F.; Hou, Z.Q.; Li, Y.; Chen, G.H.; Zhao, M.; Zhang, T.F.; Zhang, C.; Wei, J.; Kang, C. Dating the giant Zhuxi W–Cu deposit (Taqian–Fuchun Ore Belt) in South China using molybdenite Re–Os and muscovite Ar–Ar system. *Ore Geol. Rev.* **2017**, *86*, 719–733. [CrossRef]

16.   Ouyang, L.B.; Li, H.Y.; Lü, Q.T.; Li, X.F.; Jiang, G.M.; Zhang, G.B.; Shi, D.N.; Zheng, D.; Zhang, B.; Li, J.P. Crustal shear wave velocity structure and radial anisotropy beneath the Middle-Lower Yangtze River metallogenic belt and surrounding areas from seismic ambient noise tomography. *Chin. J. Geophys.* **2015**, *58*, 4388–4402, (in Chinese with English abstract).

17.   Xu, X.; Gao, S.L. The structure and formation of the Cenozoic fault basin in the Lower Yangtze region. *Earth Sci. Front.* **2015**, *22*, 148–166, (in Chinese with English abstract).

18.   Jarvis, A.; Reuter, H.I.; Nelson, A.; Guevara, E. Hole-filled seamless SRTM data V4: International Centre for Tropical Agriculture (CIAT). 2008. Available online: http://srtm.csi.cgiar.org (accessed on 7 February 2020).

19.   Wang, W.T.; Wang, X.; Meng, C.M.; Dong, S.; Wang, Z.G.; Xie, J.J.; Wang, B.S.; Yang, W.; Xu, S.H.; Wang, T. Characteristics of the seismic waves from a new active source based on methane gaseous detonation. *Earthq. Res. China* **2019**, *33*, 354–366.

20.   Wei, Z. Design of a P-wave seismic vibrator with advanced performance. *Geoarabia* **2008**, *13*, 123–136.

21.   Wang, B.S.; Zhu, P.; Chen, Y.; Niu, F.L.; Wang, B. Continuous subsurface velocity measurement with coda wave interferometry. *J. Geophys. Res.* **2008**, *113*, B12313. [CrossRef]

22.   Caldwell, J.; Dragoset, W. A brief overview of seismic air-gun arrays. *The Leading Edge* **2000**, *19*, 898–902. [CrossRef]

23.   Wang, B.S.; Tian, X.F.; Zhang, Y.P.; Li, Y.L.; Yang, W.; Zhang, B.; Wang, W.T.; Yang, J.; Li, X.B. Seismic Signature of an Untuned Large-Volume Airgun Array Fired in a Water Reservoir. *Seismol. Res. Lett.* **2018**, *89*, 983–991. [CrossRef]

24. Yang, W.; Wang, B.S.; Ge, H.K.; Song, L.L.; Yuan, S.Y.; Li, G.Y. Characteristics and signal detection method of accurately controlled routinely operated signal system. *J. China Univ. Petrol.* **2013**, *37*, 50–69, (in Chinese with English abstract).

25. Zhang, Y.P.; Wang, B.S.; Wang, W.T.; Xu, Y.H. Preliminary result of tomography from permanent station in the Anhui air-gun experiment. *Earthq. Res. China* **2016**, *32*, 3331–3342, (in Chinese with English abstract).

26. Jiang, S.M.; Wang, B.S.; Zhang, Y.P.; Chen, Y. The influence of noise on the stacking effect of airgun signal and the automatic data screening method. *J. Seismol. Res.* **2017**, *40*, 534–542, (in Chinese with English abstract).

27. Chen, Y.; Wang, B.S.; Yao, H.J. Seismic airgun exploration of continental crust structures. Sci. *China Earth Sci.* **2017**, *60*, 1739–1751. [CrossRef]

28. Thurber, C.H. Earthquake locations and three-dimensional crustal structure in the Coyote Lake Area, central California. *J. Geophys. Res. Solid Earth* **1983**, *88*, 8226–8236. [CrossRef]

29. Thurber, C.H. Local earthquake tomography: Velocities and Vp/Vs-theory. In *Seismic Tomography: Theory and Practice*; Iyer, H.M., Hirahara, K., Eds.; Chapman and Hall: London, UK, 1993; pp. 563–583.

30. Thurber, C.H.; Eberhart-Phillips, D. Local earthquake tomography with flexible gridding. *Comput. Geosci.* **1999**, *25*, 809–818. [CrossRef]

31. Evans, J.; Eberhart-Phillips, D.; Thurber, C.H. User's manual for SIMULPS12 for imaging Vp and Vp/Vs: A derivative of the 'Thurber' tomographic inversion SIMUL3 for Local Earthquakes and Explosions. USGS Open-File Report. Available online: https://pubs.usgs.gov/of/1994/0431/report.pdf (accessed on 7 February 2020).

32. Lin, G.; Thurber, C.H. Seismic velocity variations along the rupture zone of the 1989 Loma Prieta earthquake, California. *J. Geophys. Res. Solid Earth* **2012**, *117*, B09301. [CrossRef]

33. Lin, G. Three-Dimensional Seismic Velocity Structure and Precise Earthquake Relocations in the Salton Trough, Southern California. *B. Seismol. Soc. Am.* **2013**, *103*, 2694–2708. [CrossRef]

34. Lin, G.; Shearer, P.M.; Matoza, R.S.; Okubo, P.G.; Amelung, F. Three-dimensional seismic velocity structure of Mauna Loa and Kilauea volcanoes in Hawaii from local seismic tomography. *J. Geophys. Res.* **2014**, *119*, 4377–4392. [CrossRef]

35. Kissling, E.; Ellsworth, W.L.; Eberhart-Phillips, D.; Kradolfer, U. Initial reference models in local earthquake tomography. *J. Geophys. Res.* **1994**, *99*, 19635–19646. [CrossRef]

36. Kissling, E. *Velest User's Guide*; Internal Report 26; Institute of Geophysics, ETH Zurich: Zürich, Switzerland, 1995.

37. Matrullo, E.; Matteis, R.D.; Satriano, C.; Amoroso, O.; Zollo, A. An improved 1-D seismic velocity model for seismological studies in the Campania-Lucania region (Southern Italy). *Geophys. J. Int.* **2013**, *195*, 460–473. [CrossRef]

38. Eberhart-Phillips, D. Three-dimensional velocity structure in northern California Coast Ranges from inversion of local earthquake arrival times. *B. Seismol. Soc. Am.* **1986**, *76*, 1025–1052.

39. He, X.R.; Chen, G.H.; Liu, J.G.; Zhang, C. On the copper-tungsten prospecting orientation in Zhuxi region. *China Tungsten Ind.* **2011**, *26*, 9–14, (in Chinese with English abstract).

40. Meng, Z.Y.; Rao, J.F.; Ouyang, Y.P.; Wei, J.; Luo, L.C.; Pan, L. Geochemical Characteristics and Minerogenic Potential of Hongmeiling Granitic Porphyry in Taqian-Fuchun Metallogenic Belt. *Acta Mineral. Sin* **2017**, *37*, 486–494, (in Chinese with English abstract).

41. Huang, Z.G.; Li, Z.Q.; Yang, Q.W. Review and reflection of Zhuxi mine drilling construction in Jiangxi province. *Explor. Eng. (Rock & Soil Drill. Tunn.)* **2017**, *44*, 42–46, (in Chinese with English abstract).

42. Xu, H.; Wang, Q.; Ma, Z.G.; Zhou, F.; Wang, L.S. Seimic properties of typical rocks in South China. *Geol. J. China Univ.* **2012**, *17*, 469–478, (in Chinese with English abstract).

43. Liu, Z.Q.; Liu, S.B.; Chen, Y.C.; Wang, C.H.; Wan, H.Z.; Chen, G.H.; Li, S.S.; Liang, L.J. LA-ICP-MS zircon U-PB isotopic dating of lamprophyre located Zhuxi copper-tungsten mine of Jiangxi Province and geological significance. *Rock Miner. Anal.* **2014**, *33*, 758–766, (in Chinese with English abstract).

44. Zhou, X.M.; Sun, T.; Shen, W.Z.; Shu, L.S.; Niu, Y.L. Petrogenesis of Mesozoic granitoids and volcanic rocks in South China: A response to tectonic evolution. *Episodes* **2006**, *29*, 26–33. [CrossRef]

45. Mao, J.W.; Xie, G.Q.; Li, X.F.; Zhang, C.Q.; Wang, Y.T. Mesozoic large-scale mineralization and multiple lithospheric extension in South China. *Acta Geol. Sin-Engl.* **2006**, *80*, 420–431.

46. Li, Y.; Pan, X.; Zhao, M.; Chen, G.H.; Zhang, T.F.; Liu, X.; Zhang, C. LA-ICP-MS zircon U-Pb age, geochemical features and relations to the W-Cu mineralization of granitic porphyry in Zhuxi skarn deposit, Jingdezhen, Jiangxi. *Geol. Rev.* **2014**, *60*, 693–708, (in Chinese with English abstract).

47. Pan, X.F.; Hou, Z.Q.; Zhao, M.; Chen, G.H.; Rao, J.F.; Li, Y.; Wei, J.; Ouyang, Y.P. Geochronology and geochemistry of the granites from the Zhuxi W-Cu ore deposit in south china: Implication for petrogenesis, geodynamical setting and mineralization. *Lithos* **2018**, *304–307*, 155–179. [CrossRef]

48. Song, S.W.; Mao, J.W.; Xie, G.Q.; Yao, Z.Y.; Chen, G.H.; Rao, J.F.; Ouyang, Y.P. The formation of the world-class zhuxi scheelite skarn deposit: Implications from the petrogenesis of scheelite-bearing anorthosit. *Lithos* **2018**, *312–313*, 153–170. [CrossRef]

49. Su, X.Y.; Wang, D.H.; Wang, C.H.; Liu, S.B.; Liu, J.G.; Fan, X.T.; Kuan, L.H.; Wan, H.Z.; Zhang, C.; Huang, G.Q. Metallogenic element geochemistry characteristics and genesis of the Zhuxi copper-tungsten polymetallic deposit, Jiangxi Province base using XEPOS X-ray fluorescence spectrometry. *Rock Miner. Anal.* **2013**, *32*, 959–969, (in Chinese with English abstract).

50. Ouyang, Y.P.; Chen, G.H.; Rao, J.F.; Zeng, X.H.; Zhang, C.; Wei, J.; Kang, C.; Shu, L.M.; Luo, L.C. On geological features and metallogenic mechanism of Zhuxi Cu-W polymetallic deposit in Jingdezhen of Jiangxi. *J. Geol.* **2014**, *38*, 359–364, (in Chinese with English abstract).

51. Wessel, P.; Smith, W.H.F.; Scharroo, R.; Luis, J.; Wobbe, F. Generic Mapping Tools: Improved Version Released. *Eos Trans. Am. Geophys. Union* **2013**, *94*, 409–410. [CrossRef]

© 2020 by the authors. Licensee MDPI, Basel, Switzerland. This article is an open access article distributed under the terms and conditions of the Creative Commons Attribution (CC BY) license (http://creativecommons.org/licenses/by/4.0/).

Article

Magnetic Survey for Iron-Oxide-Copper-Gold (IOCG) and Alkali Calcic Alteration Signatures in Gadarwara, M.P, India: Implications on Copper Metallogeny

P.V. Sunder Raju * and K. Satish Kumar

Council of Scientific and Industrial Research-National Geophysical Research Institute, Uppal Road, Hyderabad, Telangana 500007, India; satishmarine_777@yahoo.co.in
* Correspondence: pvsraju@ngri.res.in

Received: 24 April 2020; Accepted: 6 July 2020; Published: 29 July 2020

Abstract: A government airborne geophysical survey flown in the late 1970s detected a large Magnetic anomaly at Gadarwara, Madhya Pradesh, in north-central India. Deep drilling indicates that the oval-shaped Magnetic anomaly is caused by underlying Magnetite-bearing banded iron formation belonging to the Mahakoshal Formation of Archean to Early Proterozoic age. The anomaly is hosted in a tectonic rift zone (Narmada-Son Lineament). After drilling alluvium up to 312 m thick, rocks intersected to depths of 612 m provided core samples for research. Broadly speaking, the samples contain banded hematite jaspilite (BHJ) and banded Magnetite (BM) iron formation with pervasive carbonate alterations. Three vertical diamond drill holes were drilled along a 1.4 km long N-S transect across the center of the geophysical anomaly. DDH-1, near the northern edge of the anomaly, went through 309 m of alluvium before intersecting bedrock and then cored 303 m of bedrock for a total depth of 612 m. Copper mineralization with appreciable amounts of cobalt, zinc, molybdenum, silver, rare earth elements, uranium and other elements was intersected. The litho-units are highly oxidised and intensely brecciated with hydrothermal overprinting of Na-K metasomatism alteration mineralogy. The second borehole, DDH-2 failed as the core drilling bit stuck in the alluvium and further drilling was abandoned, whereas the third borehole DDH-3 didnot intersect a Magnetite-hematite association and cored only siltstone. Two-dimensional model studies suggest that the signature of high Magnetic anomaly is at a depth of 0.4 km from the surface, with a width of 3.5 km, dipping at 45° in a northerly direction. The causative body has a Magnetic susceptibility of 0.0052 C.G.S. units, suggestive of a hematite with quartz veinlets lithology. Based on predictive Magnetic exploration models for Iron-Oxide-Copper-Gold (IOCG), such deposits can be inferred from geological observations combined with petrophysical data and forward modelling of the observed Magnetic signatures. This paper reports a prospective IOCG-like mineralization style hosted in a rift (Narmada-Son) type of tectonic environment.

Keywords: Gadarwara; central India; mineralization; IOCG; Narmada-Son-Lineament; Magnetic anomaly

---

1. Introduction

Magnetic surveys are rapid and provide clues from prospect to province. Magnetic anomalies can also provide direct signatures of certain types of ore deposit or mineralization. IOCG provinces extent across the globe and range in geological timescales from Archean to Cenozoic and include the modern examples such as Iron-Oxide Copper Gold (IOCG), the Kiruna type IO ± A(Iron Oxide Apatite) [1,2]: Examples include the Gawler craton [3,4], Olympic Dam, Clonclurry, Tennant Creek in Australia [5–8], Kiruna in Sweden and Carajas, in Brazil [9,10], and the lesser understood Wernecke [11],

Great Bear Magmatic zone [12] in the Canadian shield [12], Khetri [6,13] and Singhbhum [14] deposits in India. The majority of these key deposits and districts are Paleo- or Mesoproterozoic in age, they have vast extents <35 km × 1.5 km, occur at depths of 3–10 km in an active Magmatic setting [15].

They can vary in alteration facies (Na-Ca-Fe-K-Mg) and chemical footprints range from cm-m-km scale [16]. The salient features of IOCG include the following [6]: (1) The occurrence of copper (Cu) with or without gold (Au); (2) Hydrothermal vein, breccia or replacement ore styles characteristically in specific structural sites; (3) Abundant Magnetite or hematite; (4) Iron oxides which have low Ti contents compared to those in most igneous rocks and bulk crust, in particular nelsonites and igneous Fe-Ti deposits; (5) Absence of clear spatial association with igneous intrusions.; (6) Strong structural and stratigraphic control; (7) abundant hydrothermal-structural breccia zone. The geological architecture and evolution of India show it to have rocks from Archean to Phanerozoic ages, and suitable geological settings [6] as potential hosts for IOCG/Kiruna-type deposits. However, in India reports on IOCG are documented from the Bhukia IOCG-IOA deposit of Aravalli-Delhi Fold Belt, Rajasthan, western India [14,17], Thanewasna, Western Bastar Craton [18], Singhbhum Craton [14], and Machanur, Eastern Dharwar Craton [19] (Figure 1). The Council of Scientific and Industrial Research- National Geophysical Research Institute (CSIR-NGRI), India carried out an airborne geophysical survey flown in the late 1970s and detected a sizeable Magnetic anomaly at Gadarwara, Madhya Pradesh, in north-central India [20] with only scanty follow-up studies. For the first time, we provide the geophysical signatures using an integrated investigation with new insights on geophysical and geochemical characteristics of these rocks and offer unique constraints on the nature and significance of the IOCG mineralization at Gadarwara, Madhya Pradesh, India.

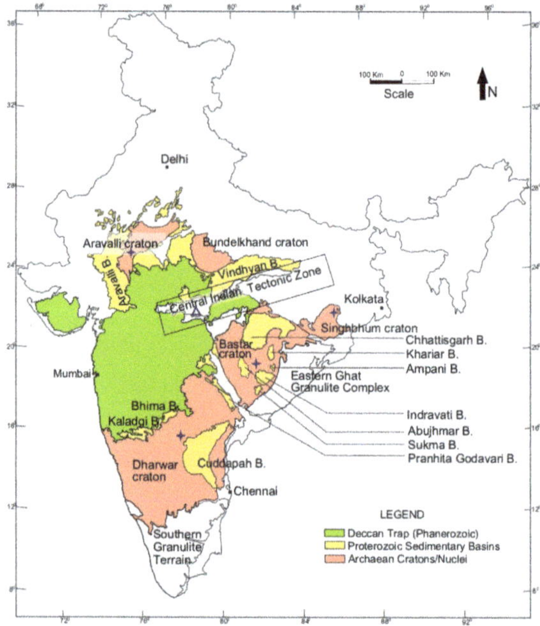

Figure 1. Simplified geological Map of India showing position of five cratonic blocks and sedimentary basins (after Mazumder et al., 2019a [21]) with reported IOCG deposits (blue stars) and study area in (triangle).

2. Regional Geology

The Son- Narmada-Tapti fault (NSF) is one of India's most prominent geomorphic features and cuts across the central part of the Indian subcontinent [1]. This east-northeast (ENE) to west-southwest (WSW) trending lineament extending in length of about ~1200 km from 72.5° E to 82.5°5″ E long and 21.5° N to 24° N latitude (Figure 2a). The NSF structural zone juxtaposes disparate and complex geological and geophysical characteristics and comprises a broad region (up to 30 km wide) of linear topographic features, contemporary seismicity, and hot springs [22]. The NSF is currently expressed as an alluvium-filled tectonic rift zone with northern and southern bounding faults. More importantly, the Son- Narmada-Tapti fault divides the Bundelkhand Craton to the north from the Dharwar Craton to the south. The Bundelkhand Craton consists of Archean amphibolitic and ultramafic slivers of greenstone belt association within a broader region of poly-metamorphosed sialic and TTG gneisses (ca. 3.5–2.7 Ga), metamorphosed volcano-sedimentary rocks, and syn- to post-tectonic granitoids (ca. 2.5–2.4 Ga), and granitoids that include a significant Paleoproterozoic component.The older rock packages are cut by at least three generations of Mafic dyke swarms, most importantly the ca. 1700 Ma dolerite dykes that are likely related to the development of rift basins of the Bijawar (or Mahakoshal Group) and Gwalior Formations. The overall tectonic trend of the Bundelkhand Craton (Figure 2b) is E to ENE flanked along its southern and northwestern Margins by the Pre-Vindhyan siliciclastic shelf sequence of the Bijawar Formation. The eastern, western and southern Margins of the BKC are covered by Vindhyan Group rocks (1400 Ma and younger), while the alluvium is covered in the northern border [23]. The stratigraphy of study area is shown in Table 1.

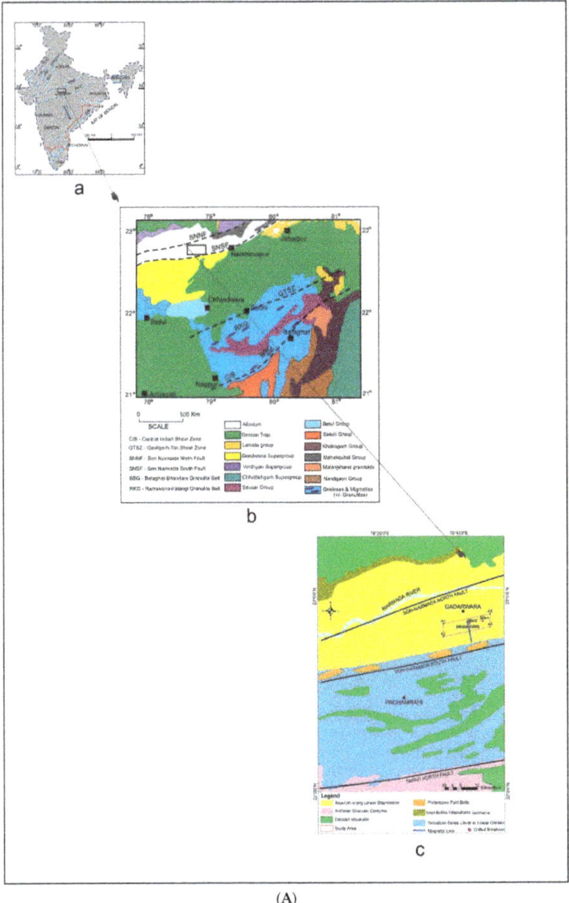

(A)

Figure 2. *Cont.*

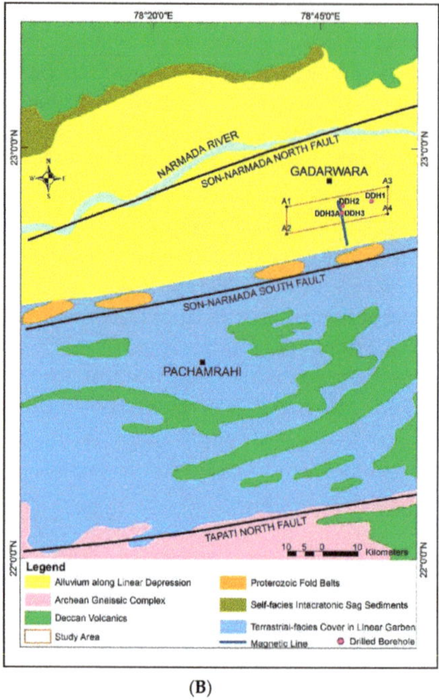

(B)

Figure 2. (**A**): (**a**) Simplified Tectonic Map of India; (**b**) Simplified geological Map of the Central Indian Tectonic Zone in central India, showing all the supracrustal belts and tectonic lineaments (modified after Yedekar et al. 1990 [24], Acharyya and Roy 2000 [25], Chattopadhyay and Khasdeo 2011 [26]; (**c**) Geological Map of Study area (after Sunder Raju., 2014 [27]. (**B**) Geological Map of the Study area.

Table 1. The stratigraphic sequence of the Narmada-Son lineament belt.

Age	Group	
Recent		Alluvium (study area)
Pleistocene		Laterite
Eocene		Trap intrusions, dolerite dykes,
		intertrappean beds, Deccan trap dikes,
		Deccan trap flows, Bagh beds, Lameta beds
		unconformity
Lower Cretaceous	Gondwana	Upper Gondwana
Upper Carboniferous		Lower Gondwanas
Lower Paleozoic	Vindhyan	Upper Vindhyans
Upper Precambrian	Super Group	Lower Vindhyans
Lower Proterozoic	Bijawar Group	Limestones, quartzites
		unconformity
Archean		Granites, gneisses, schists and phyllites

The generalise stratigraphic succession of the region (modified after Geological Survey of India publication: Mallet 1869 [28]; GSI, 1976 [29]).

The Mahakoshal Supergroup (Figure 2b) has a linear, east-north-easterly-trending zone between the two cratonic regions (Dharwar, Bundelkhand) and mostly exposed within the bounding faults of the Narmada Son Lineament [30]. The Bundelkhand craton consist of metavolcanic and metasedimentary rock sequences with significant basic/ultrabasic intrusions formed in a rift environment [30]. The older parts of the Mahakoshal Supergroup are siliciclastic and variably referred to as the Bijawar group (ca 2600–2400 Ma) [31] which at their type section in the Son Valley, lacks the deformation and associated intrusions of the Mahakoshal in the Narmada valley. The Mahakoshal Supergroup includes the Sakoli and Nandgaon bimodal volcanics with Dongargarh and Malanjkhand K-granite island-arc type intrusions from a Paleoproterozoic period (2.2–2.3 Ga) in the Bhandara craton. The Mahakoshal Group exposed in the Son valley was intruded by Alkali syenite and lamprophyre at 1796 Ma to 1610 Ma [32]. The Mahakoshal was deformed by folding and faulting resulting from Meso- to Neoproterozoic convergence [30] and is therefore locally and otherwise known as the Mahakoshal Deformed Zone. The Vindhyan Basin (Figure 2b) rocks unconformably overlap upon older Mahakoshal and Bijawar Supergroups and Bundelkhand cratonic rocks. The Vindhyan Group consist of quartzites of 5–6 km thick pile associated with metasedimentary rocks that are divisible into a Lower Group (Semri Series) of ~1.7 Ga [33] and Upper Group (Kaimur, Bhander and Rewa Series) that are 1.1 to 0.7 Ga. Pyroclastic units and felsites and sedimentological changes indicate changes in basin architecture, likely due to fault reactivations [34]. Vindhyan Group rocks are flat-lying and have not been intruded by granitoids. The northern limit of the Central Indian Tectonic Zone (CITZ) is defined by the ENE-WSW trending Son Narmada North Fault (SNNF) and is traceable from Markundi in Uttar Pradesh in the east to Hosangabad in Madhya Pradesh in the west [25]. Geologically, the lineament bounds are represented by two faults, and the Mahakoshal rocks are dominant in between these two faults, Son Narmada South Fault (SNSF) and Son Narmada North Fault (SNNF). (Figure 2c) [35] Further east, the rift faults and associated folds crop out with their subsidiary structures. Locally along the rift in the east, there are gold and base metal deposits of various types which are economically viable [30]. The BIFs found are classified as of Algoma type [30].

3. Methodology

3.1. Magnetic Survey

A ground Magnetic traverse of 15-line km was carried out in the study area by a private party. The approximate bearing of the traverse was 170–350°. The survey utilized two GEM Systems GSM-19W Magnetometers, equipped with in-built GPS, and a Geometrics G856 base station Magnetometer, recording the diurnal Magnetic variation at 20 s. intervals. The study area is demarcated in Figure 3 with the Magnetic traverse and drill hole locations.

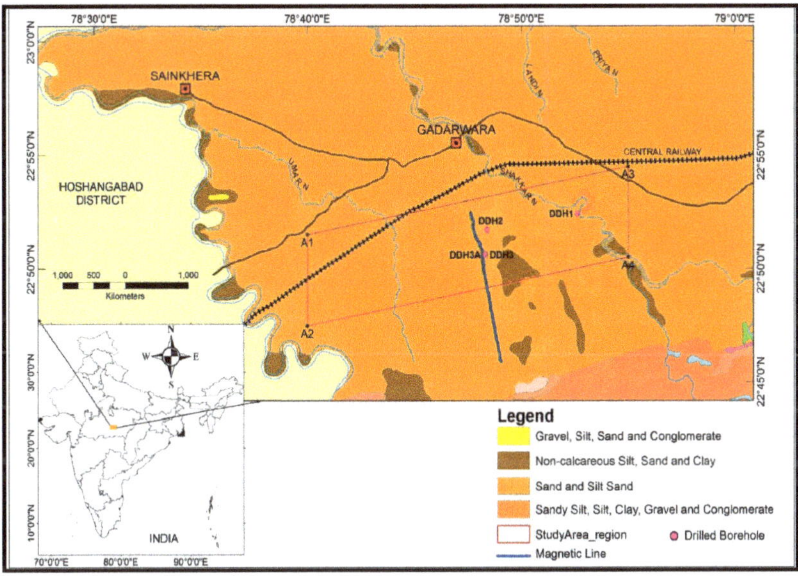

Figure 3. Study area boundaries.

3.2. The Gadarwara Magnetic Anomaly

In 1979, the CSIR-NGRI (-National Geophysical Research Institute, Hyderabad, India) carried out regional aeromagnetic surveys in three blocks along the Narmada Son Lineament for the Geological Survey of India. [36] The Magnetic data collected at a nominal line spacing of 2000 m at an altitude of 1000 m (~3200 feet). This area is mostly covered by thick alluvium and outcrops are sparse with a few exposures of the Mahakoshal Supergroup (Figure 2b). The aeromagnetic survey picked up a glaring anomaly in the central block located towards the south of Gadarwara town (22.552° N, 78.472° E) [31,37]. The magnitude of the anomaly is about a 600 nT, above background values and covers an area of approximately 40 km^2. This Magnetic anomaly is widespread andtrends to the east-northeast, and is about 1800 m wide and 20 km long. A subsequent ground Magnetic survey better constrained the anomaly to about 1200 nT amplitude [31]. The concealed anomaly is caused by Magnetite bearing "banded Iron Formations" or BIFs. BIFs assemblages are commonly associated with IOCG deposits around the globe [6]. The dominant meta-sedimentary rocks consist of cherty BIF (CBIF) with ribbon-like jasperoids, alternating bands of crypto-crystalline to microcrystalline chert and anhedral hematite with little or no clastic admixture. The thickness of hematite layers varies (1–2 mm) and hosts sericite and chlorite-ripidolite commonly. The dominance of hematite consists of specularite with replacement of silicates and Magnetite. The common gangue mineral phases include quartz, Ca(Mg-Fe) carbonates, chlorite, sericite, and apatite. Hematite rocks contain higher sulphidic mineral contents than the Magnetite rocks. REE and U enrichments occur as discrete mineral phases. Accessory zircon and apatite occur as acicular clusters. Magnetite is prominent in all the samples, and in association with hematite. Magnetite grain shapes vary from rounded, sub-rounded, euhedral, amoeboid and mesh-like with Magnetite ex-solution appear like lamellae. Magnetite-dominated assemblages contain apatite as clusters and acicular needles. Sulphides are minor with pyrite >> chalcopyrite. Varying morphologies of zircon, monazite, and xenotime are observed in SEM studies. They are typically of ~25–50 µm in size, but difficult to further quantify because of small size and altered character.

The Main sulphides mineral assemblage consists of chalcopyrite and pyrite. In particular pyrite, are round to sub-round grains and occur in interstitial spaces within fine-grained Magnetite. The sulphidic BIFs contain calcite, siderite or ferroan dolomite. Minor hercynitic spinels also occur. The essential Fe-bearing minerals in carbonate are siderite and ankerite, with lesser ferroan dolomite, calcite and Magnesium siderite.

4. Results

4.1. Drilling Results

In 2007, Adi Gold Mining Pty. Ltd. (New Delhi, India) initiated an exploration program to drill a 1.4 km fence of three vertical diamond drill holes from north to south across the center of the geophysical anomaly. DDH-1 was drilled (Figure 4 and Table 1) from near the northern edge of the anomaly, went through 309 m of alluvium coring 303 m of bedrock for a total depth of 612 m. The litho-units encountered in the drill core (Figure 4) include felsic tuff, metasedimentary rocks dominated by banded chert, and various BIF facies, such as cherty BIF, hematite BIF, Magnetite BIF and carbonate BIF. The samples contained carbonate either as a component of the rock Matrix, within coarse-grained, quartz-rich segregations parallel to the metamorphic foliation, or in cross-cutting veins. Rock samples that were taken from drill cores had coatings of travertine, iron/jaspilite and iron oxides. The sample locations and lithology are listed in Table 2. Furthermore, the XRD results suggest and confirm the presence of sheet silicates (i.e., kaolinite, montmorillonite, dickite, mica, chlorite), iron oxides, quartz, and carbonate, etc. The DDH 1 core was strongly weathered to alteration products with hematite and goethite, and the iron oxides, Making explicit recognition of banded hematized Magnetite, red jasper and white chert beds, difficult. The beds are tightly folded, and veinlets of quartz are not uncommon with pseudomorphs and cubic cavities left by oxidized pyrite. DDH-2 advanced through 288.8 m of alluvium before being abandoned. DDH 3, located near the southern edge of the anomaly, went to 307.0 m depth before being lost in alluvium. DDH-3A (drilled from a location adjacent to DDH-3) intersected bedrock at 312 m and cored until a depth of 430.8 m, retrieving 118.4 m of the core. Although boreholes had been drilled in this region during past exploration efforts looking for coal [38,39], these are presumed to be the first deep drill holes to intersect Precambrian bedrock in the Narmada Son lineament.

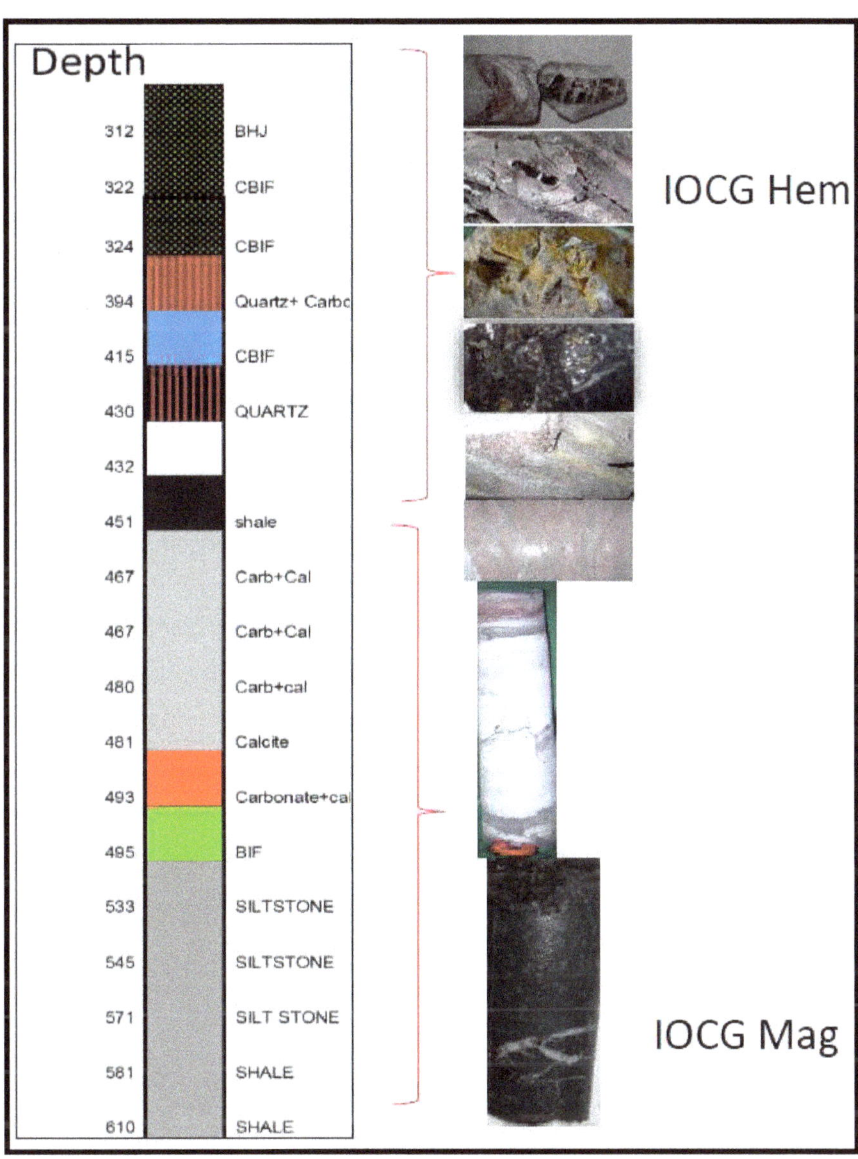

Figure 4. Core Log, Lithounits and core section of borehole DDH-1.

Table 2. Core section lithology of DDH-1.

Depth (in mts)	Litho-Unit	Description
312	BHJ	Dark brown to reddish brown in colour, intensely weathered and fractured, banded-hematite-jasper (BHJ), non-magnetic, specular hematite showing metallic lustre thin < 1 mm, fine grained jasper and hematite, soft yellow limonitic staining is noted at the fractured planes. The sulphides are pyrite, chalcopyrite and iron hydroxide like hematite.
322	Cherty IF	Weathered cherty iron bearing formation, the fine grained chert is hard and shiny lustre with sulphides and mostly pyrite, chalcopyrite and weathered staining on surface could be due leaching of iron oxides
324	Cherty IF	Weathered cherty iron bearing formation, the fine grained chert is hard and shiny lustre with sulphides and mostly pyrite, chalcopyrite and weathered staining on core surface.
394	Quartz + Carbonate	Dark brown to reddish brown in colour, intensely weathered and fractured, banded-hematite-jasper (BHJ), non-magnetic, specular hematite showing metallic lustre thin < 1 mm, fine grained. Due to intense weathering process the argillite's are altered to clay and to limonitic (yellow powdery nature) at carbonate show reactive nature with intense effervescence
415	CIF	Weathered cherty iron bearing formation, the fine grained chert is hard and shiny lustre with sulphides and mostly pyrite, chalcopyrite and weathered staining on surface could be due leaching of iron oxides.
430	Quartz	Fine grained quartz vein sample with saccharoiadal texture.
450	Shale	Grey to dark grey in colour, very fine grained, carbonate rock with calcite occurring as grey bands of sulphidic minerals and often as relict textures. Both fine bands of purple coloured minerals varying from white to purple green.
467	Carbonate + Calcite	Carbonate show reactive nature with intense effervescence with typical calcite texture perfect cleavage and conchoidal with brittle around with gentle hardness.
480	Carbonate + Calcite	Carbonate show reactive nature with intense effervescence with typical calcite texture perfect cleavage and conchoidal with brittle around with gentle hardness.
481	Calcite	Calcite typical texture with perfect cleavage and conchoidal with brittle around with gentle hardness.
493	Carbonate + Calcite	Carbonate show reactive nature with intense effervescence with typical calcite texture perfect cleavage and conchoidal with brittle around with gentle hardness.
495	Iron Formation	Dark brown to reddish brown in colour, intensely weathered and fractured, banded-hematite-jasper(BHJ), non-magnetic, specular hematite showing metallic lustre thin < 1 mm, fine grained jasper and hematite, soft yellow limonitic staining is noted at the fractured planes. The sulphides are pyrite, chalcopyrite and iron hydroxide like hematite.
533	Siltstone	Grey to dark grey in colour, very fine grained, carbonate rock with calcite occurring as bands with grey mineral and often as relict texture, fine bands of purple coloured mineral (carbonate-as it reacts with HCl), the colour of carbonates vary from white to purple and green; fractured. Magnetite (very fine-grained) crystals as fine zones, Pyrite, Chalcopyrite.
545	Siltstone	Grey to dark grey in colour, very fine grained, carbonate rock hard and fissile. The colour of carbonates vary from white to purple and green; fractured. Magnetite (very fine-grained) crystals as fine zones, Pyrite, Chalcopyrite.
571	Siltstone	Grey to dark grey in colour, very fine grained, carbonate rock with calcite occurring as bands with grey mineral and often as relict texture, fine bands of purple coloured mineral (carbonate—as it reacts with HCl), the colour of carbonates vary from white to purple and green; fractured. Magnetite (very fine-grained) crystals as fine zones, Pyrite, Chalcopyrite.
581	Siltstone	Grey to dark grey in colour, very fine grained, carbonate rock with calcite fractured. Magnetite (very fine-grained) crystals as fine zones, Pyrite, Chalcopyrite and other sulphides.
610	Shale	Dark grey to black in colour, hard, very fine grained shale with inter-layers of carbonate; well foliated, fine grains sulphides (pyrite) randomly oriented.
610	Shale	Dark grey to black in colour, hard, very fine grained shale with inter-layers of carbonate; well foliated, fine grains sulphides (pyrite) randomly oriented.

The lithologies encountered in DDH3 consist of banded chert and fine-grained felsic tuff. Some of the beds are thinly laminated, and others are folded in convoluted patterns. Most notably, the drilling intersected an approximately 50 m thick Magnetite-bearing BIF assemblage. The samples contain banded hematite jaspelite (BHJ) and banded Magnetite (BM) iron formation. Locally, both foliaform and cross-cutting quartz veinlets, with and without traces of fine-grained pyrite are present. All samples have pervasive carbonate alteration. The BIF assemblage likely belongs to the Mahakoshal Supergroup of Archean to early Proterozoic age [30]. This Magnetic anomaly could be explained by the Magnetics method response from underlying Magnetite bearing "banded iron formation" or BIF with around 80% ferruginous chert-bearing Magnetite (BM), banded hematite jaspilite (BHJ) oxidized iron formations.

4.2. Magnetic Anomaly Modelling

The Magnetic anomalies interpretation is normally carried by Matching observed and calculated values of the anomalous Magnetic field. Talwani and Heirtzler (1964) were first to examine a nonmagnetic space containing a uniformly Magnetized two-dimensional structure approximated by a polygonal prism and to suggest a numerical and computational technique of the forward modeling [40].

The modelling is based on the measurements sampled at 200 m spacing for a 15 km long profile with traverse bearing of 170–350° (Figures 5 and 6). The observed ground TMI anomaly is bipolar in nature with an amplitude of ~1200 nT (Figure 6). The basement depth is estimated and calculated using a 1D log normalized radially averaged power spectrum [41,42] (Figure 7a,b). The straight line part of the spectrum at low wave number is considered for depth estimation, ~300 m is attributed to the core litho-log of DDH-1 borehole i.e., top alluvium, outer hematite, inner hematite, alteration of sodic/potassic rocks, hematite breccia, hematite breccia with quartz veinlets and at depth a Massive Magnetite. To understand the signatures of the Magnetic anomalies, initially a synthetic modeling is carried out. (Figure 8) As, the present study area is located in the south of Gadarwara, m.P, Central India bounded by geographic longitudes 78.7931° to 78.8173° and latitudes 22.7651° to 22.8743°. To estimate the geomagnetic components such as Total Magnetic field, Inclination and Declination, we adopted International Geomagnetic Reference Field (IGRF) model-2005 which is closest to surveyed year 2006. As per the IGRF-2005 model, the Total Magnetic Intensity field, geomagnetic inclination and declination of the study area are 44,900 nT, 33° and −0.43° respectively using Geosoft, 2006 version. The main objective of the synthetic model is to understand the shape of the Magnetic anomaly at this location with respect to observed field Magnetic anomaly. The shape of synthetic Magnetic anomaly will provide the origin of the Magnetic sources at the time of formation. In general, shape of the geomagnetic anomaly varies from equator to pole with a high negative anomaly peak at equator, bi-polar anomaly at 45° latitude and high positive anomaly peak at poles from south to north [43]. The nature of Magnetic anomaly at this geomagnetic latitude is also dependent to width, depth, depth extent, susceptibility of the Magnetic source, however these parameters influence the width and amplitude of the anomaly only. Hence synthetic anomaly is generated with the known constraints such as Magnetic total field, inclination and declination (IGRF, 2005 model), depth (from DDH 1 borehole data), width [31], dip [31] of the Magnetic sources. In the present case, we observed similarities both in synthetic and observed Magnetic i.e., high positive to low negative from south to north. Hence, remanence Magnetisation might be not present in the causative sources. The shape and width of the anomaly suggests a basement source up to 3.5 km in width and Rule of thumb depth estimations and examination of the amplitude fall-off rates suggest a depth to Magnetic source of approximately 396 m depth.

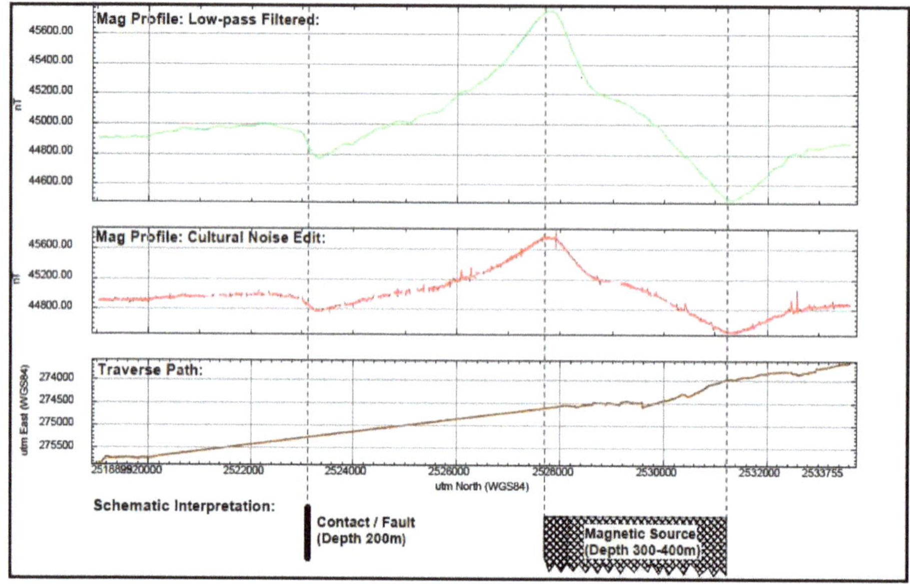

Figure 5. The Magnetic profile along traverse path, Magnetic data and low pass filter.

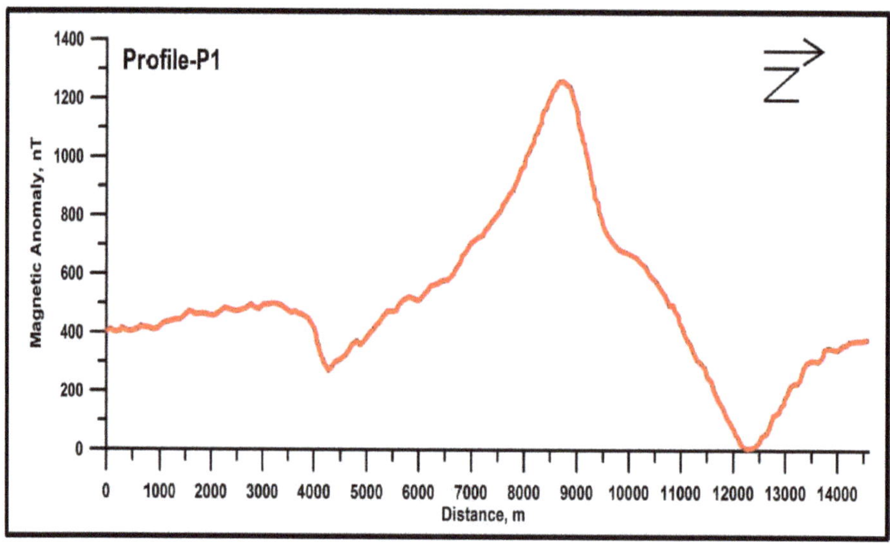

Figure 6. Observed total intensity Magnetic anomaly along profile P1.

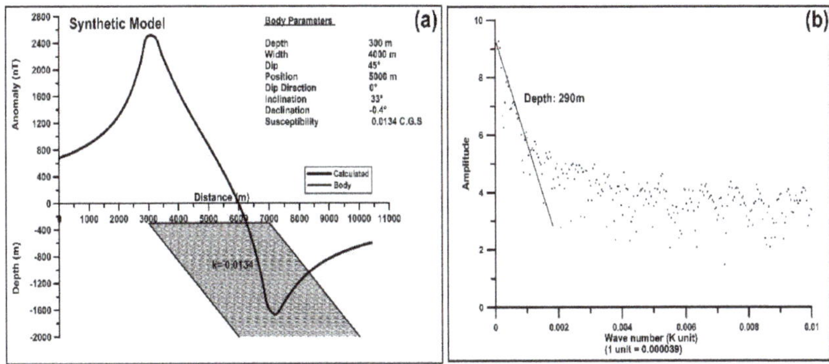

Figure 7. (**a**) synthetic Magnetic anomaly at geomagnetic latitude and (**b**) power spectrum Map showing the depth.

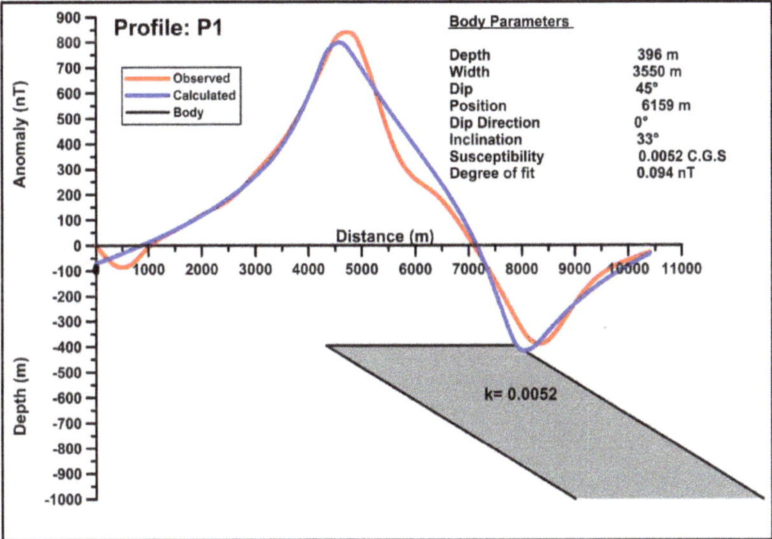

Figure 8. Interpreted 2D depth model for Magnetic anomaly.

The two dimensional model studies were carried using MaGMOD, Ltd., Toronto, Canada [31,44]. The geomagnetic Inclination and declination (33° and −0.4°), depth and width (300 m and 2000 m) of the causative source, dip (45°), Magnetization (600 nT) and strike (E-W) direction were used as initial inputs. A bipolar anomaly between 4000 and 14,500 m is considered for model studies from the profile P1 (Figure 8). The calculated Magnetic anomaly from the interpreted depth model corroborated with the observed anomaly with an R.M.S. error of 0.094 nT (Figure 8). The model studies suggest that the signature of high Magnetic anomaly is obtained from the surface to a depth of 400 m, with the mineralized body width of 3500 m at a dip angle of 45° towards north direction. The magnetic susceptibility 0.0052 C.G.S. units correspond to hematite with quartz veinlets lithology.

There is a slight possibility that an additional Magnetic source is located on the southern edge of the broader Magnetic source Manifested by a bipolar anomaly super-position on the Magnetic high coincident with the southern edge. An additional anomaly in the southern portion of the traverse is indicative of a contact or faulting. In general, the magnetic signatures of IOCG systems reflect superposed or juxtaposed Magnetic anomalies inferring a deeper source for the Magnetic anomaly.

The possible reason could be that the hematite rich zone has been heated >600 °C by regional or contact metamorphism and therefore might have acquired sufficient remanence to produce Magnetic anomalies. Hence, the observed Magnetic anomaly is due to the presence of hematite or Magnetite, weathered and oxidised/brecciated iron rich layers at 396 m depth.

5. Discussion

The drill holes DDH 1 and DDH 3 are the first deep drill holes to intersect Precambrian bedrock in the Narmada Son lineament. Unusual geological litho-units cause the geophysical anomaly at depths probably due to underlying Magnetite bearing "banded iron formation" or BIF of around 80% ferruginous chert-bearing Magnetite (BM), banded hematite jaspelite (BHJ) oxidised iron formations. The basinal fluids preserved in evaporate-bearing sedimentary successions are the important and significant contribution to the fluid budget of hematite-rich IOCG systems. The mineral footprints in the core samples investigated from Gadarwara suggest that samples have undergone hydrothermal, and oxidation alterations of primary mineralisation [45]. The hydrothermal footprint is expressed by intense silicification related to pervasive quartz veinlets. These hydrothermal alteration systems [2,46,47] are typically expressed by silicate and oxide alteration minerals and can be characterized by three zones including: (1) An upper hematite- and/or Magnetite-rich zone, ~1–3 km in horizontal diameter, associated with copper and iron sulphides (e.g., chalcopyrite, bornite and pyrite) together with sericite, chlorite and carbonate alteration. In the Olympic Dam copper-gold province it is evident that higher grade copper-gold and/or uranium/rare earth element mineralisation occurs within the upper, hematitic zones [48]. (2) A mid-depth zone rich in Magnetite with biotite or potassium feldspar ± minor pyrite, showing petrological similarities to core samples investigated typically of a few hundred meters to ~1–3 km in lateral dimensions. (3) A lower zone with sodic-calcic alteration rich in albite, actinolite ± clinopyroxene (e.g., diopside), with minor Magnetite, potentially extending laterally up to 5–10 km in Major IOGC provinces such as the Olympic copper-gold province and Cloncurry IOCG district. The major four essential components in IOCG ore systems of Olympic Dam type are (a) the crustal architecture i.e., reactivated orogenic architecture occurring at craton Margin settings, (b) The energy sources to ascent the hydrothermal fluids i.e., High temperature associated with A and I types felsic and coeval Mafic-ultramafic Magmatism. The two stage of mineralization from deep source high temperature brine of Magmatic-hydrothermal and/or basinal origins and shallow sourced lower temperature fluids of meteoric origin, (c) early high temperature iron oxides and (d) late Cu-Au, Cu from Mainly Mafic sources and U could be from felsic sources and large scale potassic alteration features. The extensive alteration zones of Magnetite, K-Feldspar, actinolite, pyrite, apatite, carbonate, quartz and chalcopyrite produce the Magnetic anomalies [7]. The similarities of oxidised alteration zones May be present above or lateral to Magnetite alteration with high grades of Cu-Au mineralisation predicted within hematitic alteration near the transition to Magnetite. Whereas, in Cloncurry Cu-Au mineralization overprinting the banded iron formation in similarity to Gadarwara type.

On a cratonic scale in all the IOCG districts the lithounits include felsic intrusions, Mafic dykes, Mafic intrusions, felsic volcanic, BIFs, clastic sediments and carbonate meta evaporates. The Gadarwara is located at the craton Margin hosting meta-sediments especially BIFs with high Magnetite contents with cross cutting dykes, especially associated with shales. The association of copper-gold mineralization with hematitic alteration and hydrothermal Magnetite with associated Cu-Au transported by late oxidizing brine solutions that had reacted with meta-sedimentary rocks such as BIFs. The shales could have supplied required sulphur for the sulphide enrichments especially copper sulphides such as chalcopyrite. The geological exploration indicators of REE-bearing meta sedimentary hosts include intraplate continental environments rifting and/faulting, alteration halos, and indicator minerals.

The key styles of copper-gold mineralisation in the Cloncurry district, to the east of Mount Isa is associated with Magnetite and potassium feldspar alteration (e.g., Ernest Henry deposit) whereas another style occurs with pyrrhotite and Magnetite (e.g., Eloise deposit. Both the Ernest Henry and

Eloise deposits are closely associated with intense Magnetite alteration. In several, Australian deposits Pressure-Temperature modelling shows that the contact zones between Magnetite- and hematite-rich alteration is highly favourable for the formation of higher grade copper-gold mineralisation [5] similar to Gadarwara (Figures 9 and 10). Such redox gradients also appear to be important in the high grade gold-copper-bismuth deposits of the Tennant Creek district, where both Magnetite- and hematite-rich deposits are present in association with chlorite ± sericite alteration [6–9]. Therefore, the most attractive copper-gold exploration targets identified in Olympic Dam, Tennant creek [4,7], are those relatively small bodies (e.g., <15 km width) where exceptionally dense and Magnetic Material (magnetite ± pyrrhotite) is laterally or vertically in contact with exceptionally dense bodies of low Magnetic susceptibility (e.g., hematite ± pyrite ± copper sulphides). At Olympic Dam, the zonation is vertically from Magnetite dominated at depth to hematite on surface i.e., upper levels. This pattern could have been altered by Major tectonic activity in the form of tilting or by faulting. It leads to debate on the Magnetic signatures of IOCG deposits which May reflect superposed or juxtaposed Magnetic anomalies [8,49]. At Gadarwara the litho-units like iron rich laminated mudstone-siltstone at depth probably formed as a submarine sedimentary facies. The retrogression and or hydrothermal alteration at upper greenschist/lower amphibolite facies led to Magnetite replacement of protolithic hematite. At lower temperature <150 °C veins of hematite develop chalcedony of quartz [1,2]. In Table 3, the comparison of IOCG deposits like Olympic Dam, Cloncurry, Kiruna, Phalborwa and Bayan Obo show similarities in ore body shape such as 1. irregular breccia, mineralization style 2. dominant minerals like Magnetite, hematite and copper sulphides and 3. enriched in copper, gold, iron ore, and REE enrichments. 4. The alteration zonation is ranging from sodic to calcic and potassic corroborated in drill core samples with Gadarwara.

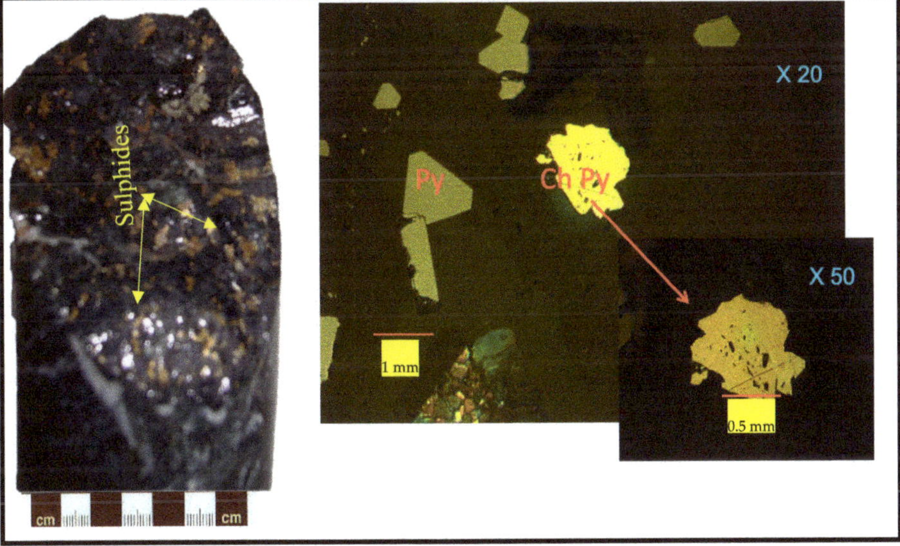

Figure 9. The drill core with sulphides, and petrography reveals Pyrite (Py) and Chalcopyrite (Chpy).

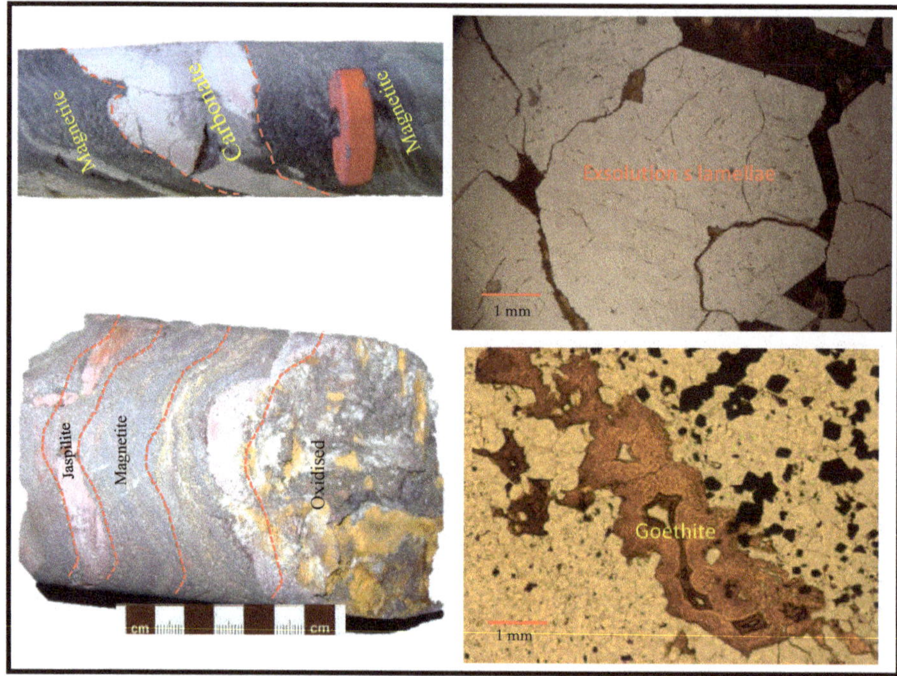

Figure 10. Oxidized brecciated iron formation with goethite and martite.

Table 3. Comparison of similarities of Gadarwara with other IOCG deposits.

Giant Ore Deposit	Type	Mineralisation	Alteration	Commodity	Orebody
Olympic Dam, Australia	Olympic Dam	Magnetite-hematite-bornite-chalcopyrite-breccia Matrix	Potassic	Fe, Cu, Au, Ag, REE, U	Pipe like and irregular breccia
Osborne, Queensland, Australia	Cloncurry	Magnetite-hematite-apatite replaced by Cu-Fe sulphides, Au etc.	Potassic	Cu, Au, Ag, Bi, Co, W	Stratabound vein, breccia
Kiirunavaara, Sweden	Kiruna	Massive Magnetite-apatite-actinolite	sodic	Fe ± Cu, Au	Tabular, pipe like, irregular
Magnitogorsk, Urals, Russia	Iron skarn	Massive Magnetite-garnet-pyroxene	Sodic	Fe ± Cu, Au	Stratabound lensoid, irregular
Phalaborwa, South Africa	Phalaborwa	Magnetite, apatite, fluorite.Cu sulphides	Sodic + potassic	Cu, Au, Ag, REE, PGE, vermiculte, Magnetite, P, U, Zr, Se, Te, Bi	Veins, layers, disseminations
Bayan Obo, Mongolia, China	Bayan Obo	Magnetite, hematite, bastnaesite, Fe-Ti-Cr-Nb oxides, monazite	Weathering of Fe oxides	REE	disseminations
Madhya Pradesh, Central India	Gadarwara	Magnetite, hematite, Cu-Fe, Au, monazite, apatite, zircon, sulphides	Sodic + potassic	Cu, Ni, Au, REE	Irregular breccia
Western Bastar Craton, India	Thaneswasna	Magnetite, hematite, pyrite, arsenopyrite, chalcopyrite, monazite	Potassic	Fe-Cu-Ba-Au-Ag-Th	breccia

The interpretation of airborne and ground Magnetic profiles using a full dike model provided the initial clues with respect to depth, width, dip, and susceptibility of the causative source as 200 m, 4000 m, 42° N, and 0.004 CGS, respectively, in the Gadarwara region [2]. In the initial stages of exploration this paved a way for further detailed ground survey and drilling data were corroborated with the 2D Magnetic modelling studies. The drill hole DDH-1 assay log (Figure 11) distinctly show enrichment of Cu, Au, Cr, U and $\sum$ REE's (total REE's) and corroborated with global model (Figure 12) as Gadarwara type (GT).

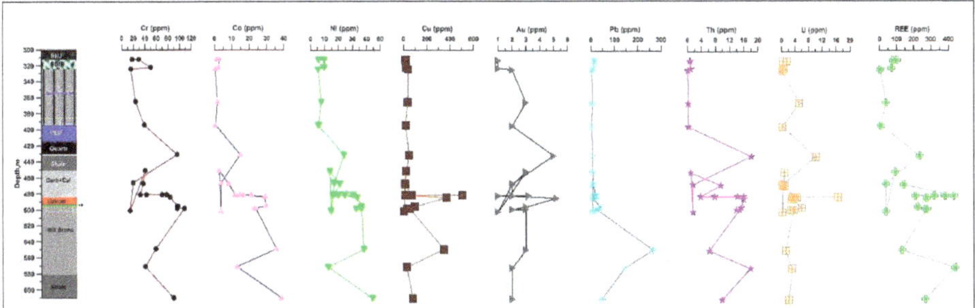

Figure 11. Lithology of DDH-1 drill core with vs assay data.

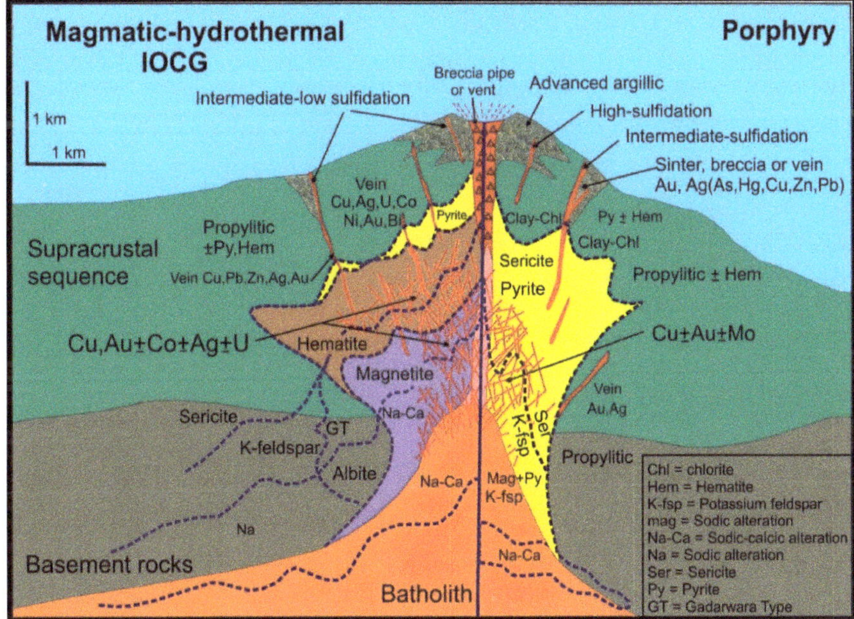

Figure 12. Modified Schematic model for Magmatic-hydrothermal systems illustrating relationship between S-rich porphyry Cu ± Mo ± Au deposits (right side) and S-poor Magmatic-hydrothermal iron oxide-copper-gold (MH-IOCG) deposits (left side). Approximate hydrothermal mineral transitions and relative spatial footprint are shown for albite–K-feldspar, K-feldspar–sericite, and magnetite-hematite transitions. Prograde and retrograde overprinting of alteration assemblages and mineralization is common. Spatial relationships for porphyry deposits after Seedorff et al. (2005) [50], Sillitoe (2010) [51], and Richards (2011) [52]; for MH-IOCG, after Hitzman et al. (1992) [48], Williams et al. (2005) [49], and Mumin et al. (2010) [53]. (Figure modified under "Fair- Use" permission from GSA) source: Magmatic-hydrothermal processes within an evolving Earth: Iron oxide-copper-gold and porphyry Cu ± Mo ± Au deposits Geology [54].

6. Conclusions

Combining our interpretations and drill core observation, we demonstrated the presence of favourable IOCG with Au-Cu mineralization, which are spatially correlated with the redox boundary. The anomaly has been confirmed by integrating the geophysical and geochemical investigations. The 2D Magnetic model suggests a depth of 396 m with a width of 3.5 km, with a north dipping body and southwards bipolar anomaly also suggesting a fault. Structural control like a fault also suggest a significant scope for further investigations. The main key variables giving rise to Gadarwara IOCG are the local host rock types which are diverse in nature. Variation in depth in different structural settings such as breccia at shallow levels and shear hosted at deeper levels. The alteration such as hematite at shallower levels and Magnetite at deeper levels identified at 396 mts in drill hole. Basinal fluids preserved in evaporate-bearing sedimentary successions are important and significant contribution to fluid budget of hematite-rich IOCG system. The mineral footprints in the core samples investigated from Gadarwara suggest that samples have undergone hydrothermal, and oxidation alterations of primary mineralisation. The Potential indicator minerals comprise sheet silicates (i.e., kaolinite, montmorillonite, dickite, mica, chlorite), iron oxides, quartz and carbonates. The PT-variations May have caused a regional transformation of kaolin group minerals and smectites to mica and chlorite. The hydrothermal footprint is expressed by intense silicification related to pervasive quartz veinlets. The oxidation of primary sulphides and iron oxides, such as hematite, Magnetite, iron carbonates (e.g., ankerite, siderite etc.) is typical of but not exclusive to IOCG type of mineralisation. The BIFS are highly oxidised with intense brecciation with hydrothermal overprinting.The EPMA analyses reflect abundant low Ti-Fe oxides, Magnetite, hematite with grunerite and Fe-bearing actinolite. There is also Na-K intense metasomatism and Gadarwara is present in a rift type of tectonic environment

Author Contributions: Data curation, P.V.S.R.; formal analysis, K.S.K.; methodology, P.V.S.R. and K.S.K.; project administration, P.V.S.R. and K.S.K.; resources, P.V.S.R; writing—original draft, P.V.S.R.; writing—review and editing, P.V.S.R. and K.S.K. All authors have read and agreed to the published version of the manuscript.

Funding: The research has no external funding

Acknowledgments: P.V.S.R. thank C. Hart for hosting him as a visiting scientist, at MDRU, UBC, Canada. We thank Andrew Nevin, CMD of Adi Gold Mining Pvt Ltd., and M. K. Sen for his support during the study, CSIR, New Delhi, for providing RRF and V.M. Tiwari, Director, CSIR-NGRI for his support. Benjamin, CSIR-TWAS fellow and Hari for preparing figures. We thank three anonymous reviewers for their excellent reviewing which helped to improvise the quality of Manuscript. P.G. Eriksson for his comments and valuable discussions during Manuscript preparation. R.K.W. Merkle for his comments and thoughtful debate on initial MS.

Conflicts of Interest: The authors declare no conflict of interest.

References

1. Hitzman, M.W.; Porter, T.M. Iron oxide-Cu-Au deposits: What, where, when, and why. *Hydrothermal Iron Oxide Copp. -Gold Relat. Depos. A Glob. Perspect.* **2000**, *1*, 9–25.
2. Hitzman, M.W.; Porter, T.M. Current Understanding of Iron Oxide Associated-Alkali Altered Mineralised Systems: Part II, A Review. In *2010 Hydrothermal Iron Oxide Copper-Gold and Related Deposits: A Global Perspective*; Porter, T.M., Ed.; PGC Publishing: Adelaide, Australia, 2010; Volume 3, pp. 33–106.
3. Bastrakov, E.N.; Skirrow, R.G. Fluid evolution and origin of Cu and Au prospect in the Olympic dam district, Gawler Craton, South Australia. *Econ. Geol.* **2007**, *102*, 1415–1440. [CrossRef]
4. Skirrow, R.G.; Bastrakov, E.N.; Barovich, K.; Fraser, G.L.; Creaser, R.A.; Fanning, C.M.; Davidson, G.J. Timing of iron oxide Cu-Au-(U) hydrothermal activity and Nd isotope constraints on metal sources in the Gawler craton, South Australia. *Econ. Geol.* **2007**, *102*, 1441–1470. [CrossRef]
5. Ehrig, K.; McPhie, J.; Kamenetsky, V.S. Geology and Mineralogical Zonation of the Olympic Dam Iron-OxideCu-U-Au-Ag Deposit, South Australia. Available online: https://www.researchgate.net/profile/Vadim_Kamenetsky/publication/291217932_Geology_and_mineralogical_zonation_of_the_Olympic_Dam_iron_oxide_Cu-U-AuAg_deposit_South_Australia/links/5c403ec1a6fdccd6b5b2dd26/Geology-and-mineralogical-zonation-of-the-Olympic-Dam-iron-oxide-Cu-U-Au-Ag-deposit-South-Australia.pdf (accessed on 6 July 2020).
6. Groves, D.I.; Bierlein, F.P.; Meinert, L.D.; Hitzman, M.W. Iron oxide copper-gold (IOCG) deposits through Earth history: Implications for origin, lithospheric setting, and distinction from other epigenetic iron oxide deposits. *Econ. Geol.* **2010**, *105*, 641–654. [CrossRef]

7. Skirrow, R.G. Gold-copper-bismuth deposits of the Tennant Creek district, Australia: A reappraisal of diverse high-grade systems. In *Hydrothermal Iron Oxide Copper-Gold & Related Deposits: A Global Perspective*; Porter, T.M., Ed.; PGC Publishing: Adelaide, Australia, 2000; Volume 1, pp. 149–160.

8. Williams, P.J.; Corriveau, L.; Mumin, A.H. Magnetite-group IOCGs with special reference to Cloncurry (NW Queensland) and northern Sweden: Settings, alteration, deposit characteristics, fluid sources, and their relationship to apatite-rich iron ores. *Geol. Assoc. Can. Short Course Notes* **2010**, *20*, 23–38.

9. Corriveau, L.; Montreuil, J.-F.; Potter, E.G. Alteration facies linkages among IOCG, IOA, and affiliated deposits in the Great Bear Magmatic zone. *Can. Econ Geol.* **2016**, *111*, 2045–2072. [CrossRef]

10. Torresi, I.; Xavier, R.P.; Bortholoto, D.F.; Monteiro, L.V. Hydrothermal alteration, fluid inclusions and stable isotope systematics of the Alvo 118 iron oxide–copper–gold deposit, Carajás Mineral Province (Brazil): implications for ore genesis. *Miner. Depos.* **2012**, *47*, 299–323. [CrossRef]

11. Hunt, J.A.; Baker, T.; Thorkelson, D.J. A review of iron oxide copper-gold deposits, with focus on the Wernecke Breccias, Yukon, Canada, as an example of a non-magmatic end member and implications for IOCG genesis and classification. *Explor. Min. Geol.* **2007**, *16*, 209–232. [CrossRef]

12. Mumin, A.H. The IOCG-Porphyry-Epithermal Continuum of Deposit Types in the Great Bear Magmatic Zone, Northwest Territories, Canada. In *Exploring for iron Oxide Copper-Gold Deposits: Canada and Global Analogues*; Corriveau, L., Mumin, H., Eds.; Geological Association of Canada: St. John's, NL, Canada, 2010; pp. 59–78.

13. Knight, J. The Khetri copper belt, Rajasthan: iron-oxide copper-gold terrane in the Proterozoic of NW India. In *Hydrothermal Iron Oxide Copper-Gold & Related Deposits: A Global Perspective*; Porter, T.M., Ed.; PGC Publishing: Adelaide, Australia, 2002; Volume 2, pp. 321–341.

14. Pal, D.C.; Barton, M.D.; Sarangi, A.K. Deciphering a multistage history affecting U-Cu (-Fe) mineralization in the Singhbhum Shear Zone, eastern India, using pyrite textures and compositions in the Turamdih U-Cu (-Fe) deposit. *Miner. Depos.* **2009**, *44*, 61–80. [CrossRef]

15. Corriveau, L.; Potter, E.G.; Montreuil, J.F.; Blein, E.K.; Fabris, A.; Reid, A.J. Alteration facies of IOA, IOCG and affiliated deposit: Understanding the similarities recognising the diversity of the ore systems GSA. In Proceedings of the GSSA Workshop, Qubéc, QC, Canada, 11–12 May 2019.

16. Corriveau, L.; Williams, P.J.; Mumin, A.H. Alteration vectors to IOCG mineralisation–From uncharted terranes to deposits. *Can. Glob. Analog.* **2010**, *20*, 89110.

17. Kirmani, I.R.; Chander, S. Petrology, geochemistry and fluid inclusion studies of Cu-Au mineralization in paleoproterozoic Salumber-Ghatol belt, Aravalli Supergroup, Rajasthan. *J. Geol. Soc. India.* **2012**, *80*, 5–38.

18. Dora, M.L.; Saha, A.K.; Randive, K.R.; Rao, K.K. Iron oxide–copper–gold mineralization at Thanewasna, western Bastar Craton. *Curr.Sci.* **2017**, *112*, 1045–1050. [CrossRef]

19. Shukla, A.K.; Behera, P.; Basavaraja, K.; Mohanty, M. Iron oxide–copper–gold-type mineralization in Machanur area, Eastern Dharwar Craton, India. *Curr. Sci.* **2016**, *5*, 1853–1858. [CrossRef]

20. NGRI. *High-Resolution Air Borne Geophysical Surveys for Mineral Exploration—Aeromagnetic Discovery of Gadarwara Gold Prospect*; NGRI Report; NGRI: Hyderabad, India, 2007; p. 59.

21. Mazumder, R.; De, S.; Sunder Raju, P.V. Archean-Proterozoic transition: the Indian perspective. *Earth Sci. Rev.* **2019**, *188*, 427–440. [CrossRef]

22. Crumansonata. Crumansonata Geoscientific studies of the Son–Narmada–Tapti Lineament Zone. *Geol. Surv. India Spec. Publ.* **1995**, *10*, 244.

23. Chakraborty, C.; Bhattacharya, A.A. The Vindhyan Basin: An overview in the light of crustal perspective. *Mem. Geol. Soc. India.* **1996**, *36*, 301–312.

24. Yedekar, D.B. The central Indian collision suture. *Geol. Surv. India, Spec. Publ.* **1990**, *28*, 1–4.

25. Acharyya, S.K.; Roy, A. Tectonothermal history of the Central Indian Tectonic Zone and reactivation of Major faults/shear zones. *J. Geol. Soc. India* **2000**, *55*, 239–256.

26. Chattopadhyay, A.; Khasdeo, L. Structural evolution of Gavilgarh-Tan Shear Zone, central India: A possible case of partitioned transpression during Mesoproterozoic oblique collision within Central Indian Tectonic Zone. *Precam. Res.* **2011**, *186*, 70–88. [CrossRef]

27. Raju, P.V.S.; Hart, C.; Kathal, P. Newly recognized IOCG-like mineralization at Gadarwara, m.P, India. *Mineral. Mag.* **2014**, *77*, 2026.

28. Mallet, F.R. On the Vindhyan Series, as exhibited in the north-western and central provinces of India. *Mem. Geol. Surv. India.* **1869**, *7*, 1–129.

29. Ghosh, D.B. The nature of Narmada-Son Lineament. *Geol. Surv. India Misc. Pub.* **1976**, *34*, 119–132.

30. Nair, K.K.K. Stratigraphy, structure and geochemistry of the Mahakoshal greenstone. *Mem. Geol. Soc. India.* **1995**, *37*, 403–432.

31. Babu, H.R. Relationship of gravity, Magnetic, and self-potential anomalies and their application to mineral exploration. *Geophysics* **2003**, *68*, 181–184. [CrossRef]

32. Bickford, M.E.; Mishra, M.; Mueller, P.A.; Kamenov, G.D.; Schieber, J.; Basu, A. U-Pb age and Hf isotopic compositions of Magmatic zircons from a rhyolite flow in the Porcellanite Formation in the Vindhyan Supergroup, Son Valley (India): implications for its tectonic significance. *J. Geol.* **2017**, *125*, 367–379. [CrossRef]

33. Ray, J.S.; Martin, M.W.; Veizer, J.; Bowring, S.A. U-Pb zircon dating and Sr isotope systematics of the Vindhyan Supergroup, India. *Geology* **2002**, *30*, 131–134. [CrossRef]

34. Kumar, S.; Schidlowski, M.; Joachimski, M.M. Carbon isotope stratigraphy of the Palaeo-Neoproterozoic Vindhyan Supergroup, central India: implications for basin evolution and intrabasinal correlation. *J. Palaeontol. Soc. India* **2005**, *50*, 65–81.

35. Jain, S.C.; Nair, K.K.K.; Yedekar, D.B. Tectonic evolution of the Son-Narmada-Tapti lineament zone. *Geol. Surv. India Spec. Pub.* **1995**, *10*, 333–371.

36. Mishra, D.C. *Gravity and Magnetic Methods for Geological Studies*; BS Publications: Hyderabad, India, 2011.

37. Rao, D.A.; Babu, H.R.; Sinha, G.S. Crustal structure associated with Gondwana graben across the Narmada-Son lineament in India: inference from aeromagnetics. *Tectonophysics* **1992**, *212*, 163–172. [CrossRef]

38. Medlicott, H.B. Notes on Satpura Coal Basin. *Mem. Geol. Surv. India* **1873**, *10*, 133–188.

39. Medlicott, H.B.; Blanford, W.T. *A Manual Geology of India*; Govt of India Press: Calcutta, India, 1879.

40. Talwani, M. Computation of Magnetic anomalies caused by two-dimensional bodies of arbitrary shape. *Comput. Miner. Ind.* **1964**, *1*, 464–480.

41. Spector, A.; Bhattacharyya, B.K. Energy density spectrum and autocorrelation function of anomalies due to simple Magnetic models. *Geophys. Prospect.* **1966**, *14*, 242–272. [CrossRef]

42. Spector, A.; Grant, F.S. Statistical models for interpreting aeromagnetic data. *Geophysics* **1970**, *35*, 293–302. [CrossRef]

43. MacLeod, I.N.; Jones, K.; Dai, T.F. 3-D analytic signal in the interpretation of total Magnetic field data at low Magnetic latitudes. *Explor. Geophys.* **1993**, *24*, 679–688. [CrossRef]

44. PGW. *Program Documentation-MAGMOD Version 1.4 Magnetic Interpretation Software Library*; Paterson, Grant and Watson Ltd.: Toronto, ON, Canada, 1982; p. 17.

45. Raju, P.V.S.; Hart, C.; Kathal, P. Integrated SWIR spectral and XRD studies on core samples from Gadarwara, Central India Craton, Madhya Pradesh, India-footprints for IOCG mineralisation. In Proceedings of the 14th SGA Biennial Meeting, Québec, QC, Canada, 20–23 August 2017; pp. 963–966.

46. Haynes, D.W.; Porter, T.M. Iron oxide copper (-gold) deposits: Their position in the ore deposit spectrum and modes of origin. *Hydrothermal Iron Oxide Copp. -Gold Relat. Depos. A Glob. Perspect.* **2000**, *1*, 71–90.

47. Hitzman, M.W.; Oreskes, N.; Einaudi, M.T. Geological characteristics and tectonic setting of proterozoic iron oxide (Cu-U-Au-REE) deposits. *Precambrian Res.* **1992**, *58*, 241–287. [CrossRef]

48. Belperio, A.; Flint, R.; Freeman, H. Prominent hill: a hematite-dominated iron oxide copper gold system. *Econ. Geol.* **2007**, *102*, 1499–1510. [CrossRef]

49. Williams, P.J.; Barton, M.D.; Johnson, D.A.; Fontboté, L.; De Haller, A.; Mark, G.; Marschik, R. Iron oxide copper-gold deposits: Geology, space-time distribution, and possible modes of origin. *Econ. Geol.* **2005**, *100*, 371–405. [CrossRef]

50. Seedorff, E.; Dilles, J.H.; Proffett, J.M., Jr.; Einaudi, M.T.; Zurcher, L.; Stavast, W.J.A.; Johnson, D.A.; Barton, M.D. Porphyry Deposits: Characteristics and Origin of Hypogene Features. *Econ. Geol.* **2005**, *100*, 251–298.

51. Sillitoe, R.H. Porphyry copper systems. *Econ. Geol.* **2010**, *105*, 3–41. [CrossRef]

52. Richards, J.P. Magmatic to hydrothermal metal fluxes in convergent and collided Margins. *Ore Geol Rev.* **2011**, *40*, 1–26. [CrossRef]

53. Mumin, A.H.; Phillips, A.; Katsuragi, C.J.; Mumin, A.; Ivanov, G. *Geotectonic Interpretation of the Echo Bay Sratovolcano Complex, Northern Great Bear Magmatic Zone*; Northwest Territories Geoscience Office: Yellowknife, NT, Canada, 2013; p. 35.

54. Richards, J.P.; Mumin, A.H. Magmatic-hydrothermal processes within an evolving Earth: Iron oxide-copper-gold and porphyry Cu ± Mo ± Au deposits. *Geology* **2013**, *41*, 767–770. [CrossRef]

© 2020 by the authors. Licensee MDPI, Basel, Switzerland. This article is an open access article distributed under the terms and conditions of the Creative Commons Attribution (CC BY) license (http://creativecommons.org/licenses/by/4.0/).

Article

Geophysical Prospecting for Groundwater Resources in Phosphate Deposits (Morocco)

Fatim-Zahra Ihbach [1], Azzouz Kchikach [1,2,*], Mohammed Jaffal [1,2], Driss El Azzab [3], Oussama Khadiri Yazami [4], Es-Said Jourani [4], José Antonio Peña Ruano [5], Oier Ardanaz Olaiz [5] and Luis Vizcaíno Dávila [5]

[1] Georesources, Geoenvironment and Civil Engineering (L3G) Laboratory, Cadi Ayyad University, Marrakesh 40000, Morocco; ihbachfatimzahra@gmail.com (F.-Z.I.); jaffalm@gmail.com (M.J.)
[2] Geology and Sustainable Mining, Mohammed VI Polytechnic University, Ben Guerir 43150, Morocco
[3] SIGER laboratory, Sidi Mohammed Ben Abdellah University, Fes 30050, Morocco; driss.elazzab@gmail.com
[4] OCP Group, Casablanca 20200, Morocco; o.khadiriyazami@ocpgroup.ma (O.K.Y.); e.jourani@ocpgroup.ma (E.-S.J.)
[5] Instituto Andaluz de Geofisica, Universidad de Granada, 18010 Granada, Spain; peruano@ugr.es (J.A.P.R.); oierarol@gmail.com (O.A.O.); luisvizcaino@hotmail.com (L.V.D.)
* Correspondence: azzouz.kchikach@um6p.ma

Received: 27 May 2020; Accepted: 10 August 2020; Published: 24 September 2020

Abstract: The Moroccan phosphate deposits are the largest in the world. Phosphatic layers are extracted in open-pit mines mainly in the sedimentary basins of Gantour and Ouled Abdoun in Central Morocco. The purpose of this study was to prospect and evaluate the water potential of aquifers incorporated in the phosphatic series using the following geophysical methods: Magnetic resonance sounding (MRS), electrical resistivity tomography (ERT), time-domain electromagnetics (TDEM), and frequency-domain electromagnetics (FDEM). The objective was, on the one hand, to contribute to the success of the drinking water supply program in rural areas around mining sites, and on the other hand, to delimit flooded layers in the phosphatic series to predict the necessary mining design for their extraction. The use of geophysical methods made it possible to stratigraphically locate the most important aquifers of the phosphatic series. Their hydraulic parameters can be evaluated using the MRS method while the mapping of their recharge areas is possible through FDEM surveys. The results obtained in two selected experimental zones in the mining sites of Youssoufia and Khouribga are discussed in this paper. The application of the implemented approach to large phosphate mines is in progress in partnership with the mining industry.

Keywords: phosphate mines; geophysical exploration; hydrogeology; phosphate extraction; gantour Basin; Ouled Abdoun Basin; Morocco

---

1. Introduction

Phosphate is a valuable mineral resource for different industries. In Morocco and all over the world, phosphate rocks are being extracted for their phosphorus content. Phosphate is used to produce phosphoric acid (PA) that have various uses in our daily lives. First, PA is mainly used in agriculture for the production of soil fertilizers. It is also used as animal feed supplements. PA is also widely solicited in laboratories because it resists to oxidation, reduction, and evaporation. It has many other industrial applications such as soaps and detergents, water treatment, dentistry, etc.

Moroccan subsoil includes more than three-quarters of the world's phosphate reserves [1,2]. These reserves exist mainly in the sedimentary basins of Gantour and Ouled Abdoun in Central Morocco (Figure 1a,b). In these basins, the age of the tabular structure of sedimentary cover ranges from Permo–Triassic (251.9 My = million years ago) to Quaternary (2.58 My) [3]; it rests unconformably on the Hercynian basement of the western part of the Moroccan Meseta [4,5]. The phosphatic series investigated in this study corresponds to the upper part of the sedimentary cover. Its age extends from the Maastrichtian (72.1 My) to Lutetian (47.8 My) [6,7]. This series outcrops in the northern part of the aforementioned basins where the phosphatic layers are exploited in open-pit mines. To the south, this series is buried under the Neogene (23 My) and Quaternary continental deposits of the Bahira and Tadla plains (Figure 1a,b).

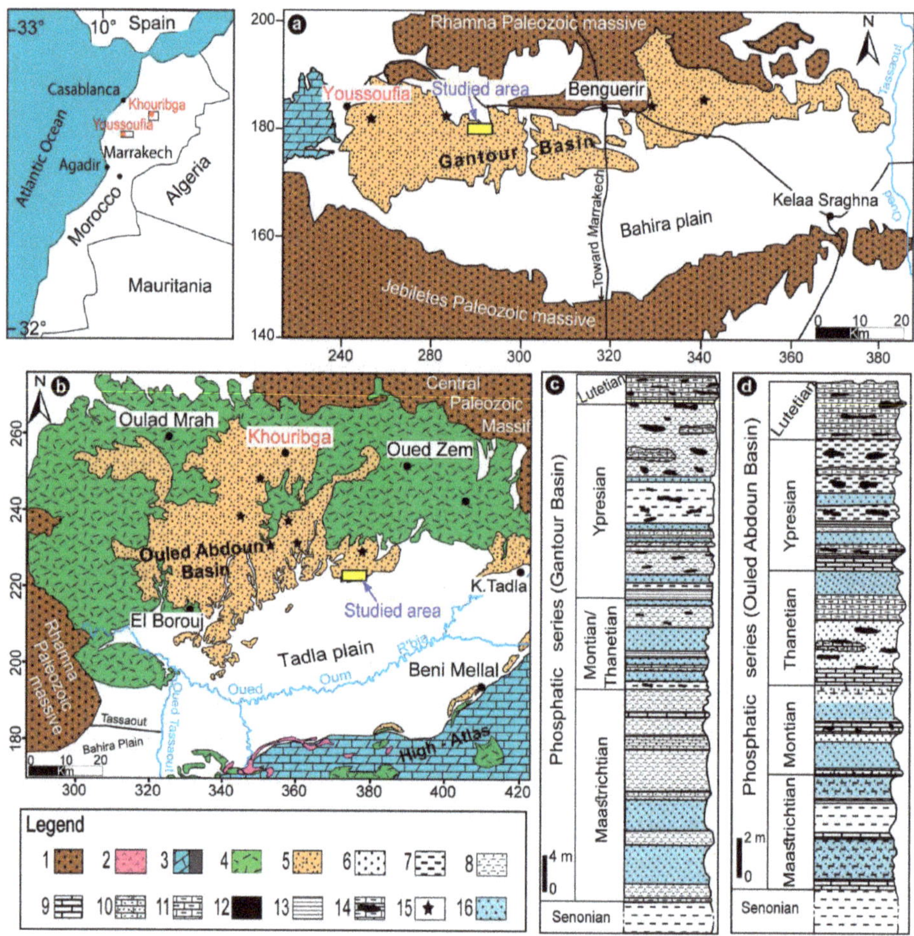

Figure 1. Location and geology sitting of study areas. (**a**) Geological map of the Gantour Basin (modified after [8]); (**b**) Geological map of the Ouled Abdoun Basin (after [9]); (**c**) Lithological columns of the phosphatic series in Gantour Basin showing the main groundwater aquifers; (**d**) Lithological columns of the phosphatic series in Ouled Abdoun Basin showing the main groundwater aquifers; 1: Paleozoic; 2: Trias; 3: Jurassic; 4: Cretaceous except Maastrichtian; 5: Maastrichtian–Eocene; 6: Uncemented phosphates, 7: Marl; 8: Phosphatic marl; 9: Limestone; 10: Phosphatic limestone; 11: Marly limestone; 12: Silex; 13: Clay; 14: Marly-siliceous limestone; 15: Open-pit mine; 16: Aquifers. (N.B. Lambert kilometric coordinates projection for the north of Morocco).

The present study was conducted to characterize the aquifers of the phosphatic series and better understand their properties and characteristics. This study is part of an ambitious sustainable development program led by the Office Chérifien des Phosphates (OCP) around phosphatic extraction sites. The drinking water supply of rural agglomerations in the vicinity of mining sites is among the main objectives of this program. In addition, OCP's geo-mining engineers are interested in determining the flooded areas in the phosphatic series which will allow them to predict all the required preparations to extract phosphates in such a hydrogeological context.

This study combined the use of at least two different geophysical methods across two selected areas, namely Youssoufia and Khouribga mining sites (Figure 1a,b). Electrical resistivity tomography (ERT) and magnetic resonance sounding (MRS) surveys were carried out in the Youssoufia area (Figure 1a); while time-domain electromagnetic (TDEM) and frequency-domain electromagnetic (FDEM) measurements were performed in the Khouribga site (Figure 1b).

The treatment and interpretation of the acquired data made it possible to locate the main horizon of the aquifer contained in the phosphatic series and map on the surface conductive corridors that would constitute potential recharging zones.

The experimental research conducted at the two aforementioned sites allowed for the establishment of a scientific approach for larger hydro-geophysical studies to be conducted on the phosphatic series. It also made it possible to define appropriate devices and sampling steps for the used geophysical methods. The MRS method is a satisfactory tool to explore the aquifers and determine their hydrodynamic parameters in the phosphatic series context. FDEM surveys, which are very easy to implement in field, are the most efficient tool for mapping aquifer recharge zones in the geological context of our studied areas. The ERT and TDEM methods can also be used in areas where MRS data are affected by natural electromagnetic noise. We discuss, in this article, the approach adopted and the main obtained results.

2. Geological and Hydrogeological Context of the Study Areas

This research was conducted at two OCP mining sites in two distinct sedimentary basins: Gantour and Oulad Abdoun. The Youssoufia site is located in the Gantour Basin. In this basin, the sedimentary series ranges from the Triassic to Quaternary period [7]. It is surrounded from the north and south, respectively, by the Paleozoic outcrops of the Rhamna and Jebiletes Massifs (Figure 1a). The phosphatic series outcrops in the northern basin where they are exploited in open-pit mines in Youssoufia and Benguerir. In the southern basin, this series is buried below the Neogene and Quaternary deposits of the Bahira Plain. In the Gantour Basin, the phosphatic series is well known owing to its exploitation and borehole recognition (Figure 1c). It starts with the Maastrichtian, mainly formed of fine sandy phosphates, phosphatic marls, and a clay–sandstone complex constituting an impermeable reference layer [7,10]. The Paleocene that is the lower epoch of the Paleogene, which includes the Montian (61.6 My) and the Thanetian (59.2 My), is mainly composed of alternating layers of uncemented phosphates, limestones and/or phosphatic marls, coprolites, and flint nodules. The Lower Eocene (Ypresian (56 My)) is commonly formed of phosphatic marls intercalated by phosphatic limestones, silexite, and siliceous marl [11,12]. Uncemented phosphatic layers become weak and often overlay impermeable clay layers, known as Ypresian clays by geo-mining engineers, in the Gantour Basin. Lutetian (47.8 My) strata mark the end of the phosphatic series. It is characterized by a fossiliferous marly limestone slab, known as the Thersitae slab [7,13–16].

The second study area is in the Ouled Abdoun Basin situated approximately 26 km south of Khouribga. This basin includes the largest phosphate deposits in the world. It extends over more than 10,000 km^2 and is bounded on the north by the Hercynian Massif of Central Morocco, on the south and east by the Jurassic escarpments of the Middle and High Atlas, and on the west by the Paleozoic outcrops of the Rhamna [14]. Geomorphologically, this basin contains two distinct units: The phosphate plateau and the Tadla Plain (Figure 1b). The phosphatic series (Figure 1d) is formed of alternating layers of uncemented phosphate, sterile limestone, and marly limestone with an average thickness of approximately 50 m in the deposits under extraction [16,17]. It begins with phosphatic marls and limestone layers that are very rich in bone debris, known as bone-bed limestones, of Maastrichtian age [18,19]. This stage marks the beginning of a phosphatogenesis that reaches its maximum during the following stages. Above, successive layers of uncemented phosphates and phosphatic limestone occur. The Montian is represented by uncemented phosphates, overlain by limestone with coprolites and flint nodules that constitute a reference layer for mining extraction. The last is also overlain by alternating regular beds of marly and phosphatic limestones, coarse-grained uncemented phosphatic layers, continuous flint horizons and, sometimes, silto-pelitic layers from Thanetian to Ypresian in age [6]. The Lower Lutetian consists mainly of alternating flint layers and slightly phosphate marls and limestones that mark the last stages of phosphatogenesis. The phosphatic series is overlain by strong fossiliferous carbonated layers, rich in gastropods, known as the Thersitae Slab [20]. The phosphatic layers outcrop or are beneath a thin quaternary cover in the Northern Ouled Abdoun Basin. They are extracted in the open-pit mines of Khouribga, Daoui, Merah, and Sidi Chennane. To the south and southeast, the phosphatic series is buried beneath the Neogene and Quaternary continental and lacustrine deposits of the Tadla Plain [21].

Hydrogeologically, the main groundwater aquifers in the sedimentary series of the Gantour and Ouled Abdoun basins correspond to karstic Turonian limestones, uncemented phosphatic horizons, and fractured limestones that cover the phosphatic series [8,21–23]. The Turonian aquifer is very deep (greater than 250 m) in the two aforementioned basins, and therefore less recognized and extracted. The aquifer corresponding to the Lutetian fractured limestones of the Thersitea slab is only important for water supply in areas where the phosphate series is buried under the Bahira and Tadla Plains. Along the perimeters of the open-pit mines in both the Gantour and Ouled Abdoun Basins, the piezometric level is generally below the Ypresian limestones.

This study is concerned with the aquifers in the uncemented phosphate horizons of the phosphatic series. Actually, this aquifer system is composed of at least eight superimposed layers of uncemented phosphates; each is overlaying a sterile impermeable interlayer of clay, marl, continuous flint bench, or siliceous phosphatic limestone. These aquifer horizons connect with each other via a network of fractures and microfaults affecting the phosphatic series. They are from bottom to top (Figure 1c,d) the following: (i) Two layers of Maastrichtian uncemented phosphates that constitute the most important aquifer of the Youssoufia and Benguerir mines; (ii) four layers of Montien–Thanetian uncemented phosphates that often overlie siliceous limestone layers. The thickness of these layers changes from one location to another in the studied basins [8]. Finally, the Ypresian phosphatic coarse layers overlie either plastic clays or siliceous marl. They correspond to the Ypresian strata indicated with blue background in Figure 1c.

3. Materials and Methods

To conduct the present hydrogeophysical study of the OCP mining sites, we chose two zones to perform experimental geophysical surveys. These zones were selected near extraction areas where geological and hydrogeological data are available to calibrate the results of the used geophysical methods. The first zone was located in the Gantour Basin at the "Panel 6" deposit according the OCP terminology (Figure 1a). The second is adjacent to the Sidi Chennane mining deposit in the Ouled Abdoun Basin (Figure 1b). During this experimental phase, we arbitrarily chose to test the MRS and ERT methods at the Youssoufia mine site, while the TDEM and FDEM methods were used at Khouribga. The same methods can be used at any of the OCP mining sites because of the similarity of the geological and hydrogeological contexts, as shown in Figure 1c,d; in addition, the two sites chosen to conduct the present study have the advantage of a low electromagnetic noise level. They are located far from any source of anthropogenic noise generated by human activities, such as power lines, radio antennae, electric fences, motors, pumps, buried conductors, etc.

3.1. ERT Profiles

ERT profiles provide two-dimensional (2D) images of subsurface resistivities. The implantation of a large number of electrodes, at regular intervals, along a rectilinear profile, enables one to obtain pseudo-sections of apparent resistivities. Using modern equipment, the acquisition is automatic by interrogating quadrupoles of increasing length, thus making it possible to obtain increasing measurement depths. A data inversion is then performed to obtain a quantitative measurement of the true electrical resistivity at real depths.

An ABEM Terrameter LS 12 with 81 electrodes was used and made available by the Andaluz Institute of Geophysics (AIG) at Granada University, Spain. We adopted an edge gradient acquisition device which allowed for a better assembly between electrodes by combining symmetrical and asymmetrical configurations. Such a configuration provides an image of subsurface resistivities at high resolution and with less artifacts, e.g., distortions of the inversion resistivity model caused by the relatively high noise contamination of the data [24]. Three ERT profiles 700 m long with 5-m electrode spacing were collected (Figure 2). A total of 6531 apparent resistivity measurements were acquired. Before the inversion of the data, their quality control was thoroughly controlled. No automatic filtering was applied but the data were edited, and all bad datum readings were manually eliminated. This resulted in slight reduction of the total corrected datum point for each profile. The data inversion was performed using the AIG's Res2dinv software (V.358-Geotomo Software). The program uses a smoothness-constrained Gauss–Newton method to obtain a 2D resistivity model [25]. The process is automatic once the inversion parameters are established and a user-defined initial model is not needed. The iterative inversion process was set at a maximum of 10 error-iterations using the L2-norm (root mean square), offering an error, between the apparent resistivities in the field and those generated by the calculated models, less than 5%. The inversion model contained 15 layers and 1692 resistivity blocks. It was refined to half the electrode spacing because of the sensitivity of the gradient array to near surface variation. The sensitivity matrix was calculated using the incomplete Gauss–Newton method that is recommended for large data sets. In general, the first 40 m of the three profiles showed good relative sensitivity values. Between 40 and 60 m depth, the sensitivity was acceptable. The ERT-3 profile showed an area of low sensitivity values due to the presence of the resistive body. Prior to the inversion process, topographic data acquired every 50 m by an OCP surveyor team were added to the resistivity measurement files.

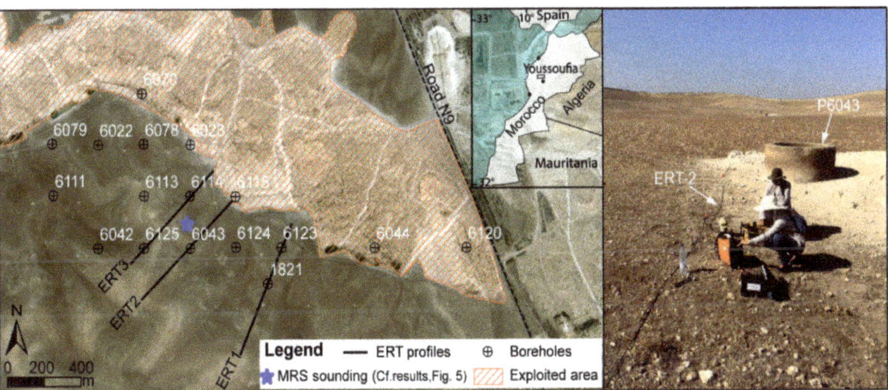

Figure 2. Location of geophysical surveys carried out in Youssoufia area and photograph showing a borehole intersected by the electrical resistivity tomography (ERT)-2 profile.

3.2. MRS Data

Schematically, the physical principle of the MRS method is based on the fact that the protons that form the hydrogen nuclei of water molecules in Earth's magnetic field have positive magnetic moments that, at equilibrium, are aligned in the direction of this main field. The excitation of these protons by emitting a disturbing magnetic field at a specific frequency, termed a "Larmor frequency," modifies this equilibrium state and causes magnetic precession moments around Earth's magnetic field. After cutting off the exciter field, and during the return to their equilibrium state, a magnetic relaxation field is generated by the protons, thus constituting the MRS measured signal. The higher the signal amplitude, the larger the number of protons in resonance; thus, this provides information regarding the subsoil water content. The importance of the precession process in the excitation and relaxation at the field cut-off is also related to the average pore size of the aquifer formation. The relaxation time (decay time of the measured signal) is longer as the solicited protons are those of water less enclosed in the rock, and therefore of an aquifer with a high hydrodynamic potential guaranteeing a sufficient pumping flow rate [26–29].

First, we measured the ambient electromagnetic noise and assessed the quality of the MRS signal in the geological context of the Youssoufia mine. Then, an MRS survey was conducted using a loop of 100 × 100 m, which is the loop size of the MRS systems employed to realize the survey. We chose this dimension in order to maximize the depth of investigation of the study area. The transmitter/receiver loop was centered on well P6043 (Figure 2) to compare and calibrate the results of this method with real hydrogeological data. Numis Pro from IRIS Instruments was used from the University Cadi Ayyad of Marrakech. MRS data acquisition and inversion were, respectively, performed using the Prodiviner and Samovar software of the aforementioned company [30]. The measured parameter is the relaxation magnetic field created by excited protons when they return to their equilibrium state. The physical parameter that we sought to identify was the water content of the different subsoil layers. The result is a plot showing the hydrogeological model corresponding to the measured signal.

3.3. TDEM Data

Electromagnetic methods are based on the diffusion of an electromagnetic field (EM) in the subsoil to determine its electrical resistivity. For the TDEM soundings conducted at the Khouribga mining site, the EM field, termed the primary field, is created by the sudden cutting off of a flowing current in a transmitting coil placed on the ground. The secondary field, related to the induced current, is measured by a receiver coil after the current cut-off. The TDEM sounding curve provides the variation in the apparent resistivity of the subsoil according to the time, in other words according to the depth, which increases with time during the secondary field measurement [31,32]. Therefore, recorded curves can be converted to pseudo-depths [33,34]. The investigation depth depends on the formation resistivities, signal–noise ratio, as well as the measurement loop size. It is classically estimated at 1.5 to 2.5 times the size of the loop, but depends on the local conditions [35]. The Tem-Fast ProSystem, from the geological service of the OCP in Khouribga, was used. The surveys were conducted using a square emission loop of 20 m side pulled on the ground. A total of 84 TDEM soundings were completed at a sampling step of 12.5 m along a profile 1050 m in length (Figure 3). Obtained data, from each survey, were inverted using the TemRes software based on the Temfast Prosystem technique [35].

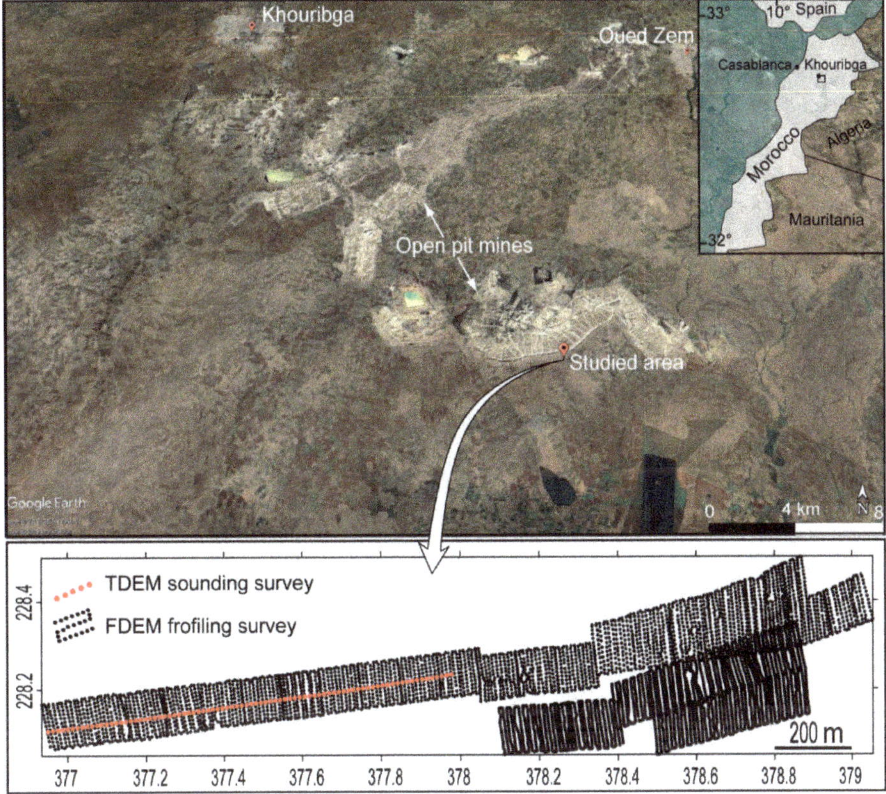

Figure 3. Location of geophysical surveys conducted in Khouribga area.

3.4. FDEM Data

The FDEM method applied in the present study was carried out using the Geonics EM-31 MK-2 system. This technique helps record the electrical conductivity whose variations reflect the facies heterogeneities in the near subsoil. The measurements were performed without any contact with the ground; this makes the field survey very easy to implement and provides a higher acquisition performance than the TDEM technique. The EM31-MK-2 system contains a transmitter and receiver coils fixed at the ends of a 3-m-long support bar and connected to the control box in the middle of the device. The transmitter generates a primary magnetic field at a given frequency. A secondary field is generated and recorded by the receiver when the primary field encounters conducting areas in the ground. The ratio of the vertical component of the secondary field in the quadrature with the primary field is proportional to the apparent conductivity of the investigated soil. The investigation depth of the EM31 method is theoretically equal to 1.5 to 2 times the distance between the transmitting and receiver coils, for a medium-conducting environment [36–38].

The EM31 method was used at the Sidi Chennane deposit (Figure 3), in the same area where the TDEM surveys were conducted. Measurements were made along profiles spaced every 10 m. These profiles cover 33 juxtaposed parcels of 100 × 100 m equaling a total length of 36.3 km and 207,746 measured values. A global positioning system (GPS) was connected and synchronized with the EM31 acquisition console, allowing for measurement geolocalization.

4. Results and Discussion

4.1. Youssoufia Mining Site

The study area belongs to the phosphate deposit of Panel 6, approximately 26 km east of Youssoufia on the main road No 9 linking Marrakech to El Jadida. Figure 4 shows the results of ERT profiles P1, P2, and P3. This figure also shows the synthetic lithological columns of the recognition wells of the phosphatic deposits created by OCP. This allowed us to constrain the models of the calculated resistivities and assign the measured values to the different lithological units of the phosphatic series. We also show on the same figure the curve of the piezometric level measured in the OCP boreholes during the geophysical field surveys.

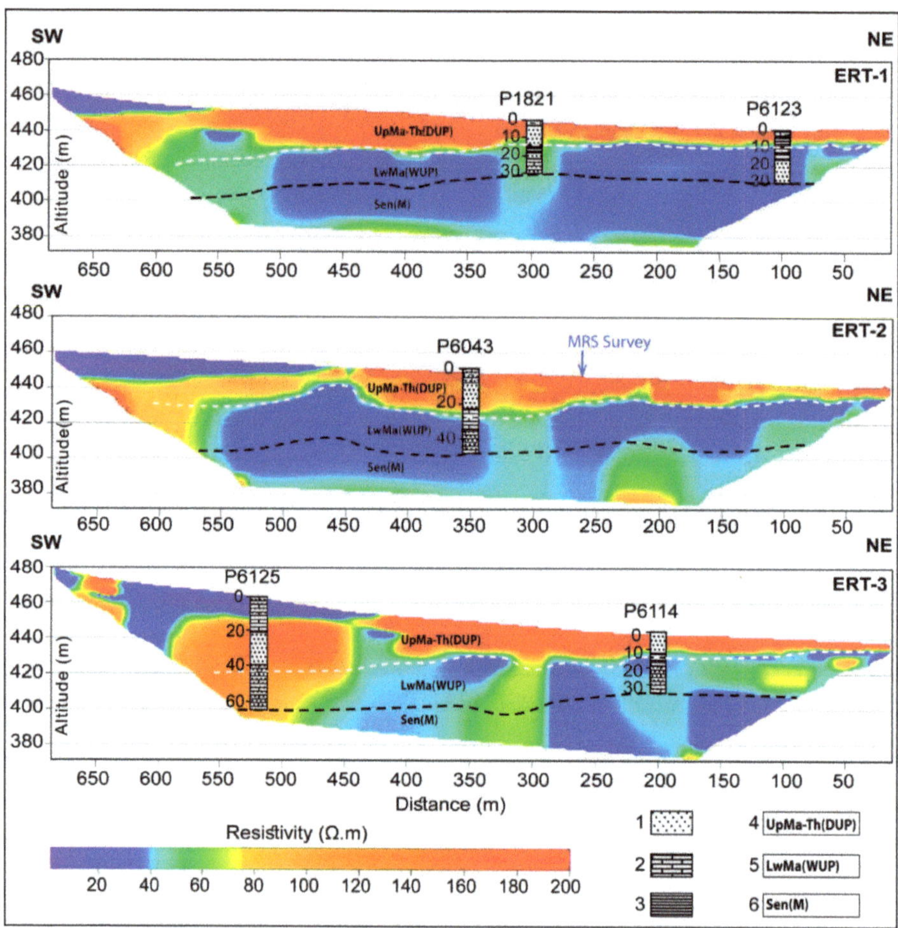

Figure 4. Model resulting from ERT data inversion calibrated to lithological boreholes columns. The dashed white line corresponds to the piezometric level measured on site. The dashed black line indicates the limit between the Maastrichtian and the Senonian formations. 1: Uncemented phosphate; 2: Limestone; 3: Clay; 4: Upper Maastrichtian–Thanetian (dry uncemented phosphate); 5: Lower Maastrichtian: (wet uncemented phosphate); 6: Senonian: (marl).

In the southwest part of the three ERT profiles, the resistivity values are less than 20 ohm·m in the sub-surface. These low values can be explained by the more or less wet marl and clay facies that we observed in this area during the geophysical surveys. The remainders of the profiles at this depth (i.e., about 10 to 20 m) are represented by higher resistivity values greater than 140 ohm·m. These values are attributed to the phosphatic limestone and uncemented phosphatic dry layers of the Upper Maastrichtian and Thanetian, between 0 and 25 m depth, in the OCP's lithological wells, but also in the surrounding mining trench. Underneath, clearly conductive areas with resistivities less than 60 ohm·m occur. They correspond to the damped or high-water content of the uncemented phosphatic layers of the Lower Maastrichtian and to Maastrichtian impermeable clays (2 m) that constitute a reference level for the extraction of the Youssoufia open-pit mining [10,17]. Conductive layers still occur beneath the Maastrichtian formations until the bottom of the ERT sections. They correspond to the Senonian marls described in the stratigraphic columns of the OCP wells since the depth of 50 m. The most conductive horizon of all the ERT sections is situated between 25 and 35 m depth. It corresponds to the uncemented phosphatic layer of the Lower Maastrichtian that overlies the aforementioned impermeable clays. At some locations, it is difficult to make delimitation according to the depth of the top and the bottom of the real aquifer horizons, corresponding to Maastrichtian uncemented phosphatic layers and marl and/or limestone intercalary, which are probably very wet. Both formations provide comparable electrical signatures, as reflected in the ERT sections by the thick blue section. Therefore, the sensitivity according to the vertical of the ERT method is more or less limited. However, laterally, the ERT sections show moderately resistive zones within the same conductive stratigraphic horizon. This lateral variation in the resistivities can be attributed to a decrease in the water content within the same conductive horizon or to local facies change. These facies variations within the same layer are actually described along the operating trenches.

The piezometric level measured in the P1821, P6123, and P6043 boreholes during the geophysical surveys exactly coincide with the interpretation provided to the ERT sections. We interpolated between the OCP boreholes and extrapolated outside of them, assigning aquifer resistivity values less than 60 ohm·m. The perfect match between the depth of the piezometric level measured in the OCP boreholes and the roof of the conductive horizon was shown by the ERT data inversion (Figure 4). Thus, this method can be used to identify and locate the aquifers in the phosphatic series at the Youssoufia mining site.

During the present study reliable MRS data were recorded after some unsuccessful attempts to perform the measurements with low stacking values, even if the survey area was located far from any electromagnetic source of noise. To obtain data of good signal-to-noise ratio (average of 5.5), it was necessary to take the measurements using 250 stacks. The emitted and the received signals had very close frequencies (difference less than 1.0 Hz). A smooth inversion of the MRS data was performed using Samovar software with a filtering window of 197.8 ms, a time constant of 15.00 ms, an average signal to noise ratio of 2.66, and a fitting error for the free induction decay 1 (FID1) of 9.05%.

Figure 5 shows MRS survey data and results including the signal relaxation curves versus time for the various pulse moments injected (Figure 5a). The hydrogeological model resulting from the inversion helps determine the depth of the aquifers and the free water content that estimates the porosity of the rock in a saturated environment. One can clearly see that the amplitude of the measured signal changed according to the excitation pulse and thus the depth. This shows the presence of subsoil water and the variation in water content from one level to another (Figure 5b). We also note that the MRS signal was clearly distinguished from the electromagnetic noise, which was fortunately very low at the studied site. Generally, two aquifer levels can be distinguished (Figure 5c): The first is between 20 and 30 m, and the second between 32 and 45 m. The water content, which reflects the porosity of the layers below the piezometric level, is, respectively, 3% and 6.7%. These values are quite comparable to those provided by pumping tests conducted on the same aquifer outside the study area [22].

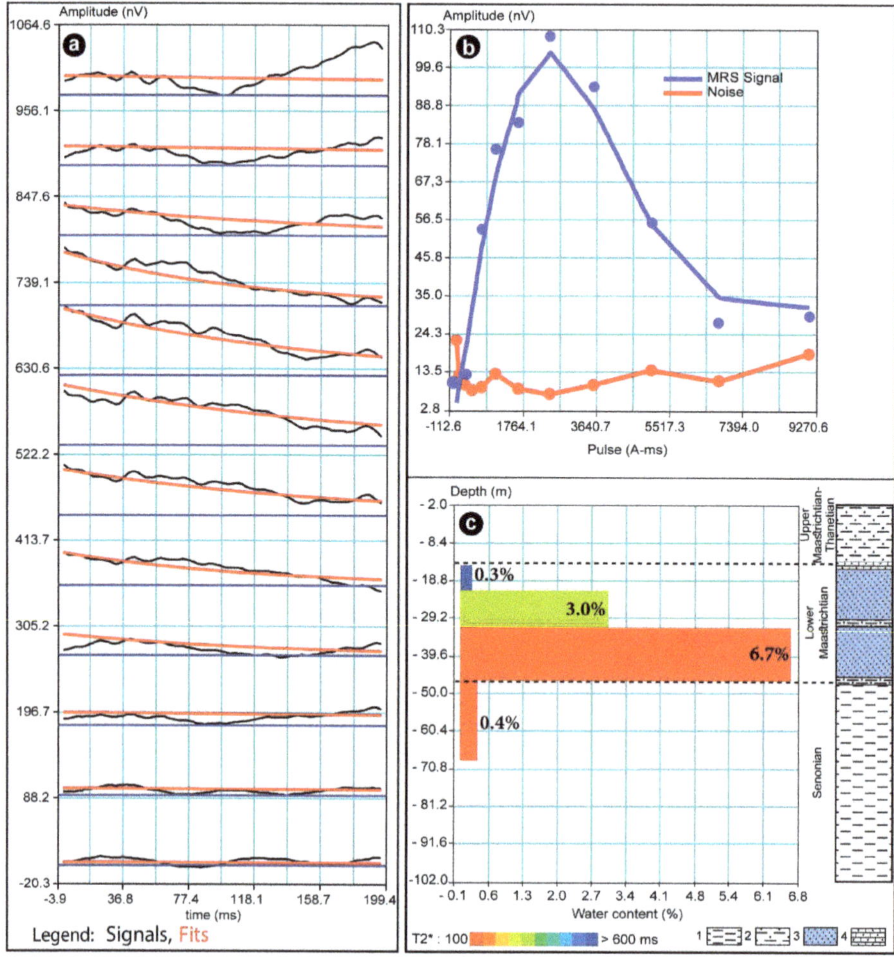

Figure 5. (**a**) Magnetic resonance sounding (MRS) signal relaxation curves versus time for the various pulse moments injected; (**b**) amplitude of measured MRS signal and noise level; (**c**) model resulting from MRS data inversion and its correlation with hydrogeological column; 1: Marl; 2: Phosphatic marl; 3: Uncemented phosphate; 4: Phosphatic limestones.

The average aquifer thickness derived from the MRS calculated model was approximately 23 m. The depth of the piezometric level measured in well n°6043 at the center of the MRS loop was 21.23 m. The MRS signal's decay time was relatively low (100–150 ms) which meant that, in general, the aquifer was moderately permeable [39–41]. The recognition well lithological sections in the phosphatic deposits showed that the aquifer delimited by the MRS data inversion corresponded to the two uncemented phosphatic layers of the Lower Maastrichtian in the study area (Figure 5c). Given that the studied area was approximately 19,242,500 m², we deduced a total water volume of approximately 22,128,875 +/- 2,251,373 m³. The uncertainty of the volume estimate was about 10% due to the error of average porosity calculated from MRS inversion and the quantification of the unsaturated part of the aquifer. This volume was quite comparable to the that obtained via the direct calculation using the OCP well piezometric data (22,829,600 +/- 2,282,960 m³). It is with this purpose that the study was conducted using MRS to better understand the water reserves and hydrodynamic parameters of the Maastrichtian aquifer. These results strongly support the use of this method in other areas of the Youssoufia open-pit mines.

4.2. Khouribga Mining Site

To facilitate interpretation of the results, we used the same color palette for both resistivity (TDEM method) and resistivity calculated from conductivity (EM31 method). The TDEM sounding data are presented as a pseudo-section obtained by interpolation of the experimental apparent resistivities from one sounding to another using the minimum curvature method [42] (Figure 6a). The same method was used to interpolate the 1D inversion of the 84 performed soundings in order to build the resistivity model illustrated in Figure 6b. This model shows a clearly conductive horizon whose top varies approximately in altitude from 495 to 510 m, and in depth from 24 to 47 m. This conductive horizon is continuous and extends for more than 1 km along the TDEM executed profile. Beneath the 70 m depth, the top of moderately resistant layers is locally observed. They correspond to the compact gray marl with gypsum of the Senonian, which can be seen in the OCP's recognition wells at this depth [43]. Above the conductive horizon, moderately resistant and resistant soils consisting of phosphatic layers with very low water content with intercalated limestone and flint layers occur. At the near-surface, the layers are even more resistant because they are dry and rich in limestone encrustation. Other resistant bodies also cut the phosphatic series. They are attributed to the disturbed areas of the phosphatic series, known as "derangements" by the OCP mining engineers [43–46]. These structures are often accompanied by a localized silicification event. The research of Kchikach et al. 2002, 2006, and 2012 [16,47,48] shows that the disturbed areas of the phosphatic series are characterized by a strongly resistive electrical signature.

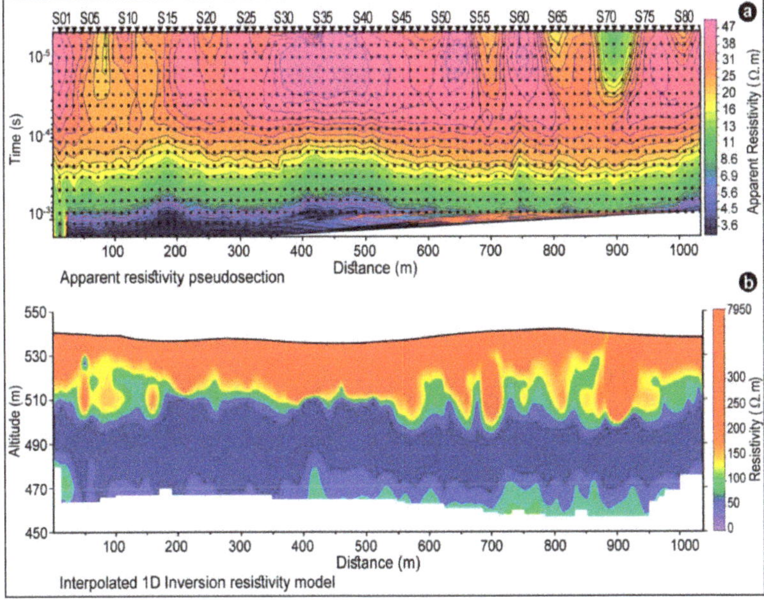

Figure 6. (**a**) Time-domain electromagnetic (TDEM) pseudo-section of experimental resistivity data from TDEM sounding survey, (**b**) interpolated 1D inversion resistivity model.

The apparent resistivity map obtained after smoothing the EM31 data, and at regular sampling steps of 1 × 1 m, shows a series of conductive anomalies in the studied area. These anomalies, with conductivity values generally greater than 20 mS/m (Figure 7), consist of altered and fractured zones of limestone encrustation. However, the areas where these encrustations are thicker and less fractured result in conductivities less than 15 mS/m. Conductive anomalies are superimposed with fractured and altered corridors that can be easily delimited in the field. Actually, during the geophysical surveys, these corridors were distinguished in the field by their wet aspect. They constitute potential zones for surface water infiltration towards the identified aquifer in the same areas as shown using the TDEM method. Despite its limited investigation depth, the EM31 technique is a very good tool for mapping groundwater aquifer recharge zones in the phosphatic series. The most productive water wells and boreholes should be, in our estimation, installed in line with these highly conductive anomalies.

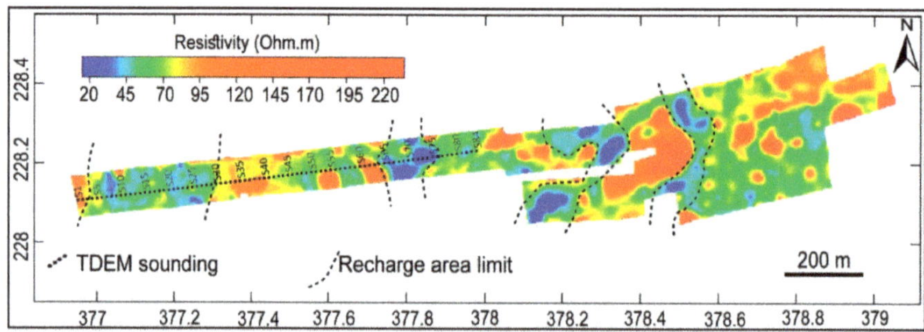

Figure 7. Apparent resistivity map obtained from EM31 data.

Finally, a summary illustration synthesizing the main outputs of the present study was established (Figure 8). The latter clearly shows that in both Youssoufia and Khouribga areas, the Lower Maastrichtian constitute a high potential water-bearing aquifer. The figure clearly depicts the convergence between the ERT and the MRS findings in Youssoufia. In fact, the conductive character of the Lower Maastrichtian is explained by the water content evidenced by the MRS survey. In the Khouribga site, TDEM data revealed that the Lower Maastrichtian is also clearly a conductor. Since this age (Lower Maastrichtian) is made with the same facies as Youssoufia (uncemented phosphate), we can assume that it may constitute the main reservoir of underground water.

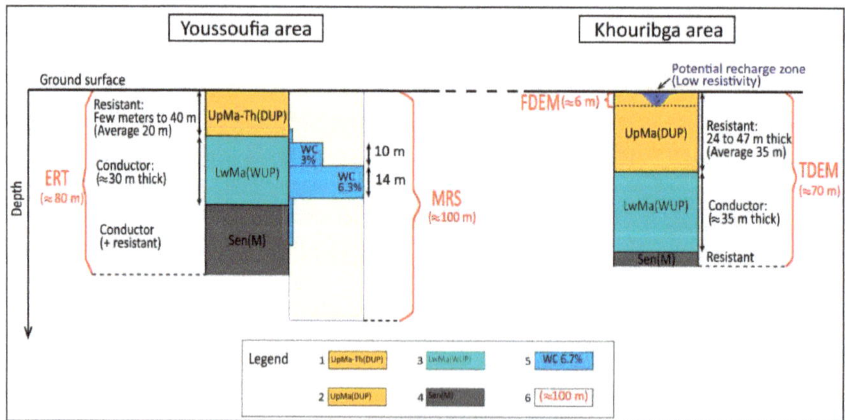

Figure 8. Summary illustration synthesizing the main outputs of the present study. 1: Upper Maastrichtian–Thanetian (Dry uncemented phosphate); 2: Upper Maastrichtian: (Dry uncemented phosphate); 3: Lower Maastrichtian: (Wet uncemented phosphate); 4: Senonian: (Compact gray marl); 5: Aquifer water content derived from MRS survey; 6: Maximal investigated depth for each geophysical method.

5. Conclusions

This study clearly shows the efficiency of the MRS method for prospecting groundwater resources and evaluating their importance in the geological context of Youssoufia open-pit mining. The hydrogeological model resulting from the MRS data inversion corroborates with the real hydrogeological data of the study area. The highlighted aquifer corresponds to two layers of high-water content Maastrichtian uncemented phosphate, as shown by the extraction works in the surrounding areas. The water volumes estimated via the piezometric data analysis and MRS method are quite comparable. Despite the average signal amplitude, the exceptional low electromagnetic noise conditions, in the geological context of the OCP open-pit mines, encourage the use of this method to survey and characterize the groundwater aquifers in the phosphatic series.

ERT profiles allowed us to delimit in detail the conductive horizons attributed to the groundwater aquifers. They clearly show that aquifers in the study area are situated below 21 m depth. Compared to the MRS method, the ERT method is less sensitive; the aquifer horizons corresponding to the uncemented phosphatic layers and intercalary of the wet marls and limestones show comparable electrical signatures, thus limiting the precise location of the top and the bottom of the supposedly saturated aquifer. However, this method is sensitive to lateral resistivity variations along the same aquifer horizon. These variations reflect the changes in water content from one zone to another along the horizon.

TDEM surveys at the Khouribga mining site highlighted a conductive horizon whose depth varies from 24 to 47 m. The interpolation between the models of the 84 surveys conducted shows that this horizon is continuous throughout the prospected zone. The integration of the lithological data sections of the surrounding boreholes enables one to attribute this conductance to the Lower Maastrichtian phosphatic layers. These layers overlie compact and impermeable Senonian marls that prevent deeper groundwater infiltration. Therefore, uncemented phosphatic layers of the Lower Maastrichtian constitute the most important aquifer in the geological context of the Khouribga open-pit mines. Conducting larger geophysical surveys, combining the TDEM and MRS surveys, which provide satisfying results in the similar hydrogeological context of the Youssoufia deposits, would enable a better understanding of the Lower Maastrichtian aquifer and evaluation of its hydrodynamic parameters.

Despite its limited investigation depth, the FDEM method can be used as a tool for mapping and delimiting the aquifer potential recharge zones in the phosphate series. The superposition of the conductivity map obtained from the FDEM data on the TDEM results, and on the spatial distribution of the aquifer hydrodynamic parameters deduced from the MRS data, could guide borehole or well installation in the most productive areas for groundwater extraction. The location of these areas would allow OCP mining engineers to implement all the necessary preparations and arrangements to extract phosphates in these flooded areas.

Author Contributions: Conceptualization: A.K. and F.-Z.I.; methodology: A.K., M.J., and D.E.A.; software: F.-Z.I., M.J., J.A.P.R., O.A.O., and L.V.D.; validation: A.K., F.-Z.I., and M.J.; investigation: F.-Z.I., A.K., M.J., D.E.A., J.A.P.R., O.A.O., and L.V.D.; resources: E.-S.J. and O.K.Y.; writing—original draft preparation: F.-Z.I. under the supervision of A.K.; writing—review and editing: A.K., M.J., D.E.A., O.K.Y., E.-S.J., J.A.P.R., O.A.O., and L.V.D.; project administration: A.K. All authors have read and agreed to the published version of the manuscript.

Funding: This research was funded by OCP-SA and CNRST Morocco project APHOS, GEO-KCH-01/2017.

Acknowledgments: This research was carried out as part of the project ID GEO-KCH 01/2017, funded by OCP. Particular thanks are addressed to the managers of the Youssoufia and Khouribga mining sites for their help and logistical support.

Conflicts of Interest: The authors declare no conflict of interest.

References

1. Bouda, A.; Salvan, M. État des connaissances sur le plateau continental marocain et ses dépôts phosphatés. *Mines Géol. Énergie* **1971**, *25*, 15–52.

2. Jasinski, S.M. *Phosphate Rock, Mineral Commodity Summaries*; U. S. Geological Survey (USGS): Reston, VA, USA, 2019; p. 122.

3. Choubert, G.; Salvan, H. Evolution du Domaine atlasique marocain depuis les temps paléozoiques. *Mém. Serv. Géol. Fr.* **1976**, *1*, 447–527.

4. Michard, A. Eléments de Géologie Marocaine. *Notes Mem. Serv. Geol. (Maroc)* **1976**, *252*, 131–146.

5. Zouhri, S.; Kchikach, A.; Saddiqi, O.; El Haïmer, F.; Baidder, L.; Michard, A. The Cretaceous-tertiary plateaus. In *Continental Evolution: The Geology of Morocco*; Springer: Berlin, Germany, 2008; pp. 331–358.

6. Azmany, M.; Farkhany, X.; Salvan, H. Gisement des Ouled Abdoum, géologie des gites minéraux marocains. *Notes Mém. Serv. Géol. Maroc* **1986**, *276*, 200–249.

7. Boujo, A. *Contribution à L'étude Géologique du Gisement de Phosphate Crétacé-Éocène des Ganntour (Maroc Occidental)*; Thèse Université de Strasbour: Strasbourg, France, 1976.

8. Karroum, M.; El Mandour, A.; Khattach, D.; Cassas, A.; Himi, M.; Rochdane, S.; Laftouhi, N.-E.; Khalil, N. Fonctionnement hydrogéologique du bassin de la Bahira (Maroc central): Apport de l'analyse des données géologiques et gravimétriques. *Can. J. Earth Sci.* **2014**, *51*, 517–526. [CrossRef]

9. Saadi, M.; Hilali, E.; Bensaïd, M.; Boudda, A.; Dahmani, M. *Carte Géologique du Maroc*; Ministere de l'Énergie et des Mines, Direction de la Géologie: Paris, France, 1985.

10. Daafi, Y.; Chakir, A.; Jourani, E.; Ouabba, S.M. Geology and mine planning of phosphate deposits: Benguerir deposit Gantour basin–Morocco. *Procedia Eng.* **2014**, *83*, 70–75. [CrossRef]

11. Belfkira, O. *Evolutions Sédimentologiques et Géochimiques de la Série Phosphatée du Maestrichtien des Ouled Abdoun (Maroc)*; Université Scientifique et Médicale de Grenoble: Grenoble, France, 1980.

12. Salvan, H. Étude préliminaire du gisement de Merah El Areh, rapport inédit. *Serv. Études Gisem. OCP* **1961**, *2*, 14.

13. Bba, A.N.; Boujamaoui, M.; Amiri, A.; Hejja, Y.; Rezouki, I.; Baidder, L.; Inoubli, M.; Manar, A.; Jabour, H. Structural modeling of the hidden parts of a Paleozoic belt: Insights from gravity and aeromagnetic data (Tadla Basin and Phosphates Plateau, Morocco). *J. Afr. Earth Sci.* **2018**, *151*, 506–522. [CrossRef]

14. El Haddi, H.; Benbouziane, A.; Mouflih, M. Geochemical Siliceous and Silicified Facies of Phosphate Series of Ouled Abdoun Basin (Morocco). *Open J. Geol.* **2014**, *4*, 295. [CrossRef]

15. Salvan, H. Phosphates. *Notes et Mémoires du Serv. Géologique du Maroc* **1952**, *87*, 283–320.

16. Kchikach, A.; Andrieux, P.; Jaffal, M.; Amrhar, M.; Mchichi, M.; Boya, B.; Amaghzaz, M.; Veyrieras, T.; Iqizou, K.J.C.R.G. Les sondages électromagnétiques temporels comme outil de reconnaissance du gisement phosphaté de Sidi Chennane (Maroc): Apport à la résolution d'un problème d'exploitation. *Comptes Rendus Geosci.* **2006**, *338*, 289–296. [CrossRef]

17. Kchikach, A.; Jaffal, M.; El Assel, N.; Chalikakis, K.; Guérin, R.; Daafi, Y.; Jourani, E.-s. Les Sondages De Resonance Magnetique Des Protons Appliques A La Caracterisation Des Aquiferes: Presentation D'une Etude Experimentale A Bouchane (Youssoufia, Maroc). *Conf. Pap.* **2013**. [CrossRef]

18. Noubhani, A.; Cappetta, H. Batoïdes nouveaux ou peu connus (Neoselachii: Rajiformes, Myliobatiformes) des phosphates maastrichtiens et paléocènes du Maroc. *Elasmobranches et Stratigr. Belg. Geol. Surv. Prof. Pap.* **1995**, *278*, 7.

19. Suberbiola, X.P.; Bardet, N.; Jouve, S.; Iarochène, M.; Bouya, B.; Amaghzaz, M. A new azhdarchid pterosaur from the Late Cretaceous phosphates of Morocco. *Geol. Soc. Lond. Spec. Publ.* **2003**, *217*, 79–90. [CrossRef]

20. Salvan, H. *Étude Complémentaire Sur le Gisement de Merah El Areh. Etude D'ensemble des Niveaux Supérieurs*; Umpublished report; Office Chérifien des Phosphates: Casablaca, Morocco, 1963; pp. 1–14.

21. Ettazarini, S.; El Mahmouhi, N. Vulnerability mapping of the Turonian limestone aquifer in the Phosphates Plateau (Morocco). *Environ. Geol.* **2004**, *46*, 113–117. [CrossRef]

22. Bougadra, A. *Synthèse Hydrogéologique de la Bahira Occidentale. Thèse de Troisième Cycle*; Université Cadi Ayyad: Marrakech, Morocco, 1991.

23. El Mokhtar, M.; Fakir, Y.; El Mandour, A.; Benavente, J.; Meyer, H.; Stigter, T. Salinisation des eaux souterraines aux alentours des sebkhas de Sad Al Majnoun et Zima (plaine de la Bahira, Maroc). *Sci. Chang. Planétaires/Sécher.* **2012**, *23*, 48–56.

24. Dahlin, T.; Zhou, B. A numerical comparison of 2D resistivity imaging with 10 electrode arrays. *Geophys. Prospect.* **2004**, *52*, 379–398. [CrossRef]

25. Loke, M.; Dahlin, T. A comparison of the Gauss–Newton and quasi-Newton methods in resistivity imaging inversion. *J. Appl. Geophys.* **2002**, *49*, 149–162. [CrossRef]

26. Chalikakis, K.; Nielsen, M.R.; Legchenko, A.; Hagensen, T.F. Investigation of sedimentary aquifers in Denmark using the magnetic resonance sounding method (MRS). *Comptes Rendus Geosci.* **2009**, *341*, 918–927. [CrossRef]

27. Legchenko, A.; Baltassat, J.M.; Bobachev, A.; Martin, C.; Robain, H.; Vouillamoz, J.M. Magnetic resonance sounding applied to aquifer characterization. *Groundwater* **2004**, *42*, 363–373. [CrossRef]

28. Legchenko, A.; Baltassat, J.-M.; Beauce, A.; Bernard, J. Nuclear magnetic resonance as a geophysical tool for hydrogeologists. *J. Appl. Geophys.* **2002**, *50*, 21–46. [CrossRef]

29. Nielsen, M.R.; Hagensen, T.F.; Chalikakis, K.; Legchenko, A. Comparison of transmissivities from MRS and pumping tests in Denmark. *Near Surf. Geophys.* **2011**, *9*, 211–223. [CrossRef]

30. IRIS Instruments. *Software for Magnetic Resonance Sounding, Numis Pro User's Manual*; IRIS Instrument: Orléans, France, 2005.

31. Galibert, P.-Y.; Guerin, R.; Andrieux, P. Structural mapping in basin-and-range-like geology by electromagnetic methods: A powerful aid to seismic. *Geophys. Prospect.* **1996**, *44*, 1019–1040. [CrossRef]

32. Descloitres, M.; Guérin, R.; Albouy, Y.; Tabbagh, A.; Ritz, M. Improvement in TDEM sounding interpretation in presence of induced polarization. A case study in resistive rocks of the Fogo volcano, Cape Verde Islands. *J. Appl. Geophys.* **2000**, *45*, 1–18. [CrossRef]

33. McNeill, J. *Principles and Application of Time Domain Electromagnetic Techniques for Resistivity Sounding*; Geonics: Mississauga, ON, Canada, 1994.

34. Schmutz, M.; Guérin, R.; Maquaire, O.; Descloitres, M.; Schott, J.-J.; Albouy, Y. Apport de l'association des méthodes TDEM (Time-Domain Electromagnetism) et électrique pour la connaissance de la structure du glissement-coulée de Super Sauze (bassin de Barcelonnette, Alpes-de-Haute-Provence, France). *Comptes Rendus l'Acad. Sci. Ser. IIA-Earth Planet. Sci.* **1999**, *328*, 797–800. [CrossRef]

35. AEMR. *Applied Electromagnetic Research: Manual for Tem-Fast ProSystem*; AEMR: Utrecht, The Netherlands, 1997; 62p.

36. McNeill, J. *Electromagnetic Terrain Conductivity Measurement at Low Induction Numbers*; Technical Note; Geonics: Mississauga, ON, Canada, 1980.

37. Nabighian, M. *Electromagnetic Methods in Applied Geophysics*; Society of Exploration Geophysicists: Tulsa, OK, USA, 1989.

38. Hördt, A.; Hauck, C. Electromagnetic methods. In *Applied Geophysics in Periglacial Environments*; Cambridge University Press: Cambridge, UK, 2008; pp. 28–56.

39. Dunn, K.-J.; Bergman, D.J.; LaTorraca, G.A. *Nuclear Magnetic Resonance: Petrophysical and Logging Applications*; Elsevier: Amsterdam, The Netherlands, 2002; Volume 32.

40. Legchenko, A.; Valla, P. Removal of power-line harmonics from proton magnetic resonance measurements. *J. Appl. Geophys.* **2003**, *53*, 103–120. [CrossRef]

41. Perttu, N. Assessment of Hydrogeological and Water Quality Parameters, using MRS and VES in the Vientiane Basin, Laos. Ph.D. Thesis, Luleå tekniska universitet, Luleå, Sweden, 2008.

42. Briggs, I.C.J.G. Machine contouring using minimum curvature. *Geophysics* **1974**, *39*, 39–48. [CrossRef]

43. El Assel, N.; Kchikach, A.; Durlet, C.; AlFedy, N.; El Hariri, K.; Charroud, M.; Jaffal, M.; Jourani, E.; Amaghzaz, M.J.E.G. Mise en évidence d'un Sénonien gypseux sous la série phosphatée du bassin des Ouled Abdoun: Un nouveau point de départ pour l'origine des zones dérangées dans les mines à ciel ouvert de Khouribga, Maroc. *Estud. Geol.* **2013**, *69*, 47–70. [CrossRef]

44. Michard, A.; Saddiqi, O.; Chalouan, A.; de Lamotte, D.F. *Continental Evolution: The Geology of Morocco: Structure, Stratigraphy, and Tectonics of the Africa-Atlantic-Mediterranean Triple Junction*; Springer: Berlin/Heidelberg, Germany, 2008; Volume 116.

45. El Kiram, N.; Kchikach, A.; Jaffal, M.; El Azzab, D.; El Ghorfi, M.; Khadiri, O.; Jourani, E.-S.; Manar, A.; Nahim, M.J.C.R.G. Phosphatic series under Plio-Quaternary cover of Tadla Plain, Morocco: Gravity and seismic data. *Comptes Rendus Geosci.* **2019**, *351*, 420–429. [CrossRef]

Minerals **2020**, *10*, 842

46. El Kiram, N.; Kchikach, A.; Jaffal, M.; Pena, J.A.; Teixido, T.; Guerin, R.; Khadiri Yazami, O.; Jourani, E. *Les Dérangements de la Série Phosphatée Dans le District Minier de Khouribga (Maroc): Une Esquisse de Leur Origine et de Leurs Méthodes de Cartographie Sous Couverture Quaternaire*; Géologues: Revue Officielle de la Société Géologique de France: Paris, France, 2017; p. 194.

47. Kchikach, A.; Elassel, N.; Gurein, R.; Teixido, T.; Pena, J.; Jaffal, M. TDEM and EM31 Methods for Detecting Sterile Bodies in the Phosphatic Bearing of Sidi Chennane (Morocco). In Proceedings of the Near Surface Geoscience 2012—18th European Meeting of Environmental and Engineering Geophysics, Paris, France, 3–5 September 2012.

48. Kchikach, A.; Jaffal, M.; Aïfa, T.; Bahi, L. Cartographie de corps stériles sous couverture quaternaire par méthode de résistivités électriques dans le gisement phosphaté de Sidi Chennane (Maroc). *Comptes Rendus Geosci.* **2002**, *334*, 379–386. [CrossRef]

© 2020 by the authors. Licensee MDPI, Basel, Switzerland. This article is an open access article distributed under the terms and conditions of the Creative Commons Attribution (CC BY) license (http://creativecommons.org/licenses/by/4.0/).

Article

Parallel Simulation of Audio- and Radio-Magnetotelluric Data

Nikolay Yavich [1,2,*], **Mikhail Malovichko** [1,2] **and Arseny Shlykov** [3]

[1]   Center for Data-Intensive Science and Engineering, Skolkovo Institute of Science and Technology, Moscow 191205, Russia; m.malovichko@skoltech.ru

[2]   Applied Computational Geophysics Lab, Moscow Institute of Physics and Technology, Dolgoprudny 141701, Russia

[3]   Institute of Earth Sciences, Saint-Petersburg University, St. Petersburg 199034, Russia; a.shlykov@spbu.ru

*   Correspondence: n.yavich@skoltech.ru

Received: 22 November 2019; Accepted: 27 December 2019; Published: 31 December 2019

Abstract: This paper presents a novel numerical method for simulation controlled-source audio-magnetotellurics (CSAMT) and radio-magnetotellurics (CSRMT) data. These methods are widely used in mineral exploration. Interpretation of the CSAMT and CSRMT data collected over an area with the complex geology requires application of effective methods of numerical modeling capable to represent the geoelectrical model of a deposit well. In this paper, we considered an approach to 3D electromagnetic (EM) modeling based on new types of preconditioned iterative solvers for finite-difference (FD) EM simulation. The first preconditioner used fast direct inversion of the layered Earth FD matrix (Green's function preconditioner). The other combined the first with a contraction operator transformation. To illustrate the effectiveness of the developed numerical modeling methods, a 3D resistivity model of Aleksandrovka study area in Kaluga Region, Russia, was prepared based on drilling data, AMT, and a detailed CSRMT survey. We conducted parallel EM simulation of the full CSRMT survey. Our results indicated that the developed methods can be effectively used for modeling EM responses over a realistic complex geoelectrical model for a controlled source EM survey with hundreds of receiver stations. The contraction-operator preconditioner outperformed the Green's function preconditioner by factor of 7–10, both with respect to run-time and iteration count, and even more at higher frequencies.

Keywords: geophysical electromagnetic modelling; CSAMT; CSRMT; CSEM

---

1. Introduction

Controlled-source audio-magnetotellurics (CSAMT) is a popular method for mineral exploration [1–5]. In principle, a non-uniform plane wave at a study area is generated by two orthogonal transmitter lines located at a sufficient distance (in the far-field zone). A typical operation frequency range is 0.1–10 kHz, which is roughly the high end of the quasi-stationary regime. Transfer functions between electric and magnetic field components, E_x, E_y, H_x, H_y, and H_z are computed from measured data and then inverted in order to reconstruct the subsurface electric conductivity. CSAMT is known to provide better data quality in the AMT dead bands (0.1–5 Hz and 700–3000 Hz) than conventional AMT surveys and it requires relatively inexpensive transmitter operations [2,5,6].

Closely connected to CSAMT is the controlled-source radio-magnetotellurics (CSRMT). This method is gaining popularity in the near-surface geophysics [7–11]. It operates at much higher frequencies (typically, up to 250 kHz–1 MHz) and thus provides excellent spatial resolution. The biggest difference with CSAMT is the displacement currents in the air [10,12] cannot be neglected to accurately compute the electric and magnetic fields, which complicates the numerical simulation dramatically. The authors of [13] demonstrated, however, that the full impedance and quasi-stationary impedances are almost identical in the 1D Earth. It suggests that the CSRMT measurements can be interpreted using an inversion software with the quasi-stationary forward simulation engine (note, though, [14,15]).

Numerical forward modeling plays a paramount role in both electromagnetic (EM) survey design and data interpretation. This demand has been driving the investigation of efficient modeling methods for nearly 60 years. Today, 1D and 2D frequency-domain modeling is relatively established, while research on 3D modeling algorithms is a hot topic.

The common problem in 3D modeling of CSAMT and CSRMT data is the solution of a large system of linear equations. This problem is present in all the discretization methods: Finite difference (FD), finite element (FE), integral equation (IE), and others; although in FD and FE modeling the system matrix is sparse, while in IE modeling the system matrix is dense. Consequently, design and implementation of an economical, scalable, and robust solver for this system is the key to successful 3D modeling and inversion.

At the frequencies from 0 to 1 MHz, the EM fields are mainly diffusive. Thus, the errors introduced by resistivity model resampling are local. Consequently, adaptation of the grid to a particular interface is a minor concern. We thus consider the second-order edge-based finite-difference discretization.

Computer simulation of both CSAMT and CSRMT data poses great computational challenges. The land surveys are frequently characterized by an exceptionally large resistivity contrast of the rocks, e.g., 1000 Ωm for granite and 0.01 Ωm for nickel ore [16], leading to very fine numerical grids needed to resolve the shortest skin depth. At the same time, the computational domain should be extended to include attenuation dictated by the longest skin depth, especially at frequencies above ~1 kHz, at which considerable amount of the electromagnetic field travels in the air and resistive rocks with little attenuation. Furthermore, in CSAMT/CSRMT setups the transmitters are grounded, making their numerical treatment more difficult comparing to the conventional audio and radio magnetotellurics (AMT/RMT), respectively. For these reasons, a CSAMT/CSRMT simulation easily results in a discrete problem with millions of discrete unknowns. Beyond being very large, the system matrix is extremely ill-conditioned. Thus, both preconditioned iterative solvers and sparse direct solvers are commonly considered. Although the use of unstructured tetrahedral grids [17] or hexahedral grids with hanging nodes [18] reduces the size of the discrete problem, an efficient solver still will be required.

Direct solvers are attractive for their universality and robustness [6,19], though they require very long initialization and large memory allocations. Thus, iterative solvers are more typically used in practical modeling and inversion. The commonly applied preconditioners are incomplete factorization completed by divergence correction [20] and multigrid [21,22].

In this paper, we assess performance of the two preconditioned iterative solvers introduced by the authors in [23]. They have been successfully tested on MT and marine controlled-source electromagnetic (CSEM) [23] and land CSEM [24] setups. Here, we applied the two preconditioners to quasi-stationary simulation in the CSRMT frequency range. Both serial and parallel modeling performance were evaluated. The solvers were applied to the secondary field formulation to avoid excessive gridding near the source.

The paper is organized as follows. We review forward modelling and briefly discuss implementation in Section 2. In Section 3, we verify modeling accuracy of our code versus an integral equation code [25]. Next, we present results of numerical simulation of a study area in Moscow syneclise. We compare performance of the two preconditioners in a single-threaded mode. Then we present a complete simulation over a realistic geologic model conducted on a cluster. Final remarks are given in Section 4.

2. Three-Dimensional Simulation of CSAMT/CSAMT Data

2.1. Efficient Finite-Difference Simulation

In CSAMT/CSRMT, the five components of the electromagnetic filed are measured (E_x, E_y, H_x, H_y, H_z) and, after preprocessing, converted to the components of the impedance and tipper. The off-diagonal components of the impedance are employed most commonly, that is [2],

$$
\begin{aligned}
Z_{xy} &= \frac{E_{x1}H_{x2} - E_{x2}H_{x1}}{H_{x1}H_{y2} - H_{x2}H_{y1}}, \\
Z_{yx} &= \frac{E_{y1}H_{y2} - E_{y2}H_{y1}}{H_{x1}H_{y2} - H_{x2}H_{y1}},
\end{aligned}
\tag{1}
$$

where subscripts 1 and 2 refer to one of the two orthogonal transmitter lines. Thus, computation of the components of the impedance or tipper requires simulation of the electric and magnetic fields E and H at several locations and frequencies.

We consider electromagnetic modeling in 3D heterogeneous isotropic media in the frequency domain. Assuming time dependence of $e^{-i\omega t}$, the electric field, $E(x, y, z)$, satisfies the following system of partial differential equations,

$$
curl\ curl\ E - i\omega\mu_0\sigma E = i\omega\mu_0 F,
\tag{2}
$$

where ω is the source angular frequency, μ_0 is the magnetic permeability of the free space, $\sigma(x, y, z)$ is the conductivity model, and $F(x, y, z)$ is the source current density. This is a linear system of three scalar equations with respect to three scalar entries of $E(x, y, z)$. We solve this partial-differential equation system in a bounded domain, completed by zero Dirichlet boundary conditions. The magnetic field H is then computed via Faraday's law,

$$
H = \frac{1}{i\omega\mu_0}\ curl\ E.
\tag{3}
$$

Since geological models predominately vary in the vertical direction, we can always split the conductivity model into background and anomalous parts,

$$
\sigma(x, y, z) = \sigma_b(z) + \sigma_a(x, y, z).
\tag{4}
$$

We further assume that the following double inequality holds,

$$
\alpha\sigma_b(z) \leq \sigma(x, y, z) \leq \beta\sigma_b(z), 0 < \alpha \leq 1 \leq \beta,
\tag{5}
$$

which controls the contrast of anomalous conductivity with respect to the background.

Given a non-uniform computational grid with $N_x \times N_y \times N_z$ cells, we apply the conventional edge-based second-order FD discretization. The FD method approximates the unknown electric field $E = \left(E_x, E_y, E_z\right)$ with a finite set of discrete values $\left\{E_{i+\frac{1}{2}\ j\ k},\ E_{i\ j+\frac{1}{2}\ k},\ E_{i\ j\ k+\frac{1}{2}}\right\}$ [26]. Each discrete value is attached to the respective edge of the FD grid, Figure 1.

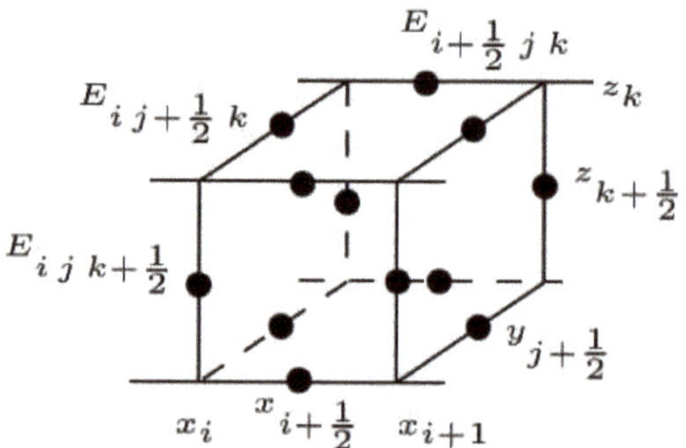

Figure 1. Discrete electric fields attached to the edges of the grid.

The actual discrete equations can be found for example in [27]. Let us form the unknown vector e of the discrete electric fields $\left\{E_{i+\frac{1}{2}\,j\,k},\,E_{i\,j+\frac{1}{2}\,k},\,E_{i\,j\,k+\frac{1}{2}}\right\}$ introduced above. Now we can write the FD discretization of (2) in a matrix form:

$$A\,e = i\omega\mu_0 f, \tag{6}$$

where A is a complex symmetric system matrix with at most 13 nonzero entries per row, corresponding to the total conductivity distribution. We denote the size of this system as n, $n \approx 3N_xN_yN_z$.

Let A_b be the FD system matrix, corresponding to the background conductivity model. Importantly, this matrix can be implicitly factorized using a fast direct algorithm, and the action of the inverse matrix can be efficiently computed [23,28]. As a result, it can be used as a preconditioner to (6):

$$A_b^{-1}A\,e = i\omega\mu_0 A_b^{-1} f. \tag{7}$$

We will refer A_b^{-1} as the Green's function (GF) preconditioner. The complexity of applying the GF preconditioner is $O\left(n^{4/3}\right)$, and auxiliary memory required is near $3n$ only. Applying the analysis presented in [23] the condition number of the GF preconditioned system can be estimated as follows,

$$cond\left(A_b^{-1}A\right) \le \beta/\alpha. \tag{8}$$

This result implies that convergence of an iterative solver applied to (7) has minor or no dependence on the grid size and cell aspect ratio, as well as the frequency, while it degrades on models with high-contrast bodies.

To minimize the impact of high-contrast bodies, another preconditioner could be constructed. Denote as Σ and Σ_b diagonal matrices, corresponding to full and background conductivity, respectively. Let us define the modified FD Green's operator and diagonal matrices according to the following formulae:

$$\mathcal{G}_b^M = 2i\omega\mu_0\Sigma_b^{\frac{1}{2}}A_b^{-1}\Sigma_b^{\frac{1}{2}} + I, \tag{9}$$

$$K_1 = \frac{1}{2}(\Sigma + \Sigma_b)\Sigma_b^{-\frac{1}{2}},\ K_2 = \frac{1}{2}(\Sigma - \Sigma_b)\Sigma_b^{-\frac{1}{2}}. \tag{10}$$

Using this operator, Equation (7) can be written in an equivalent form as follows (see Appendix A):

$$\hat{e} = \mathcal{G}_b^M K_2 K_1^{-1}\hat{e} + i\omega\mu_0\Sigma_b^{\frac{1}{2}}A_b^{-1} f, \tag{11}$$

where $\hat{e} = K_1 e$. By introducing a new operator, $C = \mathcal{G}_b^M K_2 K_1^{-1}$, we rewrite Equation (11) as follows:

$$(I - C)\hat{e} = i\omega\mu_0\Sigma_b^{\frac{1}{2}}A_b^{-1} f. \tag{12}$$

We will refer this system a contraction operator (CO) preconditioned system, since it was shown in [23] that $\|C\| < 1$. Moreover, it can be proved that this preconditioned system has a smaller or equal spectral condition number than that of the GF preconditioner, implying faster or equal convergence of the iterative solvers,

$$cond(I - C) \leq \max\{1/\alpha, \beta\}. \tag{13}$$

The complexity of applying the CO preconditioner is $O(n^{4/3})$ as well. The preconditioners in Equations (7) and (12) were incorporated into the BiCGStab iterative solver [29]. We used the complex-valued version of the solver with the standard complex-valued dot product.

Practical controlled-source modeling requires excessive gridding near the source location. To avoid this, we preferred secondary field modeling, i.e., the electric field $E(x, y, z)$ is represented as a sum $E_a(x, y, z) + E_b(x, y, z)$, where $E_b(x, y, z)$ is the response due to background conductivity model known analytically [30,31]. In this case, the secondary field $E_a(x, y, z)$ will be the unknown function and its FD discretization is performed. The actual source $F(x, y, z)$ in Equation (2) is substituted with the secondary source $\sigma_a(x, y, z)E_b(x, y, z)$.

2.2. Applicability of Quasi-Statinary Simulation

As the first step, we tested applicability of the quasi-stationary simulation in the context of controlled-source electromagnetics. The displacement current in the air must be considered when the electromagnetic field generated by a high-frequency dipole oscillator is measured at large offsets [10,12]. Otherwise observed components of the electromagnetic field cannot be matched to the computed quasi-stationary ones; this is known as the propagation effect. Formally, it is achieved by replacing $i\omega\mu_0\sigma$ term in Equation (1) with $i\omega\mu_0\sigma - \omega^2\mu_0\varepsilon_0\varepsilon^{rel}$, where ε_0 is the vacuum dielectric permittivity and $\varepsilon^{rel} \geq 1$ is the relative dielectric permittivity. It complicates the 3D numerical simulation considerably. However, in contrast to the individual components, the surface displacement-current impedance was almost identical to the quasi-stationary one, at least in a 1D Earth. To illustrate this point, we considered an X-oriented horizontal electric dipole (HED) located at the origin of Cartesian coordinate system and two receiver stations (Figure 2).

The first station corresponded to relatively short offset ($X = 450$ m, $Y = 750$ m) and the second one imitated a typical CSAMT offset ($X = 450$ m, $Y = 4500$ m). We computed the electromagnetic field for the two cases. In the first case, the model was the one in Table 1. Computations were conducted in the quasi-stationary regime, where the squared wavenumber of i-th layer was given by $i\omega\mu_0\sigma_i$. The air conductivity was set to $\sigma_0 = 10^{-8}$ S/m. In the second case, the squared wavenumbers of each i-th layer were defined as $i\omega\mu_0\sigma_i - \omega^2\mu_0\varepsilon_0\varepsilon_i^{rel}$. We set $\varepsilon_0^{rel} = 1$ in the air, and $\varepsilon_i^{rel} = 4$, $i = 1, 2, \ldots$ in the Earth. The computations were performed by the 1D code of [13]. Computed curves are presented in Figure 2. At frequencies higher than 7 kHz (the close receiver) and 27 kHz (the distant receiver), the quasi-stationary electromagnetic field (dashed lines) differed considerably from the displacement-current field (solid lines). However, ratios of the horizontal component remained essentially the same.

Table 1. The background 1D resistivity model.

#	Top, m	Thickness, m	Resistivity, Ωm
1	$-\infty$	∞	10^8
2	0	8	500
3	8	92	20
4	100	10	10^4
5	110	10	20
6	120	∞	10^4

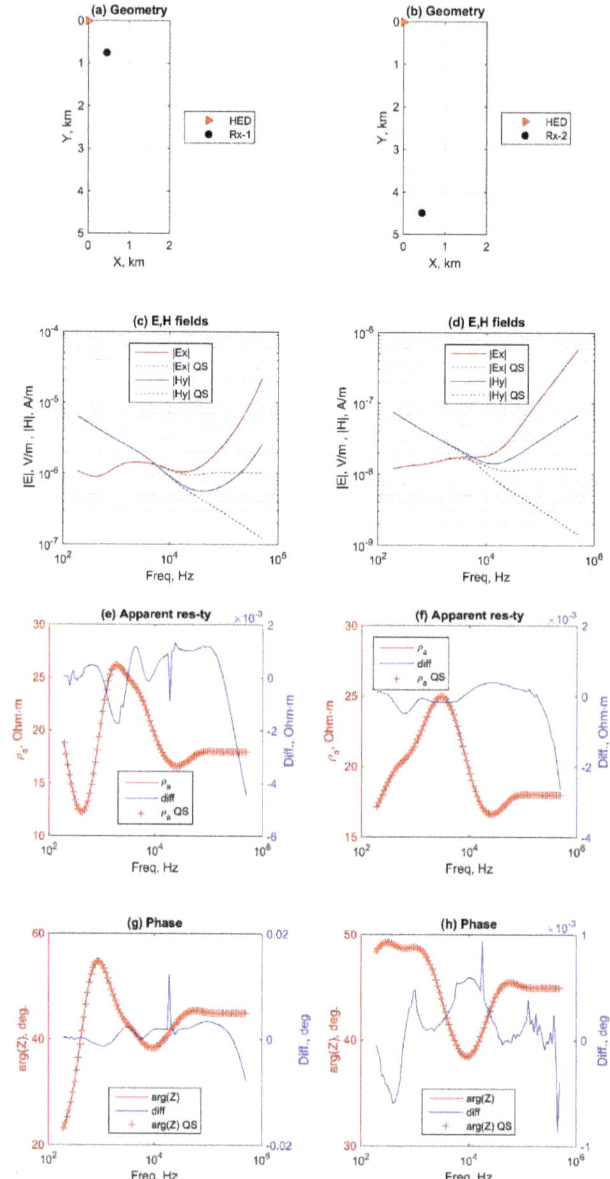

Figure 2. Impact of the displacement currents on the electromagnetic field and the surface impedance on the top of a layered Earth. The left column of panels illustrates computations relative to a short transmitter–receiver separation (receiver Rx-1). The right column of panels presents computations for a long transmitter–receiver separation (receiver Rx-2). Panels (**a,b**) depict geometry of the simulation, the top view. An X-oriented horizontal electric dipole (HED) was placed at the origin. Receiver station Rx-1 was located at (450, 750) m, whereas receiver station Rx-2 had coordinates (450, 4500) m. Panels (**c,d**) depict magnitudes of individual components of the electric field. Panels (**e,f**) depict Cagniard's apparent resistivity (ρ_a). Panels (**g,h**) show impedance phase. In the legends, QS stands for the quasi-stationary computations.

2.3. Implementation

Forward modeling of a single source involves the following steps:

- Modeling grid preparation,
- Resistivity model resampling,
- Right-hand side computation,
- Secondary electric field computation, and
- Computation of the total electric field, recovery of the magnetic field.

Our implementation was in C++. The most computationally demanding steps are the right-hand side (RHS) and secondary field computation. To reduce the computational burden, these steps were parallelized using OpenMP shared memory implementation. Simulation of multiple sources and frequencies was parallelized using the Message Passing Interface MPI. In [32], we demonstrated good scalability of this scheme.

3. Numerical Experiments

3.1. Code Verification on Simple Models

Our code in general has been validated and benchmarked against analytical layered-earth plane-wave solutions in [23]. Verification of the code in the low-frequency controlled-source electromagnetics scenario was performed in [32]. Here, we conducted several tests concerning particularities of the CSAMT/CSRMT, mainly high frequencies and high conductivity contrast.

We used Consortium for Electromagnetic Modeling and Inversion's software PIE3D for benchmarking. PIE3D was rigorously verified and has been used in production for years [25,33,34].

In this test, we used a resistivity model consisting of a 1D background model and a 100 Ωm parallelepiped body. The background model is presented in Table 1. The parallelepiped body had coordinates of the two opposite corners (110, −28, 20) m and (240, 28, 52) m; thus, that was a brick $132 \times 56 \times 32$ m^3 with its top at 20 m (Figure 3).

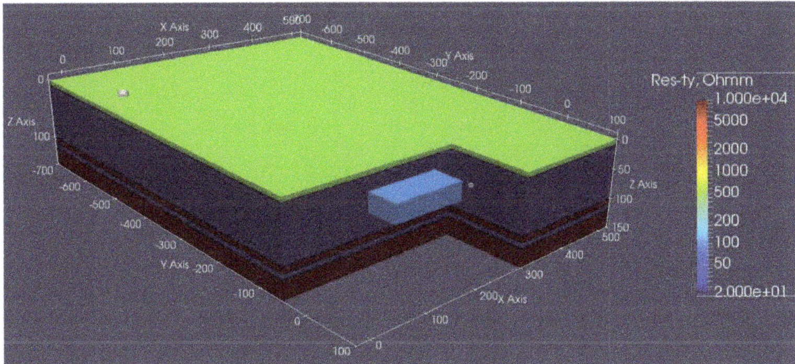

Figure 3. The 3D model used for benchmarking. The grey cube depicts position of the receiver station. The sphere is placed at transmitter position.

The transmitter was an X-orientated point electric dipole located at (32.54, −553.5) m. The receiver was at (200, 80) m. In this test, we used seven frequencies: 0.1, 0.2, 0.5, 1, 2, 5, and 10 kHz.

Finite-difference computations were performed with large and fine numerical grids. The core domain was the same at each frequency $478 \times 158 \times 150$ m^3. It included both the anomalous body and receiver location (Figure 4).

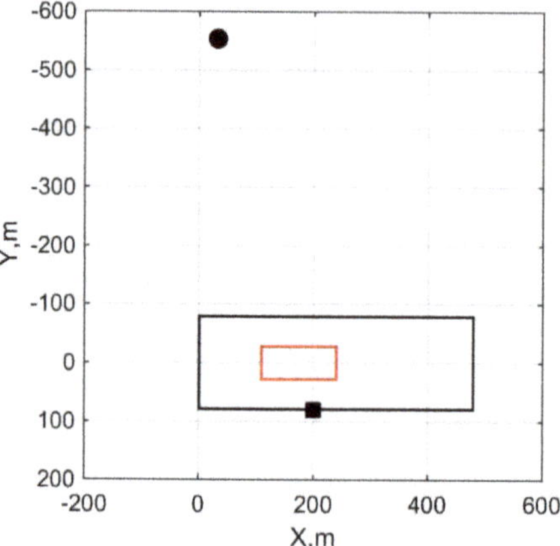

Figure 4. Geometry of the computation grid. The core domain (black rectangle) covers the anomalous body (red rectangle) and receiver position (small black square). The small black circle depicts position of the transmitter.

The grid step size was identical in all frequencies, $h = 2$ m. The padding part of the grids was different for different frequencies. The overall physical size of the grids varied from $56.0 \times 56.0 \times 28.1$ km^3 at 0.1 kHz to $6.7 \times 6.4 \times 3.3$ km^3 at 10 kHz. The number of the internal edges was between 17 M at 10 kHz and 23 M at 0.1 kHz.

PIE3D ran with grid step size $h = 2$, meaning that the body was discretized into $66 \times 28 \times 16$ cubical cells. Computed electromagnetic responses are compared in Figure 5.

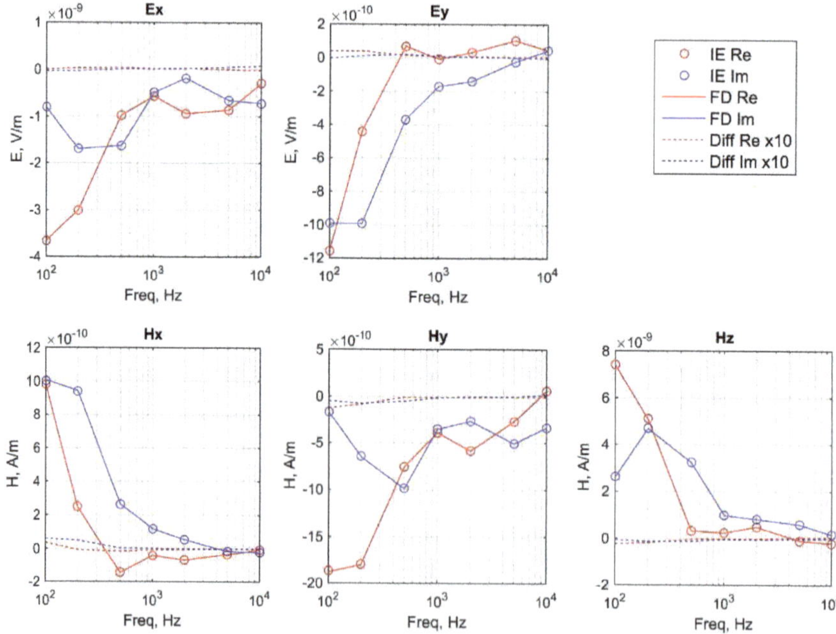

Figure 5. Benchmark against PIE3D software. "IE" and "FD" mean the integral-equation software (PIE3D) and the finite-difference software (our code), respectively. "Re" and "Im" mean the real and imaginary parts, respectively. Note that the difference curves are magnified by factor 10.

We observed a close match between individual components. The normalized misfit between IE and FD responses, that is $\|u_{IE} - u_{FD}\| / \|u_{IE}\|$, varied from 0.2% to 0.6%. This is encouraging, taking in mind that the computations were done by two very different approaches with no shared code.

3.2. Conductivity Model of Aleksadrovka

We evaluated our numerical code on a realistic 3D model using an acquisition geometry from a real geophysical survey. The model mimicked composition of Moscow State University's geophysical test camp near the village of Aleksandrovka in Kaluga Region, Russia. The acquisition geometry is presented in (Figure 6). In this work, we did not compare modelled versus measured data.

Figure 6. Map of the study area from which the acquisition geometry was taken for the numerical simulation. The boxes depict receiver stations. The receiver lines are numbered from 1 to 8.

The receiver grid consisted of 192 receiver stations and two orthogonal transmitter lines (see Figures 6 and 7). All receivers had azimuth 68.5°. Coordinates of receiver stations and transmitter lines can be found in the Supplementary Materials File S1. The five components of the electromagnetic field were simulated.

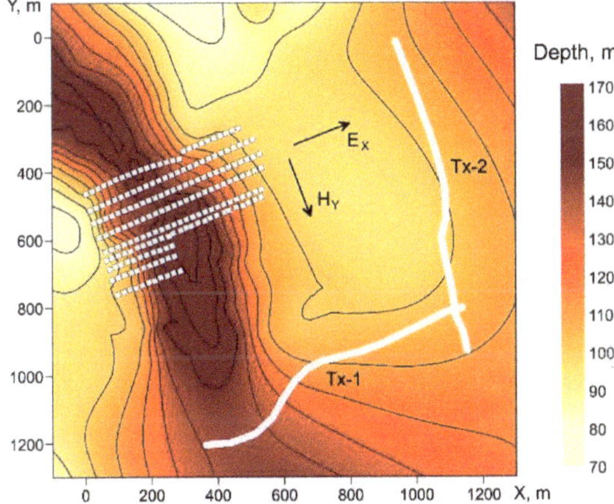

Figure 7. Topography of the top of the first layer of limestones. The white squares depict receiver stations. The white lines are grounded transmitters lines. Arrows Ex and Hy indicate orientation of the local coordinate axes at receiver stations. Coordinates of the acquisition system are attached to the electronic version of this article.

The model used in this study was compiled from various geologic studies, geophysical surveys, including seismics, and drilling data. Aleksandrovka area is located in Moscow syneclise and has relatively simple morphology of pre-Quaternary layers, but the shallow part of the subsurface has a more complicated structure due to Moscow glaciation [35]. The reference 1D geoelectrical model is based on a 300 m borehole drilled in the camp for water supply purposes. The lithology column is shown in Figure 8.

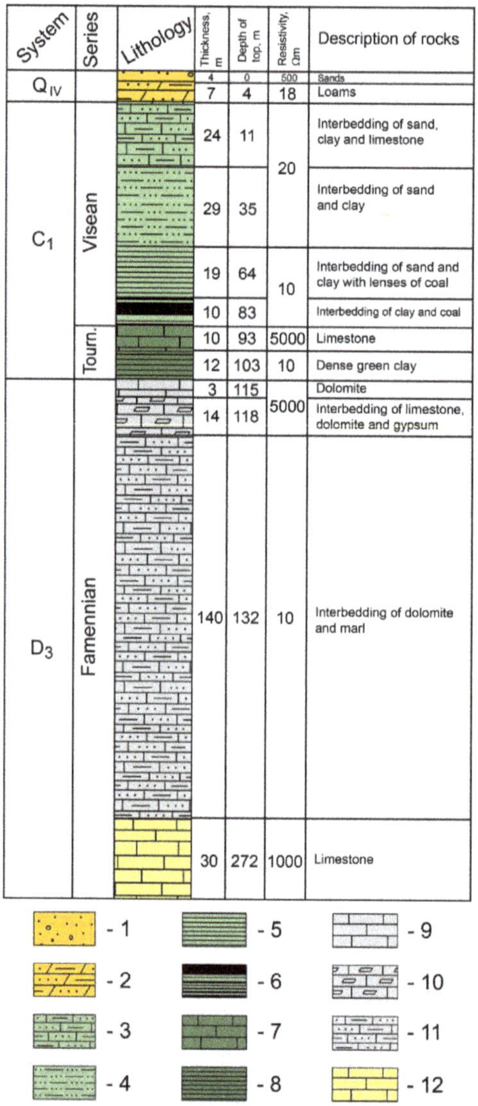

System	Series	Lithology	Thickness, m	Depth of top, m	Resistivity, Ωm	Description of rocks
Q_{IV}			4	0	500	Sands
			7	4	18	Loams
C_1	Visean		24	11	20	Interbedding of sand, clay and limestone
			29	35		Interbedding of sand and clay
			19	64	10	Interbedding of sand and clay with lenses of coal
			10	83		Interbedding of clay and coal
	Tourn.		10	93	5000	Limestone
			12	103	10	Dense green clay
D_3	Famennian		3	115	5000	Dolomite
			14	118		Interbedding of limestone, dolomite and gypsum
			140	132	10	Interbedding of dolomite and marl
			30	272	1000	Limestone

- 1
- 2
- 3
- 4
- 5
- 6
- 7
- 8
- 9
- 10
- 11
- 12

Figure 8. The lithologic column from the borehole shown in Figure 1. Lithology explanation: 1—Sands (Q_{IV}), 2—loams (Q_{IV}), 3—Interbedding of sand, clay and limestone (C_1), 4—Interbedding of sand and clay (C_1), 5—Interbedding of sand and clay with lenses of coal (C_1), 6—Interbedding of clay and coal (C_1), 7—Limestone (C_1), 8—Dense green clay(C_1), 9—dolomite (D_3), 10—Interbedding of limestone, dolomite and gypsum (D_3), 11—Interbedding of dolomite and marl (D_3), 12—Limestone (D_3).

The top part of the column (above 93 m) is composed of mostly clayey sediments with a resistivity of approximately 20 Ωm except for the shallowest Quaternary sands. The sands have a resistivity of about 500 Ωm. There are two thin (~10 m) layers of limestones and dolomites at the depth of 93 m and 115 m. Their resistivity is estimated to 5000 Ωm. The depth of the top of the first carbonate layer was clearly detected by CSRMT across the area. It is underlain by Famennian rocks that consist mostly of dolomites and marls with the overall thickness of 140 m. This formation has a resistivity of about 10 Ωm. Since this estimate came from AMT data and the Famennian formation contains several thin dolomite layers known to be resistive within the region, this resistivity estimate can be regarded as the lateral resistivity. Below this layer, the well entered hard limestones showing its upper 30 m. We estimated the thickness of this layer as 30 m and resistivity as 1000 Ωm, based on the results of the AMT inversion and numerical simulation experiments. Properties of underlying rocks down to 1 km have been obtained from 1D inversion of a single AMT curve. The depth of the Famennian rocks was estimated as 480 m. A very conductive layer (1.5 Ωm) in the range from 480 to 730 m, as well as a resistive (630 Ωm) basement below 730 m, were introduced by the 1D inversion. The final 1D reference model is presented in Table 2.

Table 2. The reference 1D resistivity model.

#	Top, m	Thickness, m	Resistivity, Ωm
1	$-\infty$	∞	10^8
2	0	12	18
3	12	32	20
4	44	26	10
5	70	23	16
6	93	10	5000
7	103	12	11
8	115	17	5000
9	132	140	11
10	272	32	1000
11	304	176	11
12	480	250	1.5
13	730	∞	670

The 1D horizontally layered reference model was deformed based on the topography of the top of the first layer of resistive carbonates. This reference surface was clearly identified from previous CSRMT survey. The most prominent feature of it is the trough oriented along the NW–SE direction, which has the maximal relative depression of about 60 m (Figure 8). Other layers with thicknesses taken from the 1D model conform to the reference surface.

Finally, we added to the model inhomogeneities that represent composition of the shallow part of the subsurface. The subsurface has a relatively complicated geoelectrical structure because of Moscow glaciation. This part has been characterized previously with a detailed CSRMT survey in the frequency range 5–1000 kHz (the far-field responses only) followed by a conventional 2D plane-wave inversion. Several superficial high-resistive sand bodies were delineated on the interpreted 2D cross-sections (Figure 9).

The biggest body was composed of Quaternary glacial or alluvial sands of 160 Ωm with its top at a depth of 20 m. The eastern part of it has an elongated shape, with approximately 100 m along its short axis and thickness of 20 m. This body was inserted into the 3D model. Finally, a 6-m-thick layer of 19 Ωm was introduced across the top of the 3D resistivity model.

The final 3D model is given in Figure 10a,b together with receivers and transmitters. We believe it represents all main features of the Aleksandrovka site. The conductivity model in the Visualization Toolkit VTK legacy file format (https://vtk.org/wp-content/uploads/2015/04/file-formats.pdf) is available in the Supplementary Materials File S1 attached to this article.

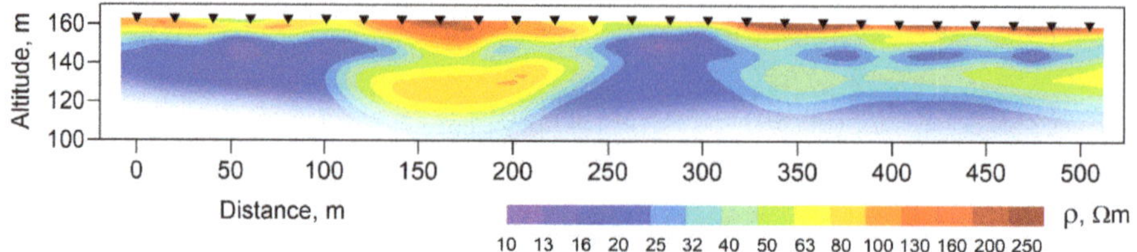

Figure 9. An example of resistivity cross-sections employed to create the shallow part of Aleksandrovka conductivity model. The cross-sections were obtained previously as a result of the 2D inversion of a CSRMT survey.

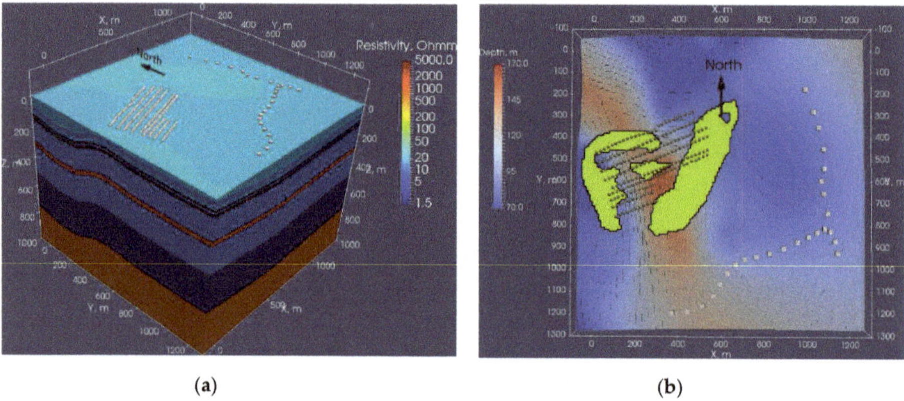

(a)   (b)

Figure 10. (**a**) The 3D resistivity model used in the numerical experiment (the shallowest 6 m were removed for visualization). The receivers are depicted as spheres, the two transmitters lines are marked by boxes. Resistivity is shown in color. (**b**) Structure of the shallow geology. The topography of the top of the limestones is shown in color (blue to red). The sand body buried at a depth of 20 m is shown in solid green. The model is attached to the electronic version of this article.

3.3. Numerical Simulation of Aleksadrovka

We compared efficiency of the sequential 3D modelling code both for GF (7) and CO (12) preconditioners. The transmitter was a point dipole located at the southern end of transmitter line Tx-2 (see Figure 3), with coordinates (1091.101, 812.101, 0.002) and azimuth 155.6°. We selected three frequencies, 192, 320, and 576 Hz. Parameters of the FD grids are presented in Table 3.

Table 3. Parameters of the finite-difference grids.

Grid Parameters	192 Hz	320 Hz	576 Hz
FD grid dimensions	$114 \times 110 \times 151$	$136 \times 128 \times 183$	$166 \times 158 \times 138$
Num of discrete unknowns	5.6 M	9.4 M	10.7 M
Grid step size in core domain, m	7.4	5.7	4.3
Core domain, m	$526 \times 497 \times 807$	$534 \times 488 \times 800$	$527 \times 492 \times 400$

The code was running on a single thread of a compute node equipped with double 12-core Intel Xeon E5-2680v3 processors running at 2.5 GHz and 128 GB of memory. The relative accuracy of the BiCGStab was set to 10^{-10}. Performance of the two preconditioned iterative solvers is presented in Table 4. The table also includes CPU time needed to prepare the RHS.

Table 4. Comparison of Green's function (GF) and contraction operator (CO) preconditioned iterative solvers.

Solver	Solver Step	192 Hz	320 Hz	576 Hz
CO	RHS computation, sec	2994	5762	9239
	Iteration count	346	389	480
	Iterative solver, sec	3658	8177	13,673
	Single iteration, sec	10.6	21.0	28.5
GF	RHS computation, sec	2993	5759	9258
	Iteration count	3042	3713	5000 (*)
	Iterative solver, sec	31,932	77,253	141,133 (*)
	Single iteration, sec	10.5	20.8	28.2

(*) Did not converge in 5000 iterations.

We observed an acceleration of 7×, 9×, and 10× at the three selected frequencies, respectively, both with respect to the iteration count and wall-clock time. Computational load per iteration for the CO preconditioner was only 7–10% larger compared to the GF one. Thus, a reduced number of iterations almost directly translated to the shorter compute time. The GF solver did not converge at 576 Hz frequency for 5000 iterations, which was set as the limit. We attributed slow converge of the GF solver to the highly conductive 1.5 Ωm layer which produced a strong horizontal resistivity contrast in the model. It should be mentioned that topography slowed the computations considerably since performance of the GF preconditioner deteriorated most severely when the air–ground interface was not a horizontal plane. In this case, the time gain of the CO over GF preconditioner was expected to be even greater than one observed in this test.

It is interesting that the time of computing the right-hand side dominated by computation of the background electric field was comparable to the time spent on the solution of a system of linear equations. This is one of the disadvantages of the secondary-field approach. However, RHS computation was easily reduced by using shorted Hankel transform filters and multithreading (see Section 2.2). In the next tests, we applied the CO preconditioned iterative solver only.

In the second set of computations, the whole survey was simulated. The electromagnetic field was computed at 192 receivers. We used 18 frequencies with range from 192 Hz to 550 kHz. It should be emphasized that the electromagnetic field above 25 kHz was not quasi-stationary for our setup, but impedance computed from individual components still can be used to interpret the real measurements (see [13]). Two transmitter lines were used to compute the two polarizations. During computations, the lines were broken into 26 straight segments (15 for Tx-1 and 11 for Tx-2), which had lengths between 40 and 185 m, and then numerically integrated along the transmitter lines. Thus, the total number of individual forward problems was 468. Parameters of the FD grids at different frequencies are given in Table 5.

Table 5. Computational grids at different frequencies.

Frequency, Hz	Core Domain, m	Grid Step in Core Domain, m	Discrete Unknowns
192	526 × 497 × 800	3.00	11.7 M
320	534 × 488 × 800	3.00	16.2 M
576	527 × 492 × 397	2.99	14.0 M
960	524 × 497 × 300	3.00	10.9 M
1500	525 × 493 × 200	2.99	11.0 M
2500	530 × 489 × 150	3.00	13.5 M
3500	531 × 490 × 150	3.00	17.1 M
5500	529 × 490 × 147	2.73	25.2 M
7500	529 × 491 × 148	2.34	34.5 M
15,000	528 × 492 × 148	1.67	70.2 M
25,000	528 × 493 × 150	1.95	47.5 M
35,000	530 × 491 × 147	3.26	17.7 M
55,000	528 × 491 × 97	2.56	23.4 M
75,000	529 × 493 × 78	2.22	29.2 M
150,000	530 × 492 × 38	1.54	45.9 M
250,000	529 × 492 × 40	1.21	76.0 M
350,000	528 × 493 × 19	1.00	86.9 M
550,000	529 × 492 × 10	0.91	93.0 M

Simulation was conducted on 26 nodes equipped with double 12-core Intel Xeon E5-2680v3 (24 cores per node) processors running at 2.5 GHz, 128 GB RAM, and InfiniBand interconnect. Each compute node was processing the 18 forward problems in the frequency range 192 Hz–550 kHz. There was message passing between nodes during computations. At each node, all 24 cores were working on each forward problem using OpenMP parallelization. The overall computations lasted for 24.5 h. Duration of the individual computation phases is given in Table 6.

Table 6. Performance of the forward simulation with OpenMP parallelization on 24 cores.

Frequency, Hz	Discrete Unknowns	RHS Computation, Sec	Iteration Count	Iterative Solver, Sec
192	11.7 M	398	459	732
320	16.2 M	572	479	1076
576	14.0 M	729	460	1043
960	10.9 M	318	502	845
1500	11.0 M	380	535	934
2500	13.5 M	561	443	1052
3500	17.1 M	765	447	1568
5500	25.2 M	1208	665	3555
7500	34.5 M	1795	553	4700
15,000	70.2 M	4489	578	12,741
25,000	47.5 M	3006	619	7264
35,000	17.7 M	768	495	1766
55,000	23.4 M	1122	436	2269
75,000	29.2 M	1362	227	1947
150,000	45.9 M	2304	64	956
250,000	76.0 M	4140	21	855
350,000	86.9 M	4120	17	1048
550,000	93.0 M	3994	14	1108

We observed that computation of the right-hand side (essentially, the background electric field) was comparable to the time allocated to the solution of a linear system. At frequencies above 150 kHz, the right-hand side dominated the other computations. It was caused by a combination of large FD grids and excellent performance of the CO preconditioned solver at the higher frequencies, at which core domains were very thin and enclosed within the top part of the model with moderate conductivity contrast. It is, however, of minor concern for inverse problems, because the background electromagnetic filed can be stored on disk. The cost of its computation is amortized across many forward problems solved over the course of the inverse problem.

Apparent resistivity and impedance phase along receive line No. 6 are presented in Figure 11. The two polarizations resulted in very similar cross-sections. Apparent resistivity approached the limit of 18 Ωm at frequencies above 25 kHz, indicating the high-frequency electromagnetic field did not penetrate below the top layer.

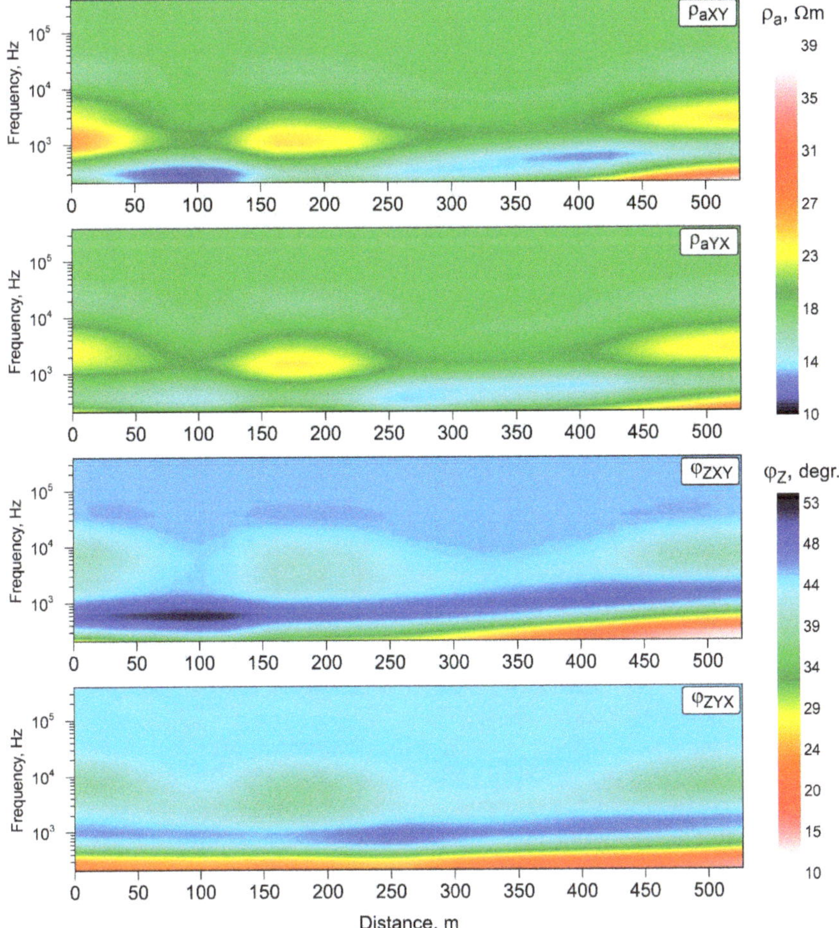

Figure 11. Results of the numerical simulation. Pseudo cross-sections along receiver line No. 6. Top two panels: XY and YX apparent resistivity. Bottom two panels: Impedance phase components, XY and YX.

Maps of the XY apparent resistivity and impedance phase are presented in Figure 12. The map at 192 Hz had a clear correlation with the topography of the first limestone layer (Figure 8), whereas the 1.5 Hz data clearly indicated the high-resistivity sand body (Figure 10b).

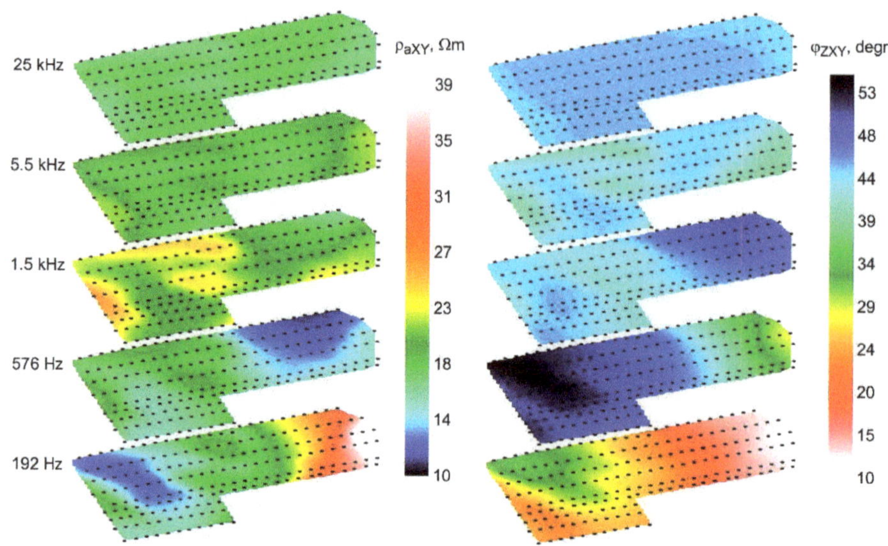

Figure 12. Results of numerical simulation. Maps of XY apparent resistivity and XY impedance component at several frequencies.

4. Discussion

In this paper, we considered an effective approach to modeling the data acquired by controlled-source audio-magnetotellurics (CSAMT) and radio-magnetotellurics (CSRMT) methods, which are widely used in mineral exploration. We assessed performance of two preconditioned iterative solvers for CSAMT/CSRMT simulation introduced earlier in [21]. Further, we presented results of numerical modeling of a real CSRMT survey performed near Aleksandrovka, Kaluga Region, Russia. The 3D model we used was based on 2D and 1D inversion as well as drilling data.

Collectively, this study demonstrated that the CO preconditioner is highly efficient even in extreme conditions of high excitation frequencies and large conductivity contrasts, encountered in CSAMT/CSRMT surveys. In contrast, the GF preconditioner degraded severely at the higher frequencies. This observation has not been published earlier as far as the authors are concerned.

From our experience, extremely large FD grids (above 50 M discrete unknowns) are easily encountered at high frequencies typical for CSRMT (above 10–50 kHz, depending on the subsurface resistivity). However, these conditions were pathological in the sense that the quasi-stationary regime is not valid anymore, and the displacement currents must be considered. The only reason such computations make sense is the quasi-stationary impedance was similar to the displacement-current one, at least in the 1D Earth, as demonstrated by [13].

From the geophysical standpoint, the data at frequencies above, perhaps, 25 kHz were redundant, because the 3D conductivity model did not contain very shallow inhomogeneities (<10 m). Our purpose was to evaluate numerical grids and simulation time if such high frequencies will be needed in the future.

Finally, we emphasize that the results of our study apply to other geophysical methods that require quasi-stationary electromagnetic modelling, specifically MT, AMT, CSMT, CSAMT, CSEM, borehole methods, and others.

Supplementary Materials: The following are available online at http://www.mdpi.com/2075-163X/10/1/42/s1, File S1: Aleksandrovka model in the VTK format.

Author Contributions: N.Y., investigation, software, and original draft preparation; M.M., software, computations, visualization, and original draft preparation; A.S., formal analysis, visualization, and original draft preparation. All authors have read and agreed to the published version of the manuscript.

Funding: This research was funded by Russian Science Foundation, grant number No. 16-11-10188.

Acknowledgments: The authors thank M.S. Zhdanov for the opportunity to benchmark our code with PIE3D software, and M. Čuma for their help in performing such calculations. This work has been carried out using computing resources of the federal collective usage center Complex for Simulation and Data Processing for Mega-science Facilities at NRC "Kurchatov Institute", http://ckp.nrcki.ru/. We would like to acknowledge the Skoltech CDISE's high-performance computing cluster, Zhores [36], for providing the computing resources that have contributed to the results reported herein.

Conflicts of Interest: The authors declare no conflict of interest.

Appendix A

In this appendix, we show equivalence of Equations (7) and (12). Before moving on, denote $\Sigma_a = \Sigma - \Sigma_b$ which is a discrete secondary conductivity. We will start with Equation (12). After switching from $\hat{e}$ to e and multiplying by $\Sigma_b^{-\frac{1}{2}}$ from the left, we receive,

$$\Sigma_b^{-\frac{1}{2}} K_1 e = \Sigma_b^{-\frac{1}{2}} \mathcal{G}_b^M K_2 e + i\omega\mu_0 A_b^{-1} f. \tag{A1}$$

Next, we substitute K_1 and $\mathcal{G}_b^M$ and employ commutativity of diagonal matrices,

$$\frac{1}{2}(\Sigma + \Sigma_b)\Sigma_b^{-1} e = \left(2i\omega\mu_0 A_b^{-1}\Sigma_b^{\frac{1}{2}} + \Sigma_b^{-\frac{1}{2}}\right)K_2 e + i\omega\mu_0 A_b^{-1} f. \tag{A2}$$

Now, we substitute K_2 and rewrite the matrices,

$$\frac{1}{2}(2\Sigma_b + \Sigma_a)\Sigma_b^{-1} e = \left(2i\omega\mu_0 A_b^{-1}\Sigma_b^{\frac{1}{2}} + \Sigma_b^{-\frac{1}{2}}\right)\frac{1}{2}\Sigma_a\Sigma_b^{-\frac{1}{2}} e + i\omega\mu_0 A_b^{-1} f. \tag{A3}$$

It is easy to note that the term $\frac{1}{2}\Sigma_a\Sigma_b^{-1} e$ cancels out, and we receive,

$$\left(I - i\omega\mu_0 A_b^{-1}\Sigma_a\right)e = i\omega\mu_0 A_b^{-1} f. \tag{A4}$$

After factoring out A_b^{-1} in the left-hand side and noting that $A = A_b - i\omega\mu_0\Sigma_a$, we obtain

$$A_b^{-1} A \, e = i\omega\mu_0 A_b^{-1} f, \tag{A5}$$

which is exactly (7).

References

1. Hu, Y.-C.; Li, T.-L.; Fan, C.-S.; Wang, D.-Y.; Li, J.-P. Three-dimensional tensor controlled-source electromagnetic modeling based on the vector finite-element method. *Appl. Geophys.* **2015**, *12*, 35–46. [CrossRef]
2. Li, X.; Pedersen, L.B. Controlled source tensor magnetotellurics. *Geophysics* **1991**, *56*, 1456–1462. [CrossRef]
3. McMillan, M.S.; Oldenburg, D.W. Cooperative constrained inversion of multiple electromagnetic data sets. *Geophysics* **2014**, *79*, B173–B185. [CrossRef]
4. Wang, T.; Wang, K.-P.; Tan, H.-D. Forward modeling and inversion of tensor CSAMT in 3D anisotropic media. *Appl. Geophys.* **2017**, *14*, 590–605. [CrossRef]
5. Zonge, K.L.; Hughes, L.J. Controlled source audio-frequency magnetotellurics. In *Electromagnetic Methods in Applied Geophysics*; Applications, Series: Investigations in geophysics, No 3; SEG Library: Tulsa, OK, USA, 1991; Volume 2, pp. 713–809.
6. Grayver, A.V.; Streich, R.; Ritter, O. Three-dimensional parallel distributed inversion of CSEM data using a direct forward solver. *Geophys. J. Int.* **2013**, *193*, 1432–1446. [CrossRef]

7. Bastani, M. EnviroMT: A new Controlled Source/Radio Magnetotelluric System. Ph.D. Thesis, ACTA Universitatis Upsaliensis, Uppsala, Sweden, 2001; p. 179.

8. Newman, G.A.; Recher, S.; Tezkan, B.; Neubauer, F.M. 3D inversion of a scalar radio magnetotelluric field data set. *Geophysics* **2003**, *68*, 791–802. [CrossRef]

9. Saraev, A.; Simakov, A.; Shlykov, A.; Tezkan, B. Controlled-source radiomagnetotellurics: A tool for near surface investigations in remote regions. *J. Appl. Geophys.* **2017**, *146*, 228–237. [CrossRef]

10. Tezkan, B. *Radiomagnetotellurics, in Groundwater Geophysics—A Tool for Hydrogeology*; Springer: Berlin/Heidelberg, Germany, 2008; pp. 295–317.

11. Turberg, P.; Müller, I.; Flury, F. Hydrogeological investigation of porous environments by radio magnetotelluric-resistivity (RMT-R 12240 kHz). *J. Appl. Geophys.* **1994**, *31*, 133–143. [CrossRef]

12. Spies, B.R.; Frischknecht, F.C. Electromagnetic Sounding. In *Electromagnetic Methods in Applied Geophysics*; Applications, Series: Investigations in geophysics, No 3; SEG Library: Tulsa, OK, USA, 1991; Volume 2.

13. Shlykov, A.; Saraev, A. Estimating the Macroanisotropy of a Horizontally Layered Section from Controlled-Source Radiomagnetotelluric Soundings. *Izv. Phys. Solid Earth* **2015**, *51*, 583–601. [CrossRef]

14. Kalscheuer, T.; Pedersen, L.B.; Siripunvaraporn, W. Radiomagnetotelluric two-dimensional forward and inverse modelling accounting for displacement currents. *Geophys. J. Int.* **2008**, *175*, 486–514. [CrossRef]

15. Persson, L.; Pedersen, L.B. The importance of displacement currents in RMT measurements in high resistivity environments. *J. Appl. Geophys.* **2002**, *51*, 11–20. [CrossRef]

16. Yavich, N.; Pushkarev, P.; Zhdanov, M. Application of a Finite-difference Solver with a Contraction Preconditioner to 3D EM Modeling in Mineral Exploration. In Proceedings of the Near Surface Geoscience 2016—First Conference on Geophysics for Mineral Exploration and Mining, EAGE, Barcelona, Spain, 4–8 September 2016.

17. Cai, H.; Hu, X.; Li, J.; Endo, M.; Xiong, B. Parallelized 3D CSEM modeling using edge-based finite element with total field formulation and unstructured mesh. *Comput. Geosci.* **2017**, *99*, 125–134. [CrossRef]

18. Grayver, A.V.; Bürg, M. Robust and scalable 3-D geo-electromagnetic modelling approach using the finite element method. *Geophys. J. Int.* **2014**, *198*, 110–125. [CrossRef]

19. Streich, R. 3D finite-difference frequency-domain modeling of controlled-source electromagnetic data: Direct solution and optimization for high accuracy. *Geophysics* **2009**, *74*, F95–F105. [CrossRef]

20. Mackie, R.L.; Smith, J.T.; Madden, T.R. Three-dimensional electromagnetic modeling using finite difference equations: The magnetotelluric example. *Radio Sci.* **1994**, *29*, 923–935. [CrossRef]

21. Mulder, W.A. A robust solver for CSEM modelling on stretched grids. In Proceedings of the 69th EAGE Conference and Exhibition, London, UK, 11–14 June 2007.

22. Yavich, N.; Scholl, C. Advances in multigrid solution of 3D forward MCSEM problem. In Proceedings of the 5th EAGE St. Petersburg International Conference and Exhibition on Geosciences, EAGE, Saint Petersburg, Russia, 2–5 April 2012.

23. Yavich, N.; Zhdanov, M. Contraction pre-conditioner in finite-difference electromagnetic modelling. *Geophys. J. Int.* **2016**, *206*, 1718–1729. [CrossRef]

24. Malovichko, M.; Tarasov, A.; Yavich, N.; Zhdanov, M. Mineral exploration with 3-D controlled-source electromagnetic method: A synthetic study of Sukhoi Log gold deposit. *Geophys. J. Int.* **2019**, *219*, 1698–1716. [CrossRef]

25. Čuma, M.; Zhdanov, M.S.; Yoshioka, K. Parallel integral equation 3d (pie3d). In Proceedings of the Consortium for Electromagnetic Modeling and Inversion Annual Meeting, Salt Lake City, UT, USA, 22–27 March 2013.

26. Monk, P.; Süli, E. A convergence analysis of Yee's scheme on nonuniform grids. *SIAM J. Numer. Anal.* **1994**, *31*, 393–412. [CrossRef]

27. Weiss, C.J.; Newman, G.A. Electromagnetic induction in a fully 3-D anisotropic earth. *Geophysics* **2002**, *67*, 1104–1114. [CrossRef]

28. Zaslavsky, M.; Druskin, V.; Davydycheva, S.; Knizhnerman, L.; Abubakar, A.; Habashy, T. Hybrid finite-difference integral equation solver for 3D frequency domain anisotropic electromagnetic problems. *Geophysics* **2011**, *76*, F123–F137. [CrossRef]

29. Van der Vorst, H.A. Bi-CGSTAB: A fast and smoothly converging variant of Bi-CG for the solution of nonsymmetric linear systems. *SIAM J. Sci. Stat. Comput.* **1992**, *13*, 631–644. [CrossRef]

30. Zhdanov, M. *Geophysical Electromagnetic Theory and Methods*; Elsevier: Amsterdam, The Netherlands, 2009.

31. Zhdanov, M. *Foundations of Geophysical Electromagnetic Theory and Methods*; Elsevier: Amsterdam, The Netherlands, 2018.

32. Yavich, N.; Malovichko, M.; Khokhlov, N.; Zhdanov, M. Advanced Method of FD Electromagnetic Modeling Based on Contraction Operator. In Proceedings of the 79th EAGE Conference and Exhibition, Paris, France, 12–15 June 2017.

33. Yoshika, K.; Zdanov, M.S. Electromagnetiv forward modeling based on the integral equation method using parallel computers. In Proceedings of the 75th Annual Inetrnational Meeting, Houston, TX, USA, 6–11 November 2005; pp. 550–553.

34. Čuma, M.; Gribenko, A.; Zhdanov, M.S. Inversion of magnetotelluric data using integral equation approach with variable sensitivity domain: Application to EarthScope MT data. *Phys. Earth Planet. Inter.* **2017**, *260*, 113–127. [CrossRef]

35. Modin, I.N.; Yakovleva, A.G. (Eds.) *Electrical Exploration: A Guide to Electrical Exploration Practices for Students of Geophysical Specialties*, 2nd ed.; PolisPRESS: Tver, Russia, 2018; Volume 1, p. 274.

36. Zacharov, I.; Arslanov, R.; Gunin, M.; Stefonishin, D.; Bykov, A.; Pavlov, S.; Panarin, O.; Maliutin, A.; Rykovanov, S.; Fedorov, M. "Zhores"—Petaflops supercomputer for data-driven modeling, machine learning and artificial intelligence installed in Skolkovo Institute of Science and Technology. *Open Eng.* **2019**, *9*, 512–520. [CrossRef]

© 2019 by the authors. Licensee MDPI, Basel, Switzerland. This article is an open access article distributed under the terms and conditions of the Creative Commons Attribution (CC BY) license (http://creativecommons.org/licenses/by/4.0/).

Article

Novel Approach to Modeling the Seismic Waves in the Areas with Complex Fractured Geological Structures

Nikolay Khokhlov *[ID] and **Polina Stognii**

Applied Computational Geophysics Lab, Moscow Institute of Physics and Technology, 141701 Dolgoprudny, Russia; stognii@phystech.edu
* Correspondence: khokhlov.ni@mipt.ru

Received: 18 November 2019; Accepted: 28 January 2020; Published: 30 January 2020

Abstract: This paper presents a novel approach to modeling the propagation of seismic waves in a medium containing subvertical fractured inhomogeneities, typical for mineralization zones. The developed method allows us to perform calculations on a structural computational grid, which avoids the construction of unstructured grids. For the calculations, the grid-characteristic method is used. We also present a comparison of the proposed method with the one described at earlier works and discuss the areas of its practical application. As an example, the numerical results for a cluster of subvertical fractures are given. A new approach for modeling fractures makes it quite easy to incorporate fractured objects into the seismic models and perform calculations without using algorithms on unstructured and curved grids.

Keywords: linear slip model; grid-characteristic method; elastic waves; modeling and inversion; seismic methods

---

1. Introduction

Many mineral deposits are associated with fractures filled by fluids occurring in volcanic rocks. The seismic method can be effectively used for detection of the fractured zones. Interpretation of seismic data collected over a mineral deposit associated with the fracture fillings requires developing effective methods of modeling the seismic waves in places with micro- and macrocracks. This is one of the challenging areas of mathematical modeling in seismic exploration for mineral deposits. Fault zones and fractures are found in many ore deposits and are important targets in the inversion and interpretation of seismic data. In addition, faults and fractures can have a large impact on the inversion and migration of seismic data (for example, anisotropy caused by fracturing). A variety of fractured structures can also be developed within hydrocarbon deposits in rock formations. They basically form a system of subvertical equally oriented fractures. In order to construct an accurate model of a deposit and develop optimal methods of its exploitation, a good general model of surrounding materials should be created, which requires an accurate reconstruction of the properties and positions of fractures [1].

Over the past few decades, several studies have been conducted on this topic. In 1980, Schoenberg [2] proposed a linear slip model, in which he described a sliding surface with special boundary conditions using approaches based on averaged medium methods. An experimental verification of this model was made in [3]. Numerical simulation of wave propagation a is well-studied area in general, with many approaches proposed to date, such as the most popular finite-difference time-marching schemes [4,5], the spectral elements [6,7], discontinuous Galerkin [8], Helmholtz finite-difference [9,10] or integral-equation solvers [11,12], etc. Still, the development of the numerical methods for modeling the fractured heterogeneities is an area of active research. For example, in [13] the authors investigated the problem of modeling fractures and faults using the finite difference schemes (FDTD). However, this approach allows us to describe and model only those fractured structures which consist of parallel fractures directed along the axes of coordinates (vertical or horizontal). This drawback can be overcome by using non-structural grids for modeling the fractured media. Examples of such approaches can be found in [5–7] where the finite-volume methods [14,15] and a grid-characteristic method [16] are used. Unstructured meshes are suitable for calculating the wave propagation in a physical region of complex shape, but they require large additional memory for storing information about the relationships of the mesh's cells. Recently, the discontinuous Galerkin (DG) method has also been widely used to solve the problems of modeling the propagation of seismic waves. For example, in [17,18], it is used to model the wave fields in the fractured media using an explicit fracture model. The method is also built on unstructured meshes. In [19], three methods were compared: The discontinuous Galerkin method [20,21], the grid-characteristic method on unstructured meshes [16], and the grid-characteristic method on structural grids [22–24]. In the work [19], it is shown that the method based on the structural grids has better performance in comparison to the unstructured meshes, in addition to a simpler implementation and operation with the grids. In the work, an anticlinar trap was used as a geological model with a large number of contact surfaces, which is a typical hydrocarbon deposit. Due to the trap structure, the seismogram involves multiple waves, which have to be separated from the useful signal, a task that can hardly be performed with low-order numerical schemes. It is shown, that three numerical methods and their software implementations are suitable for field data computations. Grid-characteristic method on structured grids is much more efficient, since it requires fewer operations per one degree of freedom and has a more efficient implementation on modern computers due to the density of data [19].

The structured grids, which are used in this paper, allow us to save the canonical structure of neighboring nodes for each grid node, which simplifies programming the method and performing the calculations. However, this formulation makes it difficult to model the inclined fractures or fractures that are not aligned with the axes of coordinates. In [25], the authors try to solve this problem using the overset grids method where a curvilinear structural grid is built around each fracture and then crosslinked with the main grid using interpolation. In this paper, we propose a novel approach to modeling the inclined fractured inhomogeneities using the grid-characteristic method on structural grids without constructing additional grids and without using the curved grids.

The approach proposed in this paper has already been briefly presented by authors at the 81st EAGE Conference and Exhibition 2019 [26]. The paper [26] gives only the idea of this approach and calculation for a cluster of subvertical fractures. In the current work, a more detailed description of the algorithm is given, verification tests on a single fracture are presented. We also investigate the effect of various positions of a single fracture with the fixed position of the source-receiver system on the resulting wave reflections. We propose rotating the source-receiver system relative to a fixed fracture to compare the results with the calculations when the source-receiver system is stationary relative to a rotatable fracture.

The results of seismic modeling for a stationary fracture in a homogeneous medium with different settings of the source-receiver system are presented. Similar results are given for a fixed source-receiver system with different fracture settings. The wave patterns of the responses for each of the series of calculations are presented. We also conduct a comparative analysis of the synthetic seismograms produced for each rotation and source-receiver system, the analysis of the influence of the frequency of the pulse of the source, as well as the step along the coordinate for these problem statements is conducted. The numerical study illustrates the effectiveness of the developed approach to modeling the seismic wave propagation in the medium with the fractured zones.

2. Mathematical Model and Method

We consider a model of ideal isotropic linear elastic material. The grid-characteristic method (GCM) uses the characteristic properties of the systems of hyperbolic equations, describing the elastic wave propagation [16]. The mathematical principles of the GCM approach are described in detail in [22]. It is based on representing the equations of motion of the linear elastic medium in the following form:

$$\mathbf{q}_t + \mathbf{A}_1 \mathbf{q}_x + \mathbf{A}_2 \mathbf{q}_y = 0 \tag{1}$$

In the last equation, $\mathbf{q}(t, x, y)$ is a vector of unknown fields, having five components and equal to

$$\mathbf{q} = \begin{bmatrix} \mathbf{v} \\ \mathbf{T} \end{bmatrix} = [\, v_1 \; v_2 \; T_{11} \; T_{22} \; T_{12} \,]^{\mathrm{T}},$$

where $\mathbf{v}$ is velocity, $\mathbf{T}$ is the stress tensor, $\mathbf{q}_t$ denotes the partial derivative of $\mathbf{q}$ with respect to t; $\mathbf{q}_x$ and $\mathbf{q}_y$ denote the partial derivatives of vector $\mathbf{q}$ with respect to x and y, respectively.

Matrices $\mathbf{A}_k$, $k = 1,\, 2$, are the 5×5 matrices given by the following expression:

$$A_1 = - \begin{bmatrix} 0 & 0 & \rho^{-1} & 0 & 0 \\ 0 & 0 & 0 & 0 & \rho^{-1} \\ \lambda + 2\mu & 0 & 0 & 0 & 0 \\ \lambda & 0 & 0 & 0 & 0 \\ 0 & \mu & 0 & 0 & 0 \end{bmatrix},$$

$$A_2 = - \begin{bmatrix} 0 & 0 & 0 & 0 & \rho^{-1} \\ 0 & 0 & 0 & \rho^{-1} & 0 \\ 0 & \lambda & 0 & 0 & 0 \\ 0 & \lambda + 2\mu & 0 & 0 & 0 \\ \mu & 0 & 0 & 0 & 0 \end{bmatrix},$$

The product of matrix $\mathbf{A}_k$ and vector $\mathbf{q}$ can be calculated as follows:

$$\mathbf{A}_k \begin{bmatrix} \mathbf{v} \\ \mathbf{T} \end{bmatrix} = - \begin{bmatrix} \rho^{-1}(\mathbf{T}\cdot\mathbf{n}) \\ \lambda(\mathbf{v}\cdot\mathbf{n})\mathbf{I} + \mu(\mathbf{n} \otimes \mathbf{v} + \mathbf{v} \otimes \mathbf{n}) \end{bmatrix},$$

where $\otimes$ denotes the tensor product of two vectors.

In the last equation $\mathbf{n}$ is a unit vector directed along the $\mathbf{x}$ or $\mathbf{y}$ directions for matrices $\mathbf{A}_1$ or $\mathbf{A}_2$, respectively.

As we discussed above, the GCM approach is based on representing the solutions of the acoustic and/or elastic wave equations at later time as a linear combination of the displaced at a certain spatial step solution at some previous time moment. This representation can be used to construct a direct time-stepping iterative algorithm of computing the wave fields at any time moment from the initial and boundary conditions. In order to develop this time-stepping formula, we represent matrices $\mathbf{A}_k$ using their spectral decomposition. For example, for matrix $\mathbf{A}_1$ we have:

$$\mathbf{A}_1 = (\mathbf{\Omega}_1)^{-1} \mathbf{\Lambda}_1 \mathbf{\Omega}_1,$$

where $\mathbf{\Lambda}_1$ is a 5×5 diagonal matrix, formed by the eigenvalues of matrix $\mathbf{A}_1$; and $(\mathbf{\Omega}_1)^{-1}$ is a 5×5 matrix formed by the corresponding eigenvectors. Note that, matrices $\mathbf{A}_1$ and $\mathbf{A}_2$ have the same set of eigenvalues:

$$\left\{ c_{\mathrm{p}}, -c_{\mathrm{p}}, c_{\mathrm{s}}, -c_{\mathrm{s}}, 0 \right\}.$$

In the last formula, c_p is a P-wave velocity being equal to $\left(\rho^{-1}(\lambda + 2\mu)\right)^{1/2}$ and c_s is an S-wave velocity being equal to $\left(\rho^{-1}\mu\right)^{1/2}$.

Let us consider some direction **x**. We assume that the unit vector **n** is directed along this direction, while the unit vectors $\mathbf{n}_1$ form a Cartesian basis together with **n**. We also introduce the following symmetric tensors of rank 2:

$$\mathbf{N}_{ij} = \frac{1}{2}\left(\mathbf{n}_i \otimes \mathbf{n}_j + \mathbf{n}_j \otimes \mathbf{n}_i\right),$$

where indexes i and j vary from 0 to 1 in order to simplify the final formulas, and $\mathbf{n}_0 = \mathbf{n}$.

It is shown in [16] that, the solution of Equation (1), vector **q**, along the x and y directions can be written as follows:

$$\mathbf{q}(t+\tau, x, y) = \sum_{j=1}^{J} \mathbf{X}_{1,j}\mathbf{q}\left(t, x - \Lambda_{1,j}\tau, y\right),$$

$$\mathbf{q}(t+\tau, x, y) = \sum_{j=1}^{J} \mathbf{X}_{2,j}\mathbf{q}\left(t, x, y - \Lambda_{2,j}\tau\right), \tag{2}$$

Here, τ is the time step of the solution, $J = 5$ corresponds to the number of eigenvalues of the matrices, and $\mathbf{X}_{1,j}$ and $\mathbf{X}_{2,j}$ are the characteristic matrices expressed through the components of matrices $\mathbf{A}_1$ and $\mathbf{A}_2$ and their eigenvalues as follows:

$$\mathbf{X}_{i,j} = \omega_{*i,j}\omega_{i,j}, \ i = 1,\ 2;$$

where $\omega_{*i,j}$ is the j's column of matrix $(\mathbf{\Omega}_i)^{-1}$, and $\omega_{i,j}$ is the j's row of matrix $\mathbf{\Omega}_i$.

Expression (1) can be used to find the solution, vector **q**, at any time moment, $t + \tau$, from the given initial conditions, thus representing a direct time-stepping algorithm of numerical modeling the elastic wave propagation in heterogeneous media.

Note that matrices $\mathbf{X}_{1,j}$ satisfy the following condition:

$$\sum_{j=1}^{J} \mathbf{X}_{1,j} = I.$$

Let us assume that matrix $\mathbf{A}_1$ has J_+ positive, J_- negative, and J_0 zero eigenvalues, respectively. Therefore, the sum of matrices, $\mathbf{X}_{1,j}$, corresponding to zero eigenvalues, can be expressed as follows:

$$\sum_{j\in J_0} \mathbf{X}_{1,j} = \mathbf{I} - \sum_{j\in J_+} \mathbf{X}_{1,j} - \sum_{j\in J_-} \mathbf{X}_{1,j}. \tag{3}$$

Considering Expression (3), Expression (2) will take the following form:

$$\mathbf{q}(t+\tau, x, y) = \mathbf{q}(t, x, y) + \sum_{j\in J_+\cup J_-} \mathbf{X}_{1,j}\left(\mathbf{q}\left(t, x - \Lambda_{1,j}\tau, y\right) - \mathbf{q}(t, x, y)\right).$$

3. Boundary and Interface Conditions

In order to obtain a unique solution of the system of hyperbolic Equation (1), one should use the corresponding boundary conditions on the surface of the modeling domain, which, in general terms, can be written as follows [16]:

$$\mathbf{D}\mathbf{q}(t+\tau, x_0, y_0) = \mathbf{d}, \tag{4}$$

where $\mathbf{D}$ and $\mathbf{d}$ are some given matrix and vector, respectively. For example, in a case of elastic media, two types of boundary conditions can be used. One is based on the given traction, $\mathbf{f}$:

$$\mathbf{T} \cdot \mathbf{n} = \mathbf{f},$$

where $\mathbf{n}$ is a unit vector of the outer normal to the boundary of the modeling domain. Another boundary condition can be based on a given velocity $\mathbf{v}_0$ at the boundary:

$$\mathbf{v} = \mathbf{v}_0.$$

The boundary conditions can be taken into account before calculating the difference scheme (using ghost nodes) or at the correction step. In this paper, we use the approach when the boundary conditions are applied after the step of the difference scheme.

At the boundary, only those characteristics that fall into the integration region are taken into account at first. We denote such a solution by the text index *in*. Then, for the case when the normal is directed to the left we get

$$\mathbf{q}(t+\tau,x,y)_{\text{in}} = \mathbf{q}(t,x,y) + \sum_{j \in J_-} \mathbf{X}_{1,j}\big(\mathbf{q}\big(t, x - \Lambda_{1,j}\tau, y\big) - \mathbf{q}(t,x,y)\big),$$

and

$$\mathbf{q}(t+\tau,x,y)_{\text{in}} = \mathbf{q}(t,x,y) + \sum_{j \in J_+} \mathbf{X}_{1,j}\big(\mathbf{q}\big(t, x - \Lambda_{1,j}\tau, y\big) - \mathbf{q}(t,x,y)\big),$$

when the normal is directed to the right. Similarly, we do this when calculating along a different spatial axis. Then, at the border, the boundary condition can be performed according to Formula (4).

A fracture can be modeled as a linear slip and contact conditions imposed independently for every fracture which allows us to avoid constructing the computational grid within the fracture itself. The double-coastal model described in [22] is used as a model of a fracture. It is a special case of the Schoenberg model [1] for a model of a fracture with zero openness. The system of equations, which models a fracture, can be written as follows:

$$\begin{aligned}
v_{1x}n_x + v_{1y}n_y &= v_{2x}n_x + v_{2y}n_y, \\
f^n_{1x} &= -f^n_{2x}, \ f^n_{1y} = -f^n_{2y}, \\
f^\tau_{1x} &= f^\tau_{2x} = 0, \ f^\tau_{1y} = f^\tau_{2y} = 0.
\end{aligned} \tag{5}$$

Here, indexes 1, 2 refer to the left and right sides of the fracture, $\mathbf{n}$ is the unit normal vector to the fracture, the superscripts n and τ are the normal and tangential components, respectively, $\mathbf{f} = (\mathbf{n}\,\mathbf{T})$ is the traction on the surface of the fracture.

4. Model of a Fracture

We propose a novel approach to modeling the inclined fractures on structural grids. The previously published methods [16,17] allow modeling in the case only, where the fracture boundary coincides with the cell boundaries (nodes) of the difference grid. In this work, we use two layers of nodes to explicitly define the fracture in order to correctly set the boundary conditions on the fracture [16].

The discretization scheme of such an algorithm is shown in Figure 1a. The fracture is indicated by a bold line, around it there is a fragmentation of the nodes of the computational grid used for the correct calculation of the boundary condition on Fracture (5). The novel approach assumes that the fracture does not coincide with the boundaries of the cells. Figure 1b schematically depicts an approach for identifying fractures in a grid. A fracture (thick black line) does not fall into the nodes of the grid. However, a discrete analogue of the fracture is considered (green line), which passes along the boundaries of the cells. Fracture is replaced by a bunch of tiny fractures scattering around it (red lines).

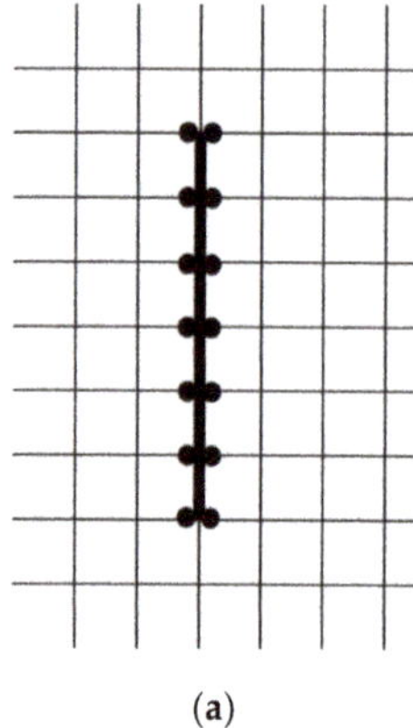 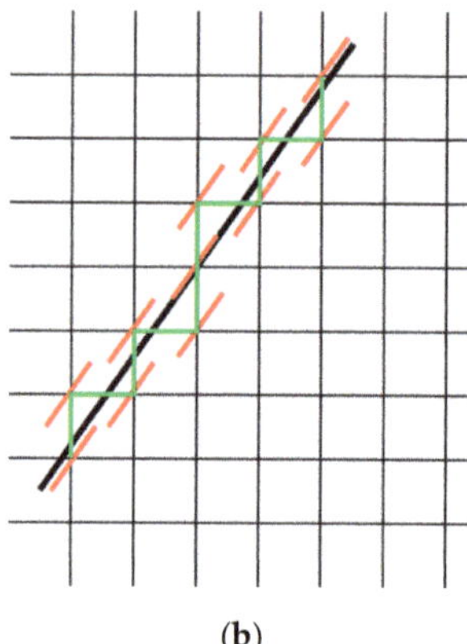

(**a**) (**b**)

Figure 1. The discretization schemes illustrating different approaches to modeling a single fracture on a structural grid. Figure (**a**) shows the approach when the fracture is aligned with the grid cells [7,8]. The new approach (**b**) implies that the fracture is inclined.

Figure 2 shows the calculation scheme of the boundary (contact) nodes near the fracture. In all cases, the nodes at the fracture boundary are duplicated.

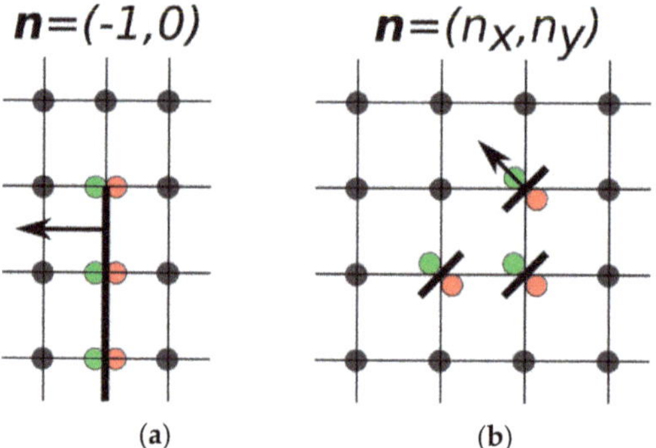

Figure 2. The layout of the contact nodes of the calculation grid. Red and green dots indicate duplicate nodes from one and the other side of the fracture. The left picture (**a**) is shown for the case when the fracture is aligned with the grid lines, the right picture (**b**) is for an inclined fracture.

In the case when the crack is aligned with the grid lines, the boundary conditions are calculated only along one axis (in the case in the figure along the x-axis). In the case of an inclined crack, the adjustment is made along two axes. In this case, information about the location of the fracture is lost, i.e., fracture turns out by a bunch of tiny fractures scattering around it and tied to grid points. This algorithm obviously has a drawback associated with the fact that the discretization of a fracture depends on the parameter of fineness of the mesh. However, it allows calculations on subvertical and inclined fractures, using the structural grids. Next, we compare these two methods.

5. Results

In this section, we present examples of calculations and a comparison of the proposed model with previously published results [16].

5.1. The Response from a Single Fracture: Fixed Fracture Position

The propagation of seismic waves from a source in a homogeneous medium with a fracture was simulated for the two-dimensional case. All calculations were carried out on rectangular structured grids. The source was located at a distance of 600 m from the center of the fracture. Five hundred receivers with a step of 1 m were located at a distance of 500 m from the center of the fracture. The homogeneous medium was a solid with a density of 2500 kg/m^3, and the velocity of longitudinal and transverse waves were 3000 and 1400 m/s, respectively. A fracture 100 m long was located vertically in a homogeneous medium 2100 × 2100 m in size. At the top, bottom, and sides of the medium, the non-reflecting boundary conditions were set for the equivalence of the problem at different positions of the source-receiver system. The time step was 10^{-4} s. A 30 Hz Ricker pulse was used as a signal source. Grid size was 2100 × 2100, the computational time for one calculation was about 1200 s at a single core of modern Intel Core i7-7700K CPU (Intel, Santa Clara, CA, USA).

The calculation was carried out for two models: A vertical fracture directed along the grid axes and coordinate axes (Figure 3a) and an inclined fracture at an angle of 30°, modeled by the method proposed in this paper (Figure 3b). In the first case, the observation system was rotated relative to the fracture so that the models completely coincided when the coordinate system was rotated.

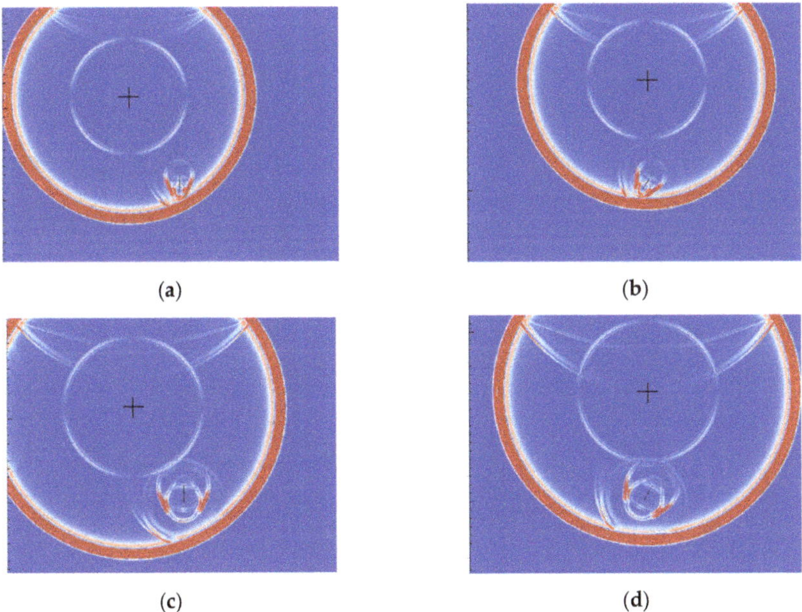

Figure 3. Fracture wave responses computed using two approaches. Figure (**a,c**) shows the wave for the initial setting of the source-receiver system relative to the fracture. On (**b,d**), the source-receiver system is located at an angle of 30° relative to its initial position.

In all Figure 3a–d, the fracture is depicted in the center of the computational domain by a black line. The source is indicated by a black cross and is located directly above the source-receiver system (the black line is directly below the propagating wave). Figure 3a shows a wave for the initial setting of the source-receiver system relative to the fracture. In Figure 3b, the source-receiver system is located at an angle of 30° relative to its initial position (Figure 3a). In all wave patterns, wave reflections from the fracture are clearly visible.

Figure 4 shows the wave field for the same calculation, but only the anomalous wave field from the fracture is displayed. The coincidence of responses is clearly seen from the figure.

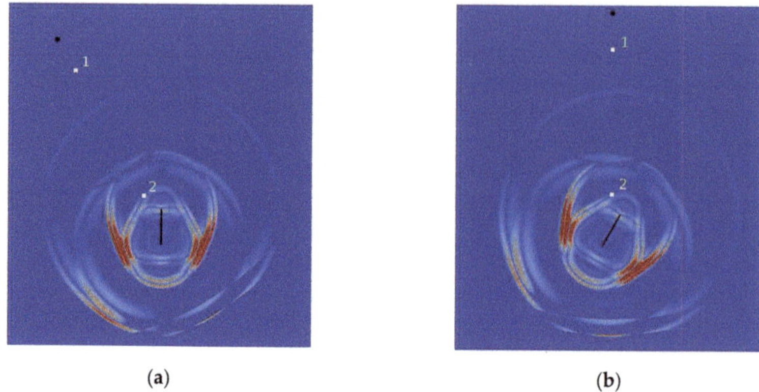

(a)   (b)

Figure 4. Anomalous wave field from the fracture for two approaches. Figure (**a**) shows the wave for the initial formulation of the source-receiver system relative to the fracture. In (**b**), the source-receiver system is located at an angle of 30° relative to its initial position. The black dot denotes the source location, white dots depict the location of the first and second receivers, respectively.

Figure 5 shows one-dimensional graphs of the values of the velocity components at one of the receivers versus time for two calculations. Only the anomalous field from the fracture is shown. The fracture is rotated at an angle of 30°. Figure 5a shows the field from the fracture on receiver 2, and Figure 5b shows the field from the fracture on receiver 2. A fairly good qualitative and quantitative agreement of the responses is seen. For the case of an inclined fracture in the time domain, t = 0.35–0.4 s "rattling" is observed due to the fact that the fracture structure is tied to discrete mesh cells. However, the main peak in the time domain is t = 0.4–0.45 s coincides and does not give a "rattle". Further passage of the wavefront also coincides well for both settings.

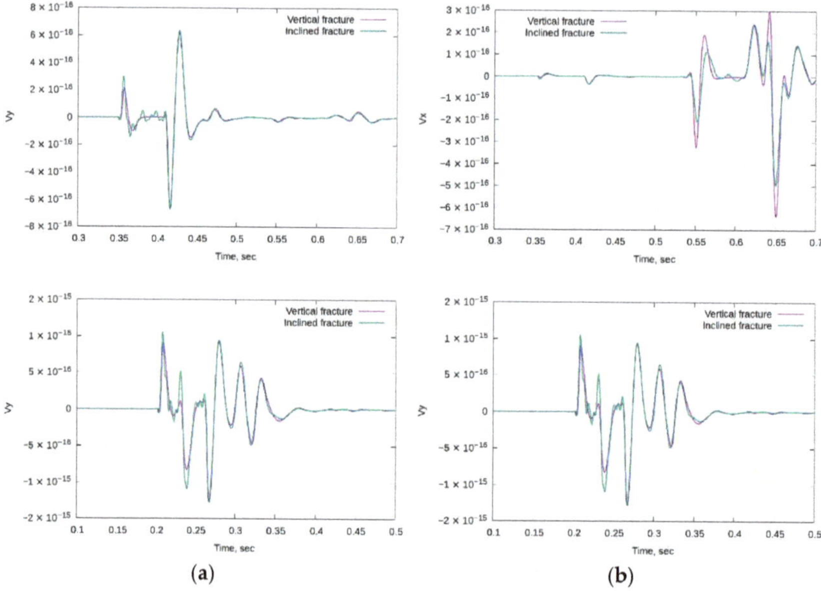

(a)   (b)

Figure 5. The value of the velocity components at one of the receivers versus time for two calculations. Only the anomalous field from the fracture is shown. Figure (**a**) shows the field from the fracture on receiver 2, and figure (**b**) shows the field from the fracture on receiver 1.

5.2. The Response from a Single Fracture: Comparison with Unstructured Mesh

In this test, a model similar to the previous one was used with the 30° inclined fracture. The source was located at a distance of 600 m from the center of the fracture, the receiver was located at a distance of 500 m from the center of the fracture. The first calculation was carried out on structural grids using the grid-characteristic method and the fracture model proposed in this paper. The second calculation was carried out on unstructured mesh using the DG method. The wave snapshots are shown in Figure 6. The source position is shown by a black dot in the figure, the receiver by white a one. A 30 Hz Ricker pulse was used as a signal source. Unlike previous calculations, a directional source was used as the source (it was selected in connection with the features of the code for DG). It can be seen from the figure that the wave snapshots coincide qualitatively well. The response from the fracture has the same structure.

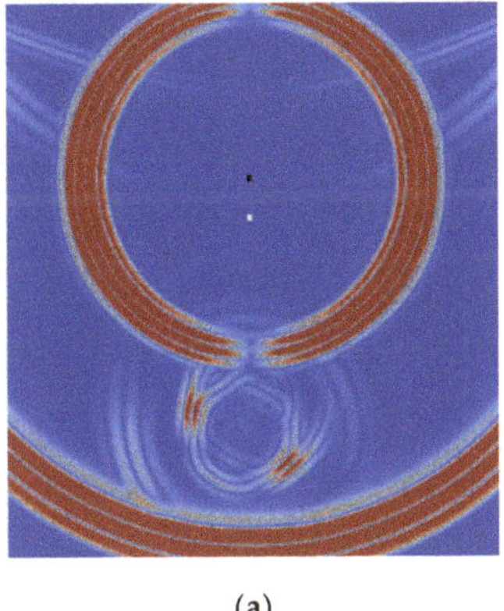
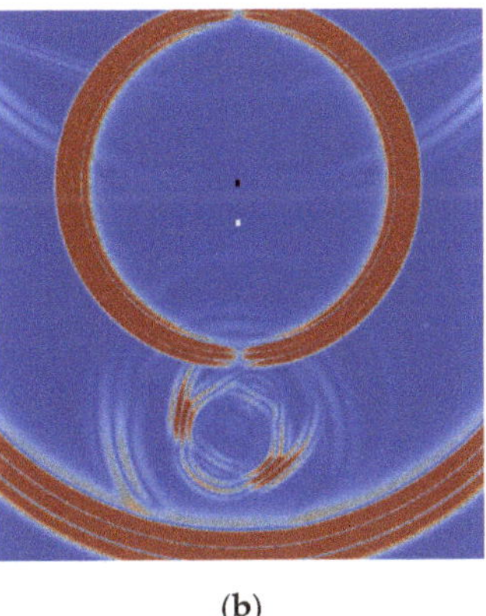

(a) (b)

Figure 6. Fracture wave responses computed using two approaches. A fracture is depicted in the center of the computational region. Figure (**a**) shows the wave for the DG (Discontinuous Galerkin) method on unstructured mesh, (**b**) for purposed method on structured mesh.

The distribution of the vertical velocity component at the receiver versus time is shown in Figure 7. Only a part of the signal is shown at time instants from 0.3 to 0.7 s, where there is a response from the fracture. The figure shows that the main response from the fracture coincides quite well. The arrival times are the same, the amplitude value is slightly different. The response from the fracture, calculated by the grid-characteristic method on structural grids is more "rattling", this is due to the discrete geometry of the fracture model.

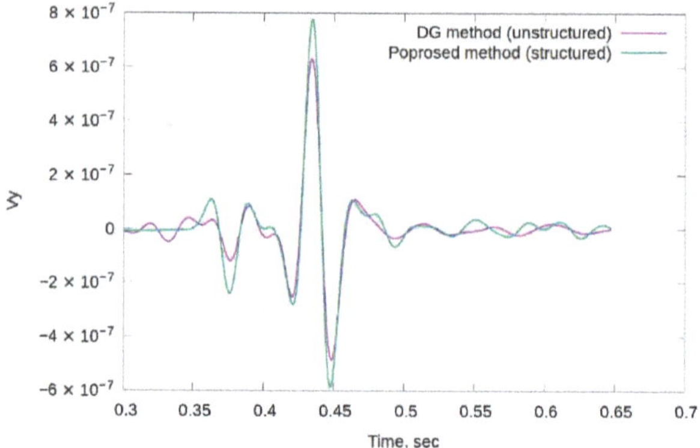

Figure 7. The value of the velocity component at one of the receivers versus time for two calculations.

5.3. The Response from a Single Fracture: Calculation Errors Due to the Angle of Inclination

In the next series of calculations, the difference in the calculated responses was studied for the case of two scenarios. In the first scenario, a vertical fracture was used, and the observation system was rotated, in the second scenario, a fixed observation system was used and the fracture was rotated. The position of the fractures and observation systems was set in such a way that, when turning, all the systems coincided with each other. The initial setting was similar to the setting from the previous section. A series of calculations was carried out for various positions of the source-receiver system with a fixed position of the fracture in space. The source-receiver system for each new calculation was rotated relative to the fracture by 10° keeping the distance to the center of the fracture constant. In total, 35 calculations were performed with a step of 10° for various orientations of the source-receiver system relative to the fixed fracture. A similar series of calculations was carried out for different positions of the fracture with a fixed orientation of the source-receiver system. With each subsequent calculation, the fracture was rotated around its center by 10°. In total, 35 calculations were performed with a step of 10° for various fracture orientations. The differences between the obtained seismograms were studied when turning the source-receiver system with a fixed fracture and when turning a fracture with a fixed source-receiver system. Moreover, the influence of the source pulse frequency on the obtained errors in the seismograms was studied. A series of calculations were carried out for the following frequencies: 15, 30, 60 Hz. Grid size for each calculation was 2100 × 2100 nodes, spacing was 1 m.

Figure 8 shows the relative deviations in the decisions for the norm L_2. Deviation was calculated by the formula

$$err = \frac{\sqrt{\sum (v(x,t) - v(x,t)_*)^2}}{\sqrt{\sum v(x,t)^2}} \cdot 100\%,$$

where $v(x,t)$ is the value of velocity at the receiver; the error was summed over all receivers. From Figure 6, it follows that with decreasing frequency, the error decreases when the source-receiver system or fractures are rotated by the corresponding angle. At a frequency of 15 Hz, the maximum error (about 6.5%) is 1–2% less than the maximum error of the similar calculations at a frequency of 30 Hz (about 8%), which, in turn, is less than the maximum calculation error at a frequency of 60 Hz (about 10%). This result correlates well with the fact that with decreasing sampling size the error falls. Due to the fact that the new model of fractures is tied to the size of the cell, the longer waves produce a better result. Moreover, for fractures with a lower slope, the solutions have a smaller error.

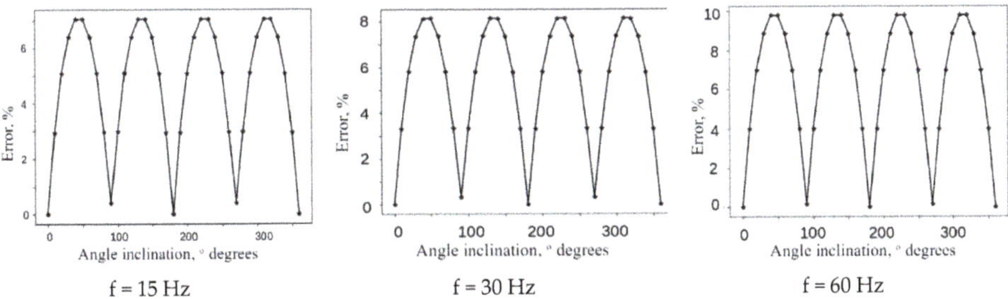

Figure 8. Relative deviation of the norm of errors for two calculations depending on the angle of rotation of the fracture and on the frequency of the signal.

Figure 9 shows a similar result for an anomalous field only (the field due to the fracture). The deviation values here are more significant, and they are associated with the discreteness of the inclined fracture. With increasing frequency, i.e., decreasing wavelength, the error increases. In Figure 5, peaks are visible that can give such a deviation in error; however, the qualitative coincidence of the waveform and the main peaks is quite good.

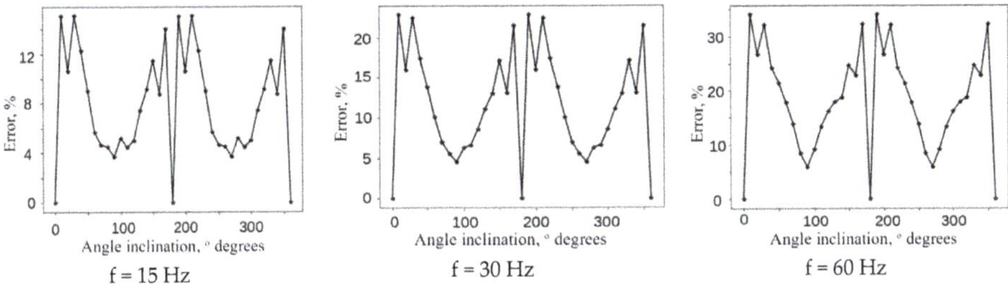

Figure 9. Relative deviation of the norm of errors for two scenarios depending on the angle of rotation of the fracture and on the frequency of the signal.

5.4. Single Fracture Response: Amplitude Study

Modern methods of interpretation increasingly take into account the absolute value of the amplitude of seismic signal, and not just the arrival times. As shown in Figure 5, the main peaks of the wave response from fractures come at the same time. Figure 10 shows the plots of the maximum amplitude of the seismic signal vs. the receiver number, based on the computations of the wave field reflections from the single fracture. We have compared the amplitudes of the seismic reflections from the inclined fracture under the angle of 30° with the amplitudes of the seismic reflections from the fracture, parallel to the boundaries of the grid. In our modeling study, 500 receivers of seismic signals were used. The blue line in Figure 10 denotes the peak amplitudes of the seismic reflections from the inclined fracture. The black line depicts the peak amplitudes of the seismic reflections from the vertical fracture, parallel to the boundaries of the grid cells. The plots show a similarity in the responses from the inclined and vertical structures.

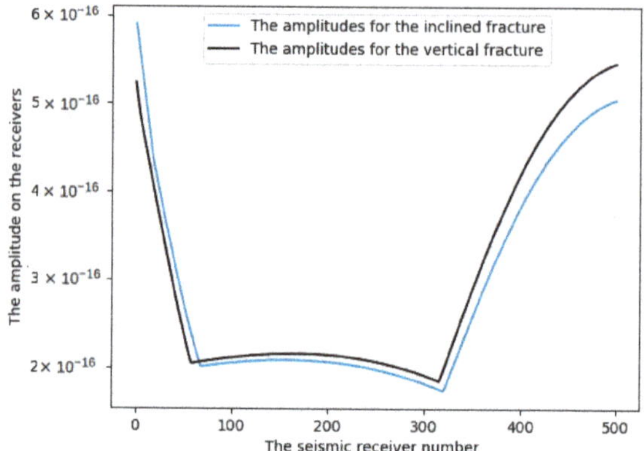

Figure 10. Peak amplitudes of the seismic signal on the receivers for the inclined fractures and for the vertical fractures.

5.5. Single Fracture Response: Error Dependence on Grid Spacing

The next series of calculations investigates the dependence of the error from the grid step. The setting is similar to the previous ones, a vertical fracture was used, and the observation system was rotated, in the second scenario, a fixed observation system was used and the fracture was rotated. A series of calculations was carried out for various positions of the source-receiver system with a fixed position of the fracture in space. The source-receiver system for each new calculation was rotated relative to the fracture by 10° keeping the distance between the source-receiver system and the center of the fracture constant. In total, 19 calculations were performed with a step of 10° for various orientations of the source-receiver system relative to the fixed fracture. Source location is the same as shown in Figures 3 and 4. The receiver is located at a distance of 100 m to the left of the center of the fracture. Source frequency is 30 Hz. A series of calculations was carried out for the grid spacing: 1, 2, 4 m. The calculation parameters are shown in Table 1.

Table 1. Calculation parameters.

Grid Spacing, m	Grid Size	Time Steps	Time Step, s	Cells Per Wavelength	Calculation Time, s
1	2100×2100	5000	10^{-4}	~46	1180 s
2	1050×1050	2500	2×10^{-4}	~23	157 s
4	525×525	1250	4×10^{-4}	~11	15 s

Figure 11 shows the relative deviations in the decisions for the norm L_2 for various grid spacings.

For calculating the error, only the anomalous field from the fracture was used. The figure shows that with decreasing spacing, the error decreases. Based on this, we can assume that there is a grid convergence of the method.

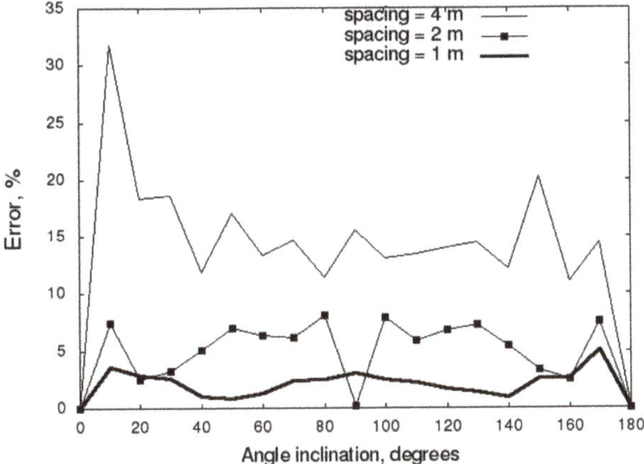

Figure 11. Relative deviation of the norm of errors for two scenarios depending on the angle of rotation of the fracture and on the frequency of the signal. The result is shown for three grid spacings: 1, 2, 4 m.

5.6. Response from a Cluster of Subvertical Fractures

The computations of the wave field reflections from a cluster of the parallel fractures were carried out. In the first case, the fractures were inclined at an angle of 20°, and the system of receivers and source of impulse were parallel to the boundaries of the grid (Figure 12a). In the second case, the fractures were parallel to the boundaries of the modeling grid, and the system of receivers and source of impulse were inclined at an angle of 20° (Figure 12b).

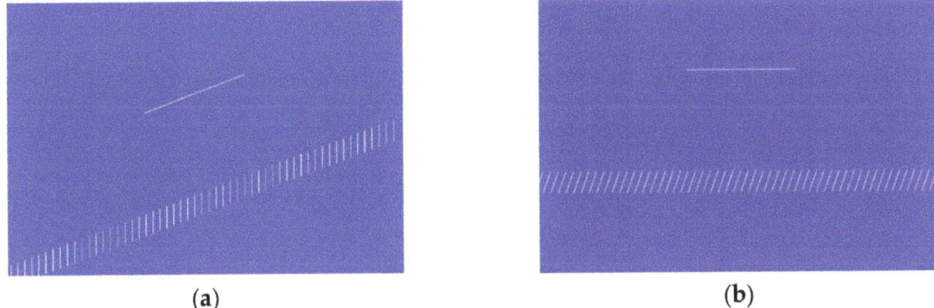

(a) (b)

Figure 12. The schematic representation of the models with the cluster of parallel fractures. Panel (a) shows a model with the inclined receivers under 20°, and a cluster of fractures parallel to the boundaries of the grid cells. Panel (b) presents the model with the cluster of the inclined fractures under 20° and the receivers parallel to the boundaries of the grid cells.

The seismic reflections from the cluster of the parallel fractures are shown in Figure 13. The overall wave reflections from the fractures are similar if we rotate the coordinate system of the model in Figure 12a under the angle of 20°, which is the angle of the inclined fractures.

Figure 14 shows the seismograms of the vertical velocity components at the receivers for two models. Only the anomaly part of the response is given. The same color scale is used. A good qualitative and quantitative agreement of responses is observed. The first response has the largest amplitude and is clearly visible on seismograms.

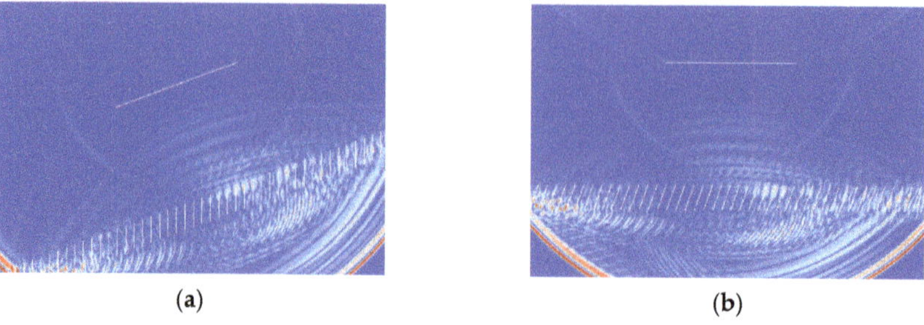

(a)   (b)

Figure 13. Wave field reflections from the cluster of parallel fractures. (**a**) The wave field from the cluster of fractures, parallel to the boundaries of the grid cells. (**b**) The wave field from the cluster of the inclined fractures under 20° boundaries.

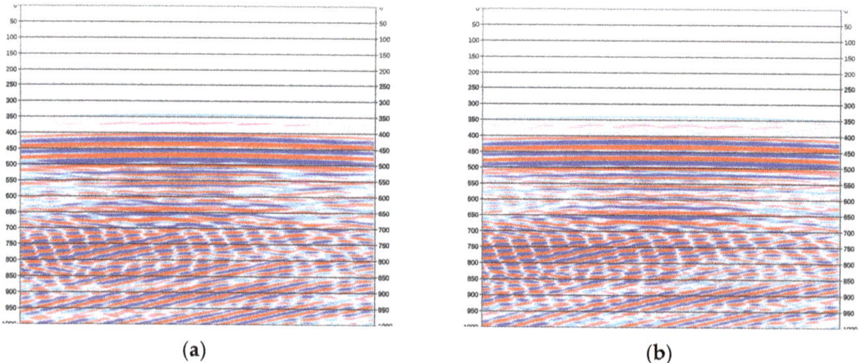

(a)   (b)

Figure 14. Vertical velocity component seismograms at receivers for a cluster of parallel fractures. An anomaly field is displayed. (**a**) The seismogram from the cluster of fractures, parallel to the boundaries of the grid cells. (**b**) The seismogram from the cluster of the inclined fractures under 20°.

Figure 15 shows the recording from the central receiver for the vertical and horizontal velocity components.

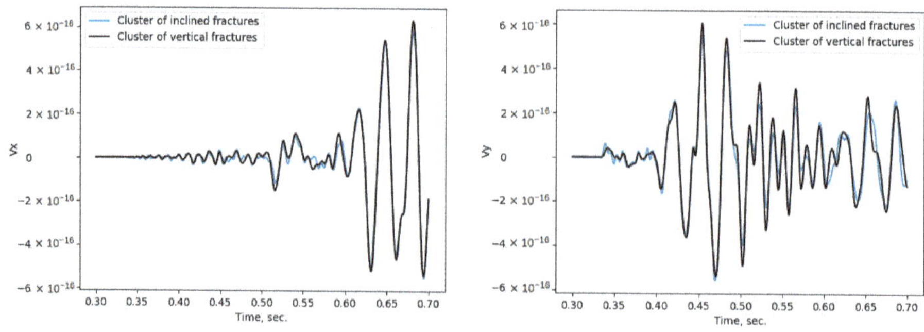

Figure 15. Velocity components of seismograms at single central receivers for a cluster of parallel fractures. An anomalous field is displayed.

A good qualitative and quantitative agreement of the result for the two methods is observed.

6. Discussion and Conclusions

This paper presents a novel approach to modeling the elastic field propagation in a medium containing the clusters of subvertical fractures. The new approach is based on the use of the grid-characteristic method [16]. Unlike previously proposed methods, this method can be used for modeling the seismic responses from subvertical fractured inhomogeneities on a structural rectangular grid. This greatly simplifies the construction of the numerical model and the use of the algorithms. The shortcomings of the proposed methodology include the fact that the crack model is discrete and does not take into account the normal of the crack. The crack is replaced by a discrete analog presented by a bunch of tiny fractures scattering around it and tied to grid points. At each point, the normal is taken into account, the total field due to interference gives a response close to real. In addition, in the case of an inclined crack, the initial grid does not change and the position of the crack can be set only up to the position of the grid node, respectively, there is an error $O(h)$, where h is the grid spacing. A comparison of the new approach with the traditional method shows that the results produced by both methods agree well with each other. The wave fields for a cluster of the inclined fractures and a cluster of the vertical fractures demonstrate a similar behavior. Thus, the results of the numerical study demonstrate that the developed method provides an effective tool for modeling the elastic field propagation in the fractured geological media. This method can be used in interpretation of the seismic data in the areas with complex fractured geological structures typical for mineral deposits and the HC reservoirs.

We proposed this method as a method for quick analysis and preliminary calculations of fractured geological models. Its advantages include:

- Fairly quick calculation and simple implementation in a structural grid. The basic calculation algorithm does not change; the method is implemented as a "corrector" step.
- The ability to calculate inclined fractures on structural grids.
- The ability to build a fractured inhomogeneity of complex shape.
- The absence of the need to build an unstructured mesh, the binding of the grid to inhomogeneity.
- Fractures can be easily added at any place in the geological models.
- The disadvantages include:
- A rather large integral error.
- Substantial grid spacing reduction required to reduce error.

The use of structural or non-structural grids to calculate dynamic wave disturbances in geological models is an open discussion. It was shown in [19] that methods on structural grids work up to 100 times faster than a method on unstructured ones. The authors of [27] also say that the use of finite-element methods and the DG method is more computationally difficult than finite difference methods. Implementation of finite-elements and DG requires an unstructured mesh, which should accurately approximate at least the main interfaces of the model. Construction of such a mesh is a problem of high complexity, which is hard to parallelize and requires manual quality control. Methods on structural grids are quite easy to parallel [23] and quite effective in both CPU [28] and GPU [29].

It should be noted that this approach is quite easily transferred to the three-dimensional case. The grid-characteristic method uses splitting in spatial coordinates, and the transition from the 2D to 3D case is quite simple. Further studies in this direction may be related to transitions to the three-dimensional case. The authors are working in this direction.

Author Contributions: Conceptualization, N.K.; methodology, N.K. and P.S.; software, N.K.; validation, P.S.; formal analysis, P.S.; investigation, N.K.; resources, P.S.; data curation, P.S.; writing—original draft preparation, N.K. and P.S.; writing—review and editing, N.K. and P.S.; visualization, P.S.; supervision, N.K.; project administration, N.K.; funding acquisition, N.K. All authors have read and agreed to the published version of the manuscript.

Funding: This research was supported by the Russian Science Foundation, project No. 16-11-10188.

Acknowledgments: This work has been carried out using computing resources of the federal collective usage center Complex for Simulation and Data Processing for Mega-science Facilities at NRC "Kurchatov Institute", http://ckp.nrcki.ru/.

Conflicts of Interest: The authors declare no conflict of interest.

References

1. Karayev, N.; Levyant, V.; Petrov, I.; Karayev, N.; Muratov, M. Assessment by methods mathematical and physical modelling of possibility of use of the exchange scattered waves for direct detection and characteristics of systems of macrocracks. *Seism. Technol.* **2015**, *1*, 22–36.
2. Schoenberg, M. Elastic Wave Behavior across Linear Slip Interfaces. *J. Acoust. Soc. Am.* **1980**, *68*, 1516. [CrossRef]
3. Hsu, C.; Schoenberg, M. Elastic Waves through a Simulated Fractured Medium. *Geophysics* **1993**, *58*, 964–977. [CrossRef]
4. Dablain, M.A. The application of high-order differencing to the scalar wave equation. *Geophysics* **1986**, *51*, 54–66. [CrossRef]
5. Moczo, P.; Robertsson, J.O.A.; Eisner, L. Advances in Wave Propagation in Heterogeneous Earth. *Adv. Geophys.* **2007**, *48*, 1–606.
6. Komatitsch, D.; Vilotte, J.P. The spectral element method: an efficient tool to simulate the seismic response of 2D and 3D geological structures. *Bull. Seismol. Soc. Am.* **1998**, *88*, 368–392.
7. Chaljub, E.; Komatitsch, D.; Vilotte, J.-P.; Capdeville, Y.; Valette, B.; Festa, G. Spectral element analysis in seismology. *Adv. Geophys.* **2007**, *48*, 365–419.
8. Dumbser, M.; Kaser, M. An Arbitrary High Order Discontinuous Galerkin Method for Elastic Waves on Unstructured Meshes II: The Three-Dimensional Isotropic Case. *Geophys. J. Int.* **2006**, *167*, 319–336. [CrossRef]
9. Pan, G.; Abubakar, A. Iterative solution of 3D acoustic wave equation with perfectly matched layer boundary condition and multigrid preconditioner. *Geophysics* **2013**, *78*, T133–T140. [CrossRef]
10. Belonosov, M.; Dmitriev, M.; Kostin, V.; Neklyudov, D.; Tcheverda, V. An iterative solver for the 3D Helmholtz equation. *J. Comput. Phys.* **2017**, *345*, 330–344. [CrossRef]
11. Malovichko, M.S.; Khokhlov, N.I.; Yavich, N.B.; Zhdanov, M.S. Parallel integral equation method and algorithm for 3D seismic modelling. In Proceedings of the 79th EAGE Conference and Exhibition, Paris, France, 12–15 June 2017. [CrossRef]
12. Malovichko, M.S.; Khokhlov, N.I.; Yavich, N.B.; Zhdanov, M.S. Acoustic 3D modeling by the method of integral equations. *Comput. Geosci.* **2018**, *111*, 223–234. [CrossRef]
13. Cui, X.; Lines, L.R.; Krebes, E.S. Seismic Modelling for Geological Fractures. *Geophys. Prospect.* **2018**, *66*, 157–168. [CrossRef]
14. Bosma, S.; Hajibeygi, H.; Tene, M.; Tchelepi, H.A. Multiscale Finite Volume Method for Discrete Fracture Modeling on Unstructured Grids (MS-DFM). *J. Comput. Phys.* **2017**, *351*, 145–164. [CrossRef]
15. Franceschini, A.; Ferronato, M.; Janna, C.; Teatini, P. A Novel Lagrangian Approach for the Stable Numerical Simulation of Fault and Fracture Mechanics. *J. Comput. Phys.* **2016**, *314*, 503–521. [CrossRef]
16. Kvasov, I.; Petrov, I. *Numerical Modeling of Seismic Responses from Fractured Reservoirs by the Grid-Characteristic Method*; Leviant, V., Ed.; Society of Exploration Geophysicists: Tulsa, OK, USA, 2019. [CrossRef]
17. Cho, Y.; Gibson, R.L.; Lee, J.; Shin, C. Linear-Slip Discrete Fracture Network Model and Multiscale Seismic Wave Simulation. *J. Appl. Geophys.* **2019**, *164*, 140–152. [CrossRef]
18. Vamaraju, J.; Sen, M.K.; De Basabe, J.; Wheeler, M. Enriched Galerkin Finite Element Approximation for Elastic Wave Propagation in Fractured Media. *J. Comput. Phys.* **2018**, *372*, 726–747. [CrossRef]
19. Biryukov, V.A.; Miryakha, V.A.; Petrov, I.B.; Khokhlov, N.I. Simulation of Elastic Wave Propagation in Geological Media: Intercomparison of Three Numerical Methods. *Comput. Math. Math. Phys.* **2016**, *56*, 1086–1095. [CrossRef]
20. Mercerat, E.D.; Glinsky, N. A Nodal High-Order Discontinuous Galerkin Method for Elastic Wave Propagation in Arbitrary Heterogeneous Media. *Geophys. J. Int.* **2015**, *201*, 1101–1118. [CrossRef]
21. Miryaha, V.A.; Sannikov, A.V.; Petrov, I.B. Discontinuous Galerkin Method for Numerical Simulation of Dynamic Processes in Solids. *Math. Model. Comput. Simulations* **2015**, *7*, 446–455. [CrossRef]

22. Favorskaya, A.V.; Zhdanov, M.S.; Khokhlov, N.I.; Petrov, I.B. Modelling the Wave Phenomena in Acoustic and Elastic Media with Sharp Variations of Physical Properties Using the Grid-Characteristic Method. *Geophys. Prospect.* **2018**, *66*, 1485–1502. [CrossRef]

23. Ivanov, A.M.; Khokhlov, N.I. Efficient Inter-Process Communication in Parallel Implementation of Grid-Characteristic Method. Springer: Cham, Switzerland, 2019; pp. 91–102. [CrossRef]

24. Golubev, V.I.; Khokhlov, N.I. Estimation of Anisotropy of Seismic Response from Fractured Geological Objects. *Comput. Res. Model.* **2018**, *10*, 231–240. [CrossRef]

25. Ruzhanskaya, A.; Khokhlov, N. Modelling of Fractures Using the Chimera Grid Approach. In Proceedings of the 2nd Conference on Geophysics for Mineral Exploration and Mining, Belgrade, Serbia, 9–13 September 2018. [CrossRef]

26. Stognii, P.; Khokhlov, N.; Zhdanov, M. Novel Approach to Modelling the Elastic Waves in a Cluster of Subvertical Fractures. In Proceedings of the 81th EAGE Conference and Exhibition, London, UK, 3–6 June 2019. [CrossRef]

27. Lisitsa, V.; Cheverda, V.; Botter, C. Combination of the discontinuous Galerkin method with finite differences for simulation of seismic wave propagation. *J. Comput. Phys.* **2016**, *311*, 142–157. [CrossRef]

28. Furgailo, V.; Ivanov, A.; Khokhlov, N. Research of Techniques to Improve the Performance of Explicit Numerical Methods on the CPU. In Proceedings of the 2019 Ivannikov Memorial Workshop (IVMEM), Velikiy Novgorod, Russia, 13–14 September 2019; pp. 79–85. [CrossRef]

29. Khokhlov, N.; Ivanov, A.; Zhdanov, M.; Petrov, I.; Ryabinkin, E. *Applying OpenCL Technology for Modelling Seismic Processes Using Grid-Characteristic Methods*; Springer: Cham, Switzerland, 2016; pp. 577–588. [CrossRef]

© 2020 by the authors. Licensee MDPI, Basel, Switzerland. This article is an open access article distributed under the terms and conditions of the Creative Commons Attribution (CC BY) license (http://creativecommons.org/licenses/by/4.0/).

Article

Three-Dimensional Regularized Focusing Migration: A Case Study from the Yucheng Mining Area, Shandong, China

Yidan Ding [1,*] , Guoqing Ma [1,*], Shengqing Xiong [2] and Haoran Wang [1]

1   College of Geo-Exploration Sciences and Technology, Jilin University, Changchun 130000, China; hrw17@mails.jlu.edu.cn
2   China Aero Geophysical Survey and Remote Sensing Center for Land and Resources, Beijing 100083, China; xsq@agrs.cn
*   Correspondence: dingyd18@mails.jlu.edu.cn (Y.D.); maguoqing@jlu.edu.cn (G.M.)

Received: 23 March 2020; Accepted: 21 May 2020; Published: 22 May 2020

Abstract: Gravity migration is a fast imaging technique based on the migration concept to obtain subsurface density distribution. For higher resolution of migration imaging results, we propose a 3D regularized focusing migration method that implements migration imaging of an entire gravity survey with a focusing stabilizer based on regularization theory. When determining the model parameters, the iterative direction is chosen as the conjugate migration direction, and the step size is selected on the basis of the Wolfe–Powell conditions. The model tests demonstrate that the proposed method can improve the resolution and precision of imaging results, especially for blocky structures. At the same time, the method has high computational efficiency, which allows rapid imaging for large-scale gravity data. It also has high stability in noisy conditions. The developed novel method is applied to interpret gravity data collected from the skarn-type iron deposits in Yucheng, Shandong province. Migration results show that the depth of the buried iron ore in this area is 750–1500 m, which is consistent with the drilling data. We also provide recommendations for further mineral exploration in the survey area. This method can be used to complete rapid global imaging of large mining areas and it provides important technical support for exploration of deep, concealed deposits.

Keywords: 3D focusing migration; regularization; conjugate migration direction; fast imaging; skarn-type iron deposits

1. Introduction

Gravity exploration is an effective method for obtaining information regarding subsurface density distribution; hence, it is an important tool in geophysical exploration and it is widely used to estimate target mineralized zones. Subsurface density distribution can be obtained via inversion of gravity data. However, from the perspective of classical theory, all geophysical inversions, including gravity inversion are ill-posed because their solutions are either non-unique or unstable. Nevertheless, geophysicists can solve this problem to some extent by adopting different types of regularization algorithms. Tikhonov [1] established regularization theory (also see [2–4]). The regularized inversion problem can be solved by finding a solution to the optimization problem that minimizes the objective functional, which can be expressed as [5–7]

$$P(\boldsymbol{m}) = P_d(\boldsymbol{m}) + \lambda P_m(\boldsymbol{m}) \tag{1}$$

where $P_d(m)$ is the misfit functional, $P_m(m)$ is the stabilizing functional that ensures the inversion result contains a priori information of the model parameters, and λ is the regularization parameter that balances the effects of $P_d(m)$ and $P_m(m)$.

On the basis of this equation, a variety of 3D regularized inversion methods for gravity data have been developed with different regularization terms [6,8], such as regularized smooth inversion based on the L2 norm (e.g., [9,10]), sparse inversion based on the L1 norm (e.g., [11]), total variation inversion (e.g., [12,13]) and focusing inversion (e.g., [14,15]). The focusing inversion method obtains relatively convergent results with sharp boundaries, which makes it is very suitable for mineral interpretation. Several classical techniques have also been developed to automatically estimate an appropriate regularization parameter [16], such as the generalized cross-validation criteria [17], L-curve criteria [18] and adaptive regularization method [6]. For the adaptive regularization method, the regularization parameter λ successively decreases during the process of the iteration, thus accelerating convergence and improving robustness.

Inversion methods can be inefficient in actual data processing due to the amount and time of calculation they require, and they are not able to complete the overall 3D density imaging of large regions [19]. These inefficiencies have restricted the study of regional structures and deep geological problems. Therefore, there is an urgent need to improve computing efficiency. Zhdanov introduced the concept of potential field migration and demonstrated its application for rapid 2D and 3D imaging of gravity and gradiometry surveys [6,20–22]. For gravity and its gradient fields, migration imaging is based on the direct integral transformation of gravity and its gradient data (which can also be mathematically described as the action of the adjoint operator on the observed data). The potential field migration method is fast and stable, and can be used for real-time imaging. However, the imaging results appear divergent [23]. This method was later extended to iterative migration [24], in which case adjoint operators are used iteratively. The forward model was introduced to estimate the accuracy of the migration results by computing the residual of the observed data and calculated data during each iteration. When the residual is less than a given value, this migration imaging model is adopted as the final result; thus, the imaging accuracy is improved. Multi-component joint iterative migration was developed subsequently and further improved the vertical resolution [25].

The potential field migration method instantaneously provides an estimation of the density distribution in a single step, while iterative migration improves the spatial resolution. However, the introduction of an iterative process reduces the computational efficiency. In order to maintain the high efficiency of migration and improve the resolution and stability, we propose a new method called 3D regularized focusing migration. We introduce an objective functional similar to that of regularized inversion and use the focusing stabilizer, which helps to generate a sharp and focused image of the model. When solving the optimization problem of minimizing the objective functional, the search direction of our algorithm is the conjugate migration direction instead of the conjugate gradient direction that is used for conventional regularized inversion. Good results can be obtained at the first migration and using the conjugate migration direction results in faster convergence than using the conjugate gradient direction in inversion. Thus, we can obtain relatively convergent results with fewer iterations. Unfortunately, the line searches are still performed multiple times while minimizing the objective functional. In order to improve the efficiency of a line search under the premise of ensuring algorithm convergence, the iterative step size is selected based on the inexact line search step size criterion proposed by Wolfe and Powell [26,27]. Adopting the Wolfe–Powell conditions allows a misfit to take less time to reach the given threshold when the model scale is relatively large.

The proposed method is efficient and stable, and can provide high-resolution subsurface density distribution imaging in real-time. It is well suited for evaluating mining targets because our efficient algorithm allows for fast imaging without being constrained by millions of cells from large mining areas. Our regularization parameter selection criterion is applicable to the case where the observed data are contaminated with Gaussian noise of uniform but unknown standard deviation, just like the real data. The focusing stabilizer can distinguish the boundary of each geological body, especially dense ore bodies.

In the following sections, we briefly introduce the principle of the regularized focusing migration method, and apply it to several synthetic models to verify the rationale and performance of our method. Then it is applied to the interpretation of gravity data collected in Yucheng, Shandong province. This province has multiple potential mining targets for skarn-type iron ore deposits that have no surface expression. We discuss the geological background of this area, interpret recently acquired gravity data, and provide a road map for further mineral exploration.

2. Methodology

2.1. Forward Modeling of Gravity Anomalies

The purpose of gravity forward modeling is to calculate the effect of the subsurface density distribution on the observation surface. The modeling domain is discretized into rectilinear cuboids of constant density (Figure 1). Discrete forward modeling for gravity anomalies produced by arbitrary density distribution can be expressed in general matrix notation as:

$$d = Am, \tag{2}$$

where d is a vector of the observed data of order n; m is a vector of the model parameters of order m; A, which is called the sensitivity matrix, is a matrix of size n by m formed by the corresponding gravity field kernels. If the subsurface modeling region is divided into $n_x \times n_y \times n_z$ geometric cuboids, then $n = n_x \times n_y$, $m = n_x \times n_y \times n_z$.

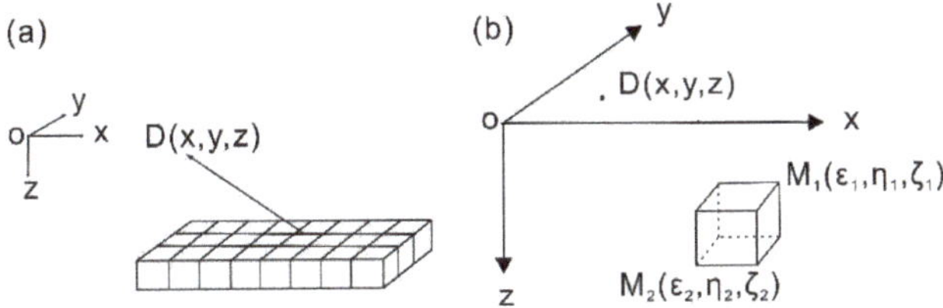

Figure 1. Representation of the subsurface density distribution. (**a**) A layer of rectilinear cuboids; (**b**) a single cuboid. The observation position is described by coordinate (x, y, z). The coordinates of cuboid corners can be completely described by two vertices $(\varepsilon_1, \eta_1, \zeta_1)$ and $(\varepsilon_2, \eta_2, \zeta_2)$.

The gravity anomalies generated by any geometric cuboid M at observation point D can be expressed as [28]

$$d = \gamma \rho_M \left\{ (\varepsilon - x)In[R - (\eta - y)] + (\eta - y)In[R - (\varepsilon - x)] - (\zeta - z)\tan^{-1} \frac{(\varepsilon - x)(\eta - y)}{(\zeta - z)R} \right\} \Big|_{\varepsilon_1}^{\varepsilon_2} \Big|_{\eta_1}^{\eta_2} \Big|_{\zeta_1}^{\zeta_2}, \tag{3}$$

where γ is the gravitational constant; ρ_M is the density of the geometric cuboid; (x, y, z) are the coordinates of surface observation point D; and $\varepsilon_1, \varepsilon_2, \eta_1, \eta_2, \zeta_1, \zeta_2$ are coordinates to determine the subdivision cuboid space, $R = \sqrt{(\varepsilon - x)^2 + (\eta - y)^2 + (\zeta - z)^2}$.

2.2. Principles of of Gravity Field Migration

Given that gravity field dataset d^{obs} is observed on the observation surface S, the objective is to determine the subsurface density distribution, m. Zhdanov proposed a fast migration imaging method on the basis of the Newton method. Its imaging principle is described briefly below [6].

Within the framework of the Newton method, a solution to the problem of parameter functional minimization can be found with a single step. When both the initial model and the prior model of anomalous density distribution are equal to zero, the solution can be expressed as:

$$m = -\frac{1}{\lambda}(W_m^* W_m)^{-1} l_0,\tag{4}$$

where λ is the regularization parameter; W_m is a model weighting matrix; W_m^* is the adjoint matrix of W_m; and l_0 is the direction of steepest descent that is introduced as a result of applying adjoint gravity operator, A^*, to the observed gravity data, d^{obs}:

$$l_0 = -A^* d^{obs},\tag{5}$$

where adjoint gravity operator, A^*, is specified as

$$A^* = \gamma \iint \frac{d}{|r'-r|^3}(z'-z)ds.\tag{6}$$

Let $\frac{1}{\lambda} = k$, and substitute Equation (5) into Equation (4), then, migration density can be rewritten as

$$m = k(W_m^* W_m)^{-1} A^* d^{obs} = k\omega_z^2 A^* d^{obs}.\tag{7}$$

Weighting operator, ω_z, is constructed based on the square root of the integrated sensitivity of the data to the density, S_z:

$$\omega_z = \sqrt{S_z}.\tag{8}$$

Unknown coefficient, k, can be determined by linear line search:

$$k = \frac{\| A^{\omega^*} d^{obs} \|^2}{\| A^\omega A^{\omega^*} d^{obs} \|^2},\tag{9}$$

where A^ω is the weighted sensitivity matrix and A^{ω^*} is the weighted adjoint gravity operator.

Migration gives us a method to analytically determine the model parameters, so rather than running an iterative algorithm, we find the optimal value for m in a single step. It is obvious that migration cannot obtain very accurate imaging results, although it has many advantages in terms of efficiency. Therefore, the regularized focusing migration method is introduced in this paper.

2.3. Regularized Focusing Migration

We can consider k in Equation (7) as a step size and $\omega_z^2 A^*$ as the migration direction. In this case, the migration is similar to the first iteration of a general inversion problem.

In regularization theory, the solution of linear inverse Equation (2) is substituted by a minimization of Tikhonov objective functional [6,7]:

$$P(m) = \| W_d(Am - d) \|^2 + \lambda \| W_m(m - m_{apr}) \|^2 \to min,\tag{10}$$

where m_{apr} is a prior model and here, m_{apr} is equal to zero; W_d is the data weighting matrix; and W_m is the model weighting matrix.

We introduce a focusing weighting matrix W_e, and the reweighted objective functional can be rewritten as:

$$P(m^\omega) = \| A^\omega m^\omega - d^\omega \|^2 + \lambda \| m^\omega \|^2,\tag{11}$$

where

$$A^\omega = W_d A W_m^{-1} W_e^{-1},\tag{12}$$

$$d^\omega = W_d d, \tag{13}$$

$$m^\omega = W_e W_m m. \tag{14}$$

There are different options for weighting matrices W_e, W_d, W_m. Here, we chose the following forms [6]:

$$W_e = diag \frac{1}{\sqrt{(m)^2 + e^2}}, \tag{15}$$

$$W_d = diag\left(AA^T\right)^{\frac{1}{2}}, \tag{16}$$

$$W_m = diag\left(A^T A\right)^{\frac{1}{2}}, \tag{17}$$

Here e is a focusing parameter and it is selected as 0.08 by trial and error [29].

For the n th iteration ($n = 1, 2, 3...$), the objective functional can be written as:

$$P(m_n^{\omega_n}) = \| A^{\omega_n} m_n^{\omega_n} - d^{\omega_n} \|^2 + \lambda_n \| m_n^{\omega_n} \|^2 . \tag{18}$$

When determining the model parameters, unlike the conjugate gradient method used in the general inversion problem, we use the conjugate migration direction method, which implies that the search direction of each iteration is along the conjugate migration direction. Then for regularized migration, according to Equation (5), we define the migration direction as

$$I_m = I(m) = A^*(Am - d) + \lambda m, \tag{19}$$

and the conjugate migration direction is defined as the linear combination of the migration direction for this step and the conjugate migration direction for the previous step. Hence the conjugate migration direction for the current model parameter $m_n^{\omega_n}$ is:

$$\widetilde{I}_n^{\omega_n} = -I_n^{\omega_n} + \beta_n \widetilde{I}_{n-1}^{\omega_{n-1}}, \tag{20}$$

where

$$I_k^{\omega_n} = I^{\omega_n}(m_n^{\omega_n}) = A^{\omega_n *}(A^{\omega_n} m_n^{\omega_n} - d^\omega) + \lambda_n(m_n^{\omega_n}), \tag{21}$$

$$\beta_n = \frac{\| I_n^{\omega_n} \|^2}{\| I_{n-1}^{\omega_{n-1}} \|^2}. \tag{22}$$

Specifically, the conjugate migration direction is defined as the migration direction when $n = 0$, namely $\widetilde{I}_0^{\omega_0} = -I_0^{\omega_0}$.

The successive line search in the conjugate migration direction can be written as:

$$m_{n+1}^{\omega_n} = m_n^{\omega_n} + k_n \widetilde{I}_{\alpha_n}^{\omega_n}. \tag{23}$$

Here, k_n is the iteration step. We take the Wolfe–Powell inexact line search conditions instead of the exact line search criterion shown in Equation (9) of the conventional iterative migration to obtain the step size [30]:

$$P\left(m_n^{\omega_n} + k_n \widetilde{I}_n^{\omega_n}\right) \le P(m_n^{\omega_n}) + \delta k_n I_n^{\omega_n T} \widetilde{I}_n^{\omega_n}, \tag{24}$$

$$\left| I^{\omega_n}\left(m_n^{\omega_n} + k_n \widetilde{I}_n^{\omega_n}\right)^T \widetilde{I}_n^{\omega_n} \right| \le \sigma \left| I_n^{\omega_n T} \widetilde{I}_n^{\omega_n} \right|, \tag{25}$$

where $0 < \delta < \sigma < 1$. Furthermore, the sufficient descent property, namely,

$$I_n^{\omega_n T} \widetilde{I}_n^{\omega_n} \le -c \| I_n^{\omega_n} \|^2, \tag{26}$$

where c is a positive constant, is crucial to ensure the global convergence of the optimization problem with the inexact line search techniques.

The actual model parameters can be converted as

$$m_{n+1} = W_m^{-1} W_{e_n}^{-1} m_{n+1}^{\omega_n},\tag{27}$$

where W_{e_n} is the focusing weighting matrix updated according to the m_n obtained on the $(n-1)$ st iteration.

Reweight m_{n+1} to get $m_{n+1}^{\omega_{n+1}}$:

$$m_{n+1}^{\omega_{n+1}} = W_m W_{e_{n+1}} m_{n+1}.\tag{28}$$

To ensure that the misfit functional converges to the global minimum, the following procedures are adopted to reduce the regularization factor [31]. First, let

$$p_n = \frac{\| m_{n+1}^{\omega_{n+1}} \|^2}{\| m_n^{\omega_n} \|^2},\tag{29}$$

and then update the regularization factor as follows:

$$\lambda_{n+1} = \begin{cases} \lambda_n & p_n < 1 \\ \lambda_n / p_n & p_n > 1 \end{cases}.\tag{30}$$

If the misfit functional is not shrinking fast enough, namely $\| r_n^{\omega_n} \|^2 - \| r_{n+1}^{\omega_{n+1}} \|^2 < 0.01 \| r_n^{\omega_n} \|^2$, the value of λ_{n+1} calculated by Equation (30) should be reduced to q times of the original value. We selected $q = 0.6$ by trial and error. It is clear that for the solution obtained in each iteration, if the value of regularization parameter is continuously reduced, then the value of the corresponding objective functional also continues to decrease. This guarantees that the objective functional will converge faster on the premise of obtaining a stable solution.

When the misfit reaches a given range:

$$Misfit = \| \frac{A m_N - d}{d} \| .100\% \leq \varepsilon_0,\tag{31}$$

the iteration process terminates and outputs m_N, otherwise it proceeds to the next iteration.

3. Synthetic Examples

We examined the feasibility and effectiveness of 3D regularized focusing migration with several typical synthetic models. First, we designed a single cube with a density contrast of 1 g/cm^3, with the center located at a depth of 150 m. The three-dimensional view of the subsurface model and its surface gravity anomalies are shown in Figure 2a,b. The subsurface inversion domain was divided into $50 \times 50 \times 25$ regular cells.

Figure 2c,d show the cross-sections through the density volume resulting from the regularized focusing migration at $y = 250$ m and $z = 150$ m, respectively. The density extremum of the imaging results corresponded well with the center of the anomalous body, and the high-density distribution precisely outlined the anomalous body, which demonstrated the feasibility of the method.

Our second synthetic example includes three anomalous bodies representing a dike complex and a prism (Figure 3a). Figure 3b shows the gravity anomaly computed by that model.

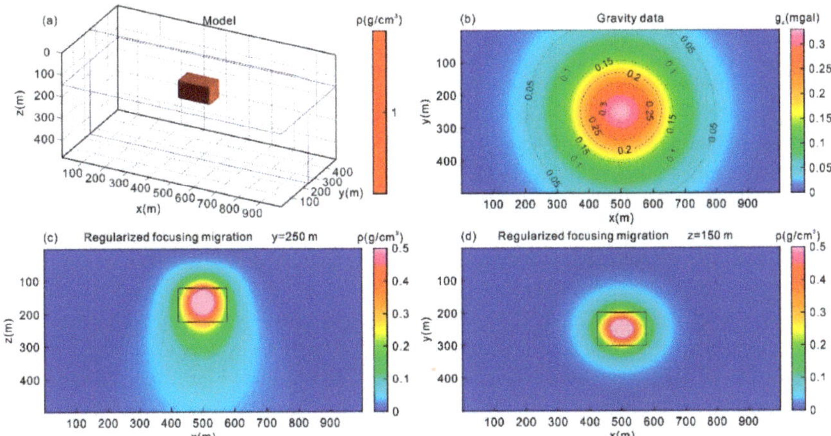

Figure 2. (**a**) Three-dimensional view of a single cube model; (**b**) gravity anomalies at $z = 0$ m generated from the model; (**c**) vertical cross-section of $y = 250$ m obtained from 3D regularized focusing migration; (**d**) horizontal cross-section of $z = 150$ m obtained from 3D regularized focusing migration.

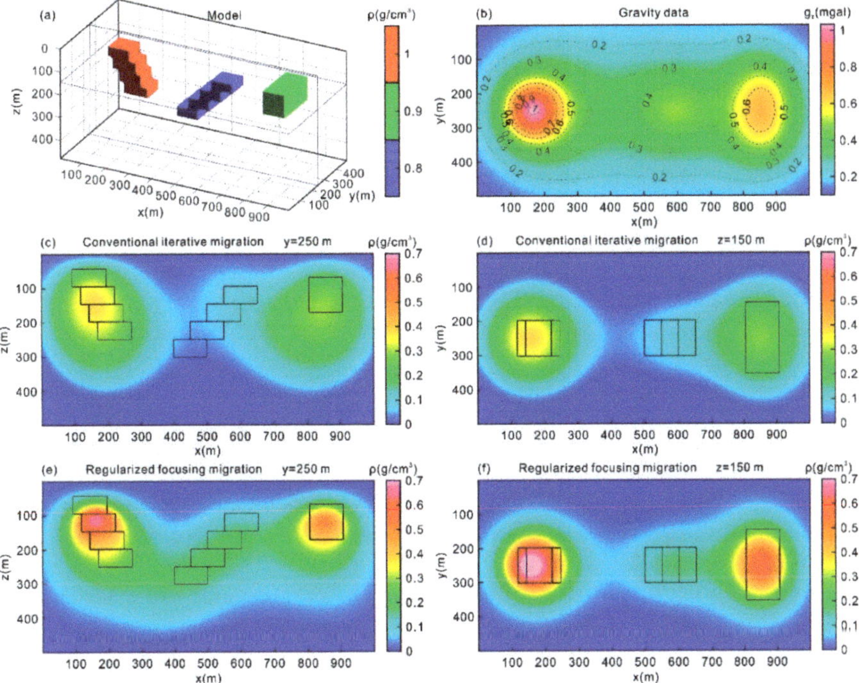

Figure 3. (**a**) Three-dimensional view of the subsurface model; (**b**) gravity anomalies at $z = 0$ m generated from the model; (**c**) vertical cross-section of $y = 250$ m obtained from conventional iterative migration; (**d**) horizontal cross-section of $z = 150$ m obtained from conventional iterative migration; (**e**) vertical cross-section of $y = 250$ m obtained from 3D regularized focusing migration; and (**f**) horizontal cross-section of $z = 150$ m obtained from 3D regularized focusing migration.

Figure 3c,f show the cross-sections of density distribution obtained by conventional iterative migration and regularized focusing migration. Regularized focusing migration method was able to distinguish the trends in the individual dikes, resulting in better convergence and a more realistic output, while better imaging the deep part of the model. However, the conventional iterative migration method barely distinguished three anomalous bodies and underestimated the dips in the modeled dikes, showing divergent density distribution that was far beyond the actual extent. In contrast, 3D regularized focusing resulted in a better spatial resolution and preserved sharper and finer boundaries.

Next, we investigated how regularized focusing migration performs in noise conditions. Figure 4a shows the three-dimensional diagram for two isolated cubes with a density contrast of 1 g/cm^3, located at depths of 70 m and 100 m. Figure 4b shows the surface gravity anomalies for this model. We added 5% random Gaussian noise to the observed gravity data, as shown in Figure 4c to simulate the data collected in actual surveys. The imaging results for noiseless and noise-corrupted data are shown in Figure 4d,g. The imaging results for both noiseless and noise-corrupted data were in good agreement with the true model. Although the two bodies were close in distance, which was equal to their side length, our method still distinguished the discrete bodies. The extremum of the derived density distribution corresponded well to the center of each geological body, and the horizontal positions and buried depths could be clearly delineated. However, the recovered density values were somewhat different from the actual conditions, especially for the deeper bodies. This is because the gravity field declines rapidly with depth, as shown in Equation (3). It is clear that the direct imaging of noise-containing data was still able to produce stable and smooth results. The volume images with a density cut-off greater than 0.3 g/cm^3 of noiseless and noise-containing data, are also shown in Figure 4h,i, respectively. The result of the direct migration of noise-containing data is very similar to that of noiseless data. Based on the results of this test, we concluded that the regularized focusing migration method has good noise resistance since the introduction of a regularization term prevents local overfitting. This method shows good performance and is able to obtain the global optimal results.

We also analyzed the efficiency of the proposed method. All tests were run on an Intel Xeon E5-2650 (Intel, Santa Clara, CA, USA) with 2.2 Ghz, 64GB of main memory. The model shown in Figure 2a was adopted for this purpose. Figure 5 shows the misfit curves as a function of the number of iterations for the regularized inversion [32], the conventional iterative migration method and our proposed method. The migration method obtains a result with a lower misfit in the first iteration, which gives rise to convergence and less calculation time. Numerical results show that the migration method can decrease the iterative number by 35% compared with the inversion method and greatly reduce the computation time.

Figure 6 shows the computation time of the regularized focusing migration method as a function of the model scale. The iteration process stopped when the misfit calculated by Equation (31) reached 5%. For models less than $6 \times 6 \times 6$ cells, the computation time for both line search criterion is constant because the model scale is relatively small and most of the time is spent on the initialization of the work. For models larger than $6 \times 6 \times 6$ cells, computation time is linearly dependent on the scale of the model. For computations based on the Wolfe–Powell conditions, the time that is consumed is higher than that of the exact line search with models under $16 \times 16 \times 16$ cells, but this is the opposite for the models exceeding $16 \times 16 \times 16$ cells. This is because the exact line search for each step involves the calculation of $A^\omega A^{\omega*}$, which means that the time complexity grows roughly as $\left(n_x.n_y\right)^2\left(n_x.n_y.n_z\right)^2$, while for the Wolfe–Powell conditions, which calculate $A^\omega m$ multiple times, each time the step size is determined, the time complexity is approximately $n_x^2 n_y^2 n_z$. The solution to the underdetermined system of equations does not depend on the exact search process. When the model scale is too large, the cost of using the exact line search is relatively high. According to the Wolfe–Powell conditions, the global convergence is expected to be fast. Although the number of steps increases, the convergence speed is accelerated.

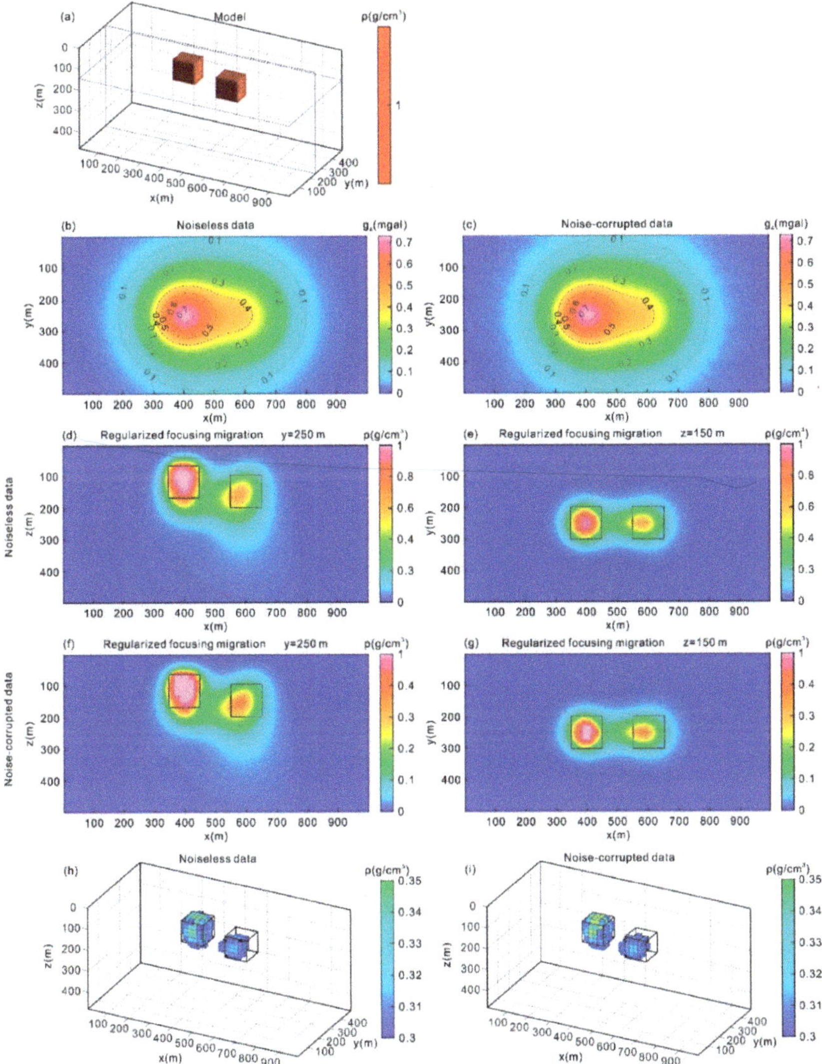

Figure 4. (**a**) Three-dimensional view of the subsurface model; (**b**) gravity anomalies at $z = 0$ m generated from the model; (**c**) gravity anomalies at $z = 0$ m generated from the model with 5% random Gaussian noise added; (**d**) vertical cross-section of $y = 250$ m for noiseless gravity data; (**e**) horizontal cross-section of $z = 150$ m for noiseless gravity data; (**f**) vertical cross-section of $y = 250$ m for noise-corrupted data; (**g**) vertical cross-section of $z = 150$ m for noise-corrupted data; (**h**) volume image with a density cut-off greater than 0.3 g/cm^3 for noiseless data; (**i**) volume image with a density cut-off greater than 0.3 g/cm^3 for noise-corrupted data.

To process the model with $50 \times 50 \times 25$ cells, the runtime for our method was only 224 s compared to 2278 s for the conventional migration method and 1.8 h for the regularized inversion method. This efficiency is due to the adoption of the conjugate migration direction and the step size is based on the Wolfe–Powell conditions when the model parameters are updated during each iteration. Besides, the continuous decrease in the regularization parameter and the introduced focusing weight can also accelerate the convergence and obtain a solution that minimizes the objective functional more quickly.

Therefore, when studying the regional structures and the deep geological problems, regularized focusing migration imaging can be used for global analysis without the need for data thinning. This is especially applicable to the investigation of large-scale mining areas, as in the following case study of the Yucheng mining area, Dezhou, Shandong, China.

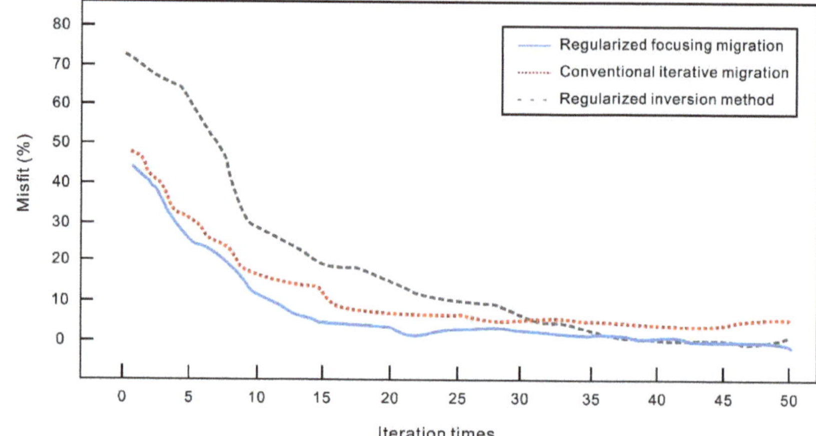

Figure 5. Misfit of three different methods: regularized focusing migration, conventional iterative migration and regularized inversion.

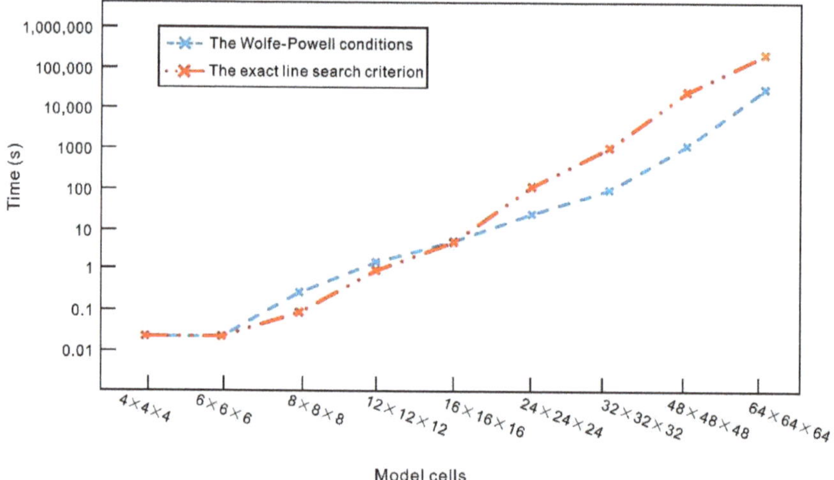

Figure 6. Computation time of regularized focusing migration method for models of different sizes. The blue line represents use of the Wolfe–Powell line search conditions as shown in Equations (24)–(26), and the red line represents use of the exact line search criterion as shown in Equation (9).

4. Field Example

In 2011, the Shandong Province Mineral Resources Potential Evaluation Project revealed that the Yucheng area had metallogenic conditions that are indicative of skarn-type ore deposits. In the following years, a variety of geological surveys showed that the area had good prospecting potential, and the iron ore bodies were confirmed during drilling. The traditional interpretation of the geophysical surveys in this area was based on the analysis of gravity or magnetic anomalies in contour maps [33–35], which provided the major geophysical characteristics of the area and was used to determine the approximate extent of the deposits. However, they alone could not determine the exact locations and depths of the mineral deposits.

In the following sections, we first summarize the metallogenic geological conditions of this area, then use our high-resolution method to invert the results of the entire airborne gravity data into 3D density models, and finally, we analyze the density imaging results to identify potential mining targets. Our work provides a reference for follow-up studies and further exploration of skarn-type iron ore under a thick overburden.

4.1. Geological Background of the Survey Area

The survey area is located near the Qihe county and Yucheng county borders to the south of Dezhou city in Shandong province, and belongs to the northwestern Shandong plain. It is located in the Luxi uplift area of the North China tectonic plate (Figure 7) [36]. The area is covered by strata of Cambrian, Ordovician, Carboniferous, Permian, Neogene and Quaternary from the base to the top [33]. The surface is completely covered by thick Quaternary and Neogene strata without any bedrock outcrops [37]. All kinds of deep geological information can be inferred from geophysical data. Due to regional tectonic movements, fault structures are widespread, including NE-striking faults, NNW-striking faults and SN-striking faults, and local developed fold structures (Figure 8).

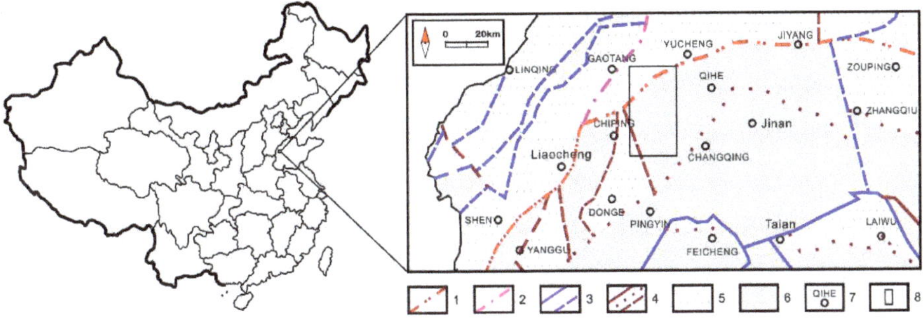

Figure 7. Tectonic background map of Yucheng area. The survey area is detailed in the black rectangle. 1 = the secondary structural unit; 2 = the tertiary structural unit; 3 = the quaternary structural unit; 4 = five-level structural unit; 5 = (latent) sag zone; 6 = (latent) high zone; 7 = toponym; 8 = survey area.

The carbonate rocks of the Ordovician Majiagou group are widely distributed and serve as host rocks for skarn-type iron ores. In addition, some iron-ore bodies also occur in the clastic rocks of the Carboniferous-Permian group. According to available borehole data, the depth of the Ordovician strata in the northwest reaches 3440 m or more, whereas it varies greatly in the southeastern area, ranging from 1165 m and 610–650 m in the east of Litun and in the northwest of Dazhang, respectively. There is no borehole data for the northwest of Pandian, and the depth of the Ordovician rocks is unknown [36]. In the survey area, magmatic activity was intense during the late Yanshanian period, and a significant volume of igneous rock mass was added at that time. The investigation of iron ores indicates that the ore bodies occur near the contact zone between the body mass of the late Yanshanian intermediate-basic intrusive rocks that are dominated by diorite and gabbro and the state of Ordovician limestone. In addition, a small number of ore bodies occur near the intersection of fault structures and in the xenoliths near the rock mass [35].

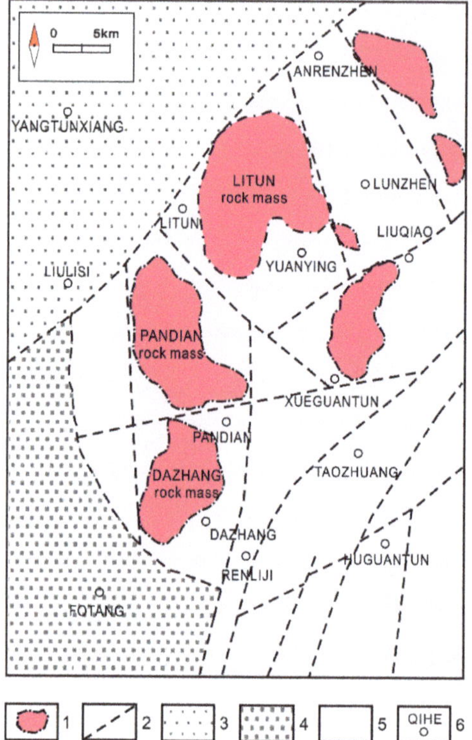

Figure 8. Geological map of the survey area. The figure is from [36]. 1 = Inferred range of concealed diorite rocks; 2 = inferred concealed fault; 3 = Yucheng-Jiyang latent sag; 4 = Lepingpu latent sag; 5 = Qihe latent high; and 6 = toponym.

4.2. Data Acquisition

The airborne gravity survey that collected gravity data in the Yucheng area, Shandong was performed by the China Aero-Geophysical Survey and Remote Sensing Center for Land and Resources (AGRS, Beijing, China). During the survey, a gravimeter was installed in the cabin of a CESSNA208 aircraft. The flight altitude was fixed at 450 m above sea level. The 45×35 km^2 survey area was flown at line spacings of 1500 m. The flight survey lines were designed to be perpendicular to the regional geological strike to ensure effective sampling. Figure 9 shows the gravity survey lines in the Yucheng area and the black rectangle delineates the region where we perform regularized focusing migration.

4.3. Migration Models of Gravity Data

We performed a terrain correction, latitude correction, and Bouguer correction on the collected gravity data to obtain Bouguer gravity anomalies (Figure 10a). The characteristics of regional gravity anomalies generally reflected the distribution of late Yanshanian intrusive rocks and the Ordovician strata uplift in the region. The iron-ore deposits of interest are shallower than 2 km because the iron ores below that have no economic mining value. The regional field below 2 km was separated from the Bouguer gravity anomaly by matched filtering [38], thus weakening the influence of deep structures on the imaging results for shallow density migration. Figure 10b shows the radial logarithmic power spectrum of the Bouguer gravity anomalies. Long-wavelength anomalies were removed to obtain the shallow local gravity anomalies. Figure 10c shows the local gravity anomalies, where three gravity highs can be seen. Four wells were drilled, including three holes in the Litun area and one hole in the

Yucheng area, and the ZK1 well in the Litun area and the ZK003 well in the Dazhang area discovered bodies of rich magnetite (Tables 1 and 2). According to borehole data, the density of magnetite was significantly higher than other rocks.

Figure 9. Gravity survey lines in the Yucheng area. The black rectangle shows the region used for regularized focusing migration.

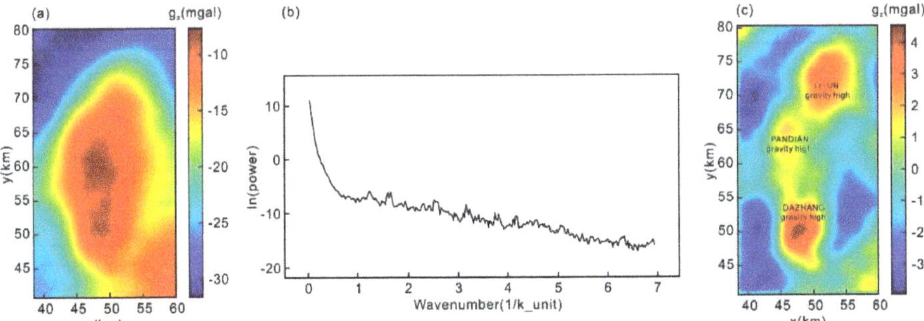

Figure 10. (**a**) Regional gravity anomaly; (**b**) radial logarithmic power spectrum; (**c**) local gravity anomaly.

Table 1. Dazhang ZK003 drilling data.

Lithology	Top Surface Depth (m)	Thickness (m)	Density (g/cm^3)
Sand/mealy sand/clay	0	571.9	
Mudstone/sandstone/limestone	571.9	18.4	2.69
Limestone	590.3	19.1	2.68
Limestone	609.4	103.7	2.72
Diorite	713.1	86.3	2.64
Skarn	799.4	2.6	3.15
Magnetite	802	15.3	4.41
Skarn	817.3	2.4	3.18
Magnetite	819.7	0.6	4.45
Diorite	820.3	105.7	2.63

Table 2. Litun ZK1 drilling data.

Lithology	Top Surface Depth (m)	Thickness (m)	Density (g/cm^3)
Sandstone/mudstone	0	1126.19	
Diorite	1126.19	19.03	2.57
Skarn dolomite	1145.22	5.02	2.66
Skarn	1150.24	7.14	2.87
Magnetite	1157.38	9.34	4.19
Skarn	1166.72	3.15	2.82
Magnetite	1169.87	8.77	4.26
Crushed zone	1178.64	5.31	
Skarn	1183.95	14.39	2.84
Magnetite	1198.34	98.15	4.49
Skarn	1296.49	16.35	2.87
Diorite	1312.64	123.46	2.68

Based on the characteristics of the stratigraphic, magmatic rocks distribution and the drilling data in the survey area, the high value of the local gravity anomalies were mainly caused by the skarn-type magnetite deposits. The positive density contrast of the migration imaging results could be used as a prospecting criterion. We performed regularized focusing migration imaging of local gravity anomalies with an initial model based on drilling data, and then delineated the spatial distribution of the iron ore under the range of positive density contrast of the imaging results. The local gravity data were inverted in a region designed as 21,500 m × 39,600 m in the X and Y directions and 2 km in the Z direction, with 88 × 98 × 50 small cells. Figure 11 shows the convergence curve of the migration. Figure 12 shows a perspective view of the migration density model with abnormal parts with a density greater than 0.7 g/cm^3, which gives a clear view of the anomalous bodies. For a better representation, several horizontal cross-sections with different depths are shown in Figure 13.

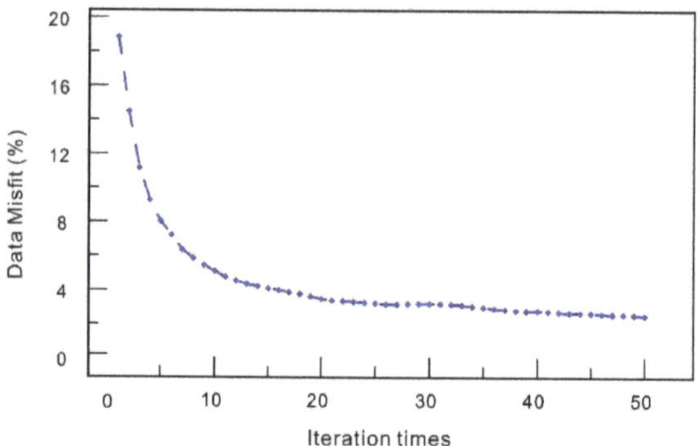

Figure 11. Convergence curve of the migration.

4.4. Geological Interpretation of the Mining Area

The results of migration imaging suggest that there are considerable mineral resources underneath the Litun, Pandian and Dazhang anomaly areas. Ore bodies derived from our analysis are mainly distributed between a depth of 700 m and 1500 m.

Figure 14 shows the ore-body distribution superimposed on a simplified geological map of the survey area. One can see that the target mineralized zones are mainly concentrated in the vicinity of fault structures since faults can provide channels for magmatic upwelling. The iron ore in Litun (I) is buried at a depth of about 1050–1350 m, that in Pandian (II, III) is about 1150–1550 m deep, and that in Dazhang (IV) is about 750–950 m. It should be noted that the Pandian area (II, III) is rich in iron ore resources underground, but these resources are deeper and less extensive, and drilling would be more difficult than in the Litun (I) and Dazhang areas (IV). The iron ores in the Litun and Dacheung areas are characterized by their great thickness, shallow burial and easy mining, and these should be regarded as key prospecting drilling areas. When these targets are exhausted and more geological information is obtained, exploration can proceed outward, including under the Pandian area along with other nearby high-density distributions that have not yet been measured.

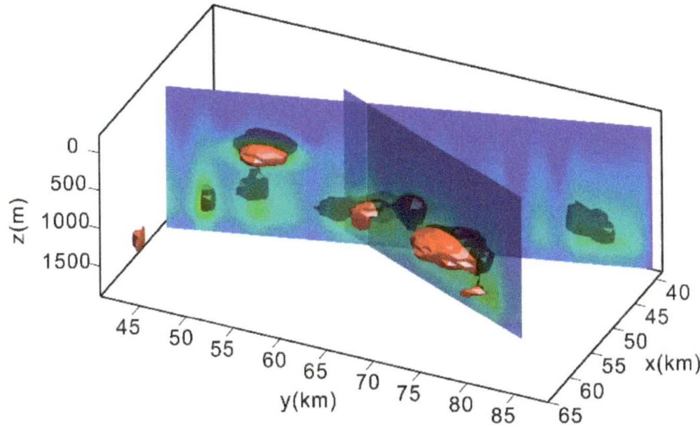

Figure 12. Perspective view of the migration density model. The red anomalous bodies show a density distribution with density greater than 0.7 g/cm^3.

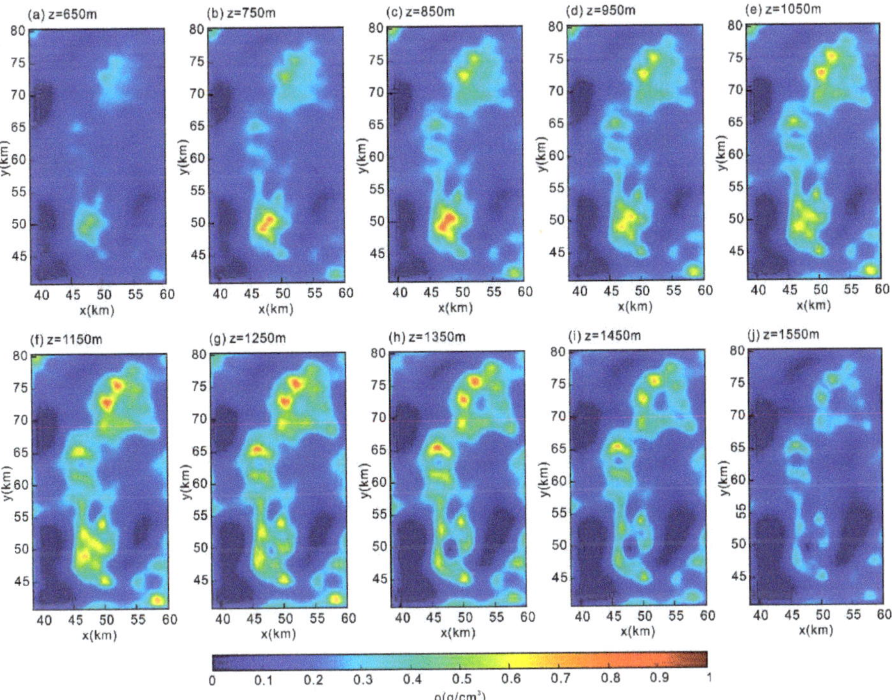

Figure 13. Horizontal cross-sections at (**a**) z = 650 m, (**b**) z = 750 m, (**c**) z = 850 m, (**d**) z = 950 m, (**e**) z = 1050 m, (**f**) z = 1150 m, (**g**) z = 1250 m, (**h**) z = 1350 m, (**i**) z = 1450 m and (**j**) z = 1550 m of imaging results.

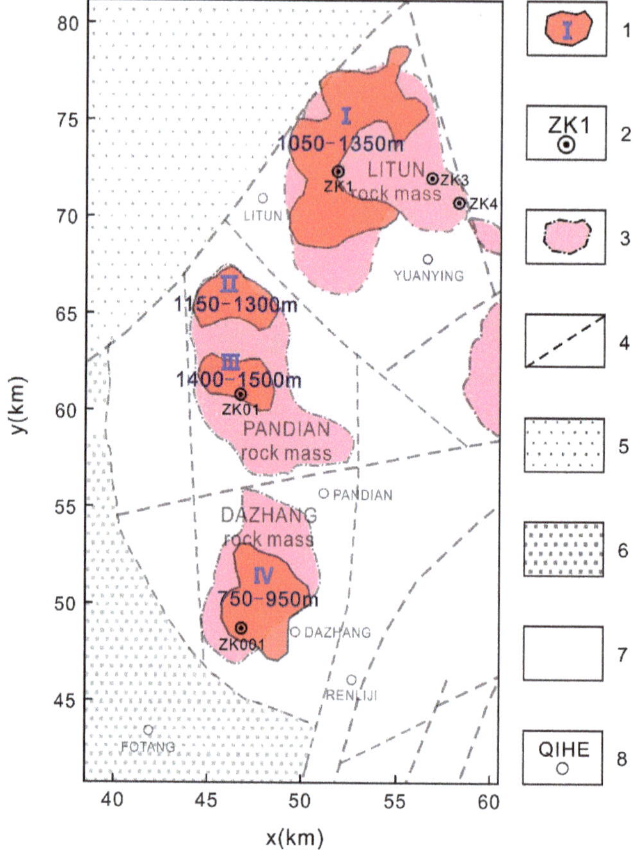

Figure 14. Map of iron ore range co-rendered with the corresponding geological map. 1 = scope of prospecting target area; 2 = drill hole; 3 = inferred range of concealed diorite rocks; 4 = inferred concealed fault; 5 = Yucheng-Jiyang latent sag; 6 = Lepingpu latent sag; 7 = Qihe latent high; 8 = toponym.

The recent well (ZK01) in this area found a body of ore that is 54.4 m thick at a depth of 1444.4 m, which is consistent with our interpretation. Therefore, the regularized focusing migration method has the ability to delineate the horizontal range of the ore body, predict its mineral depth and summarize prospecting targets.

5. Conclusions

Here, we developed 3D regularized focusing migration of large-scale gravity data with higher horizontal and vertical resolution than conventional gravity migration imaging. As validated by synthetic models, this method can determine complicated geological structures and it has high stability in noise conditions. Furthermore, the proposed algorithm based on the conjugate migration direction and the Wolfe–Powell conditions inherits the high efficiency of conventional migration and can produce real-time imaging with high accuracy. This provides a method that can delineate deep mineral resources.

We have shown how this method was efficiently used for the interpretation of the airborne gravity data collected over the Yucheng mining area, Dezhou, Shandong, China. High-resolution imaging results were obtained for the high-density body distribution, which helped to identify several potential mining targets with no surface expression. Currently, one of the predicted targets outlined in our study is undergoing further drilling exploration based on these airborne gravity analyses and well results so far coincide with our predictions. This method offers a basis for further exploration in this area to map the skarn-type iron-ore bodies under a thick overburden.

Gravity and magnetic combination migration will be used further in this area to reduce the non-uniqueness of the results. Joint migration of multiple geophysical data is the focus of our future research.

Author Contributions: Conceptualization, Y.D. and G.M.; data curation, Y.D. and G.M.; formal analysis, Y.D.; funding acquisition, G.M.; investigation, Y.D.; methodology, Y.D.; project administration, G.M.; resources, S.X.; software, Y.D. and H.W.; supervision, G.M.; validation, Y.D.; visualization, Y.D.; writing—original draft, Y.D.; writing—review & editing, G.M. All authors have read and agreed to the published version of the manuscript.

Funding: This research was funded by the National Key Research and Development Program of China, grant numbers 2017YFC0602203 and 2017YFC0602000.

Conflicts of Interest: The authors declare no conflict of interest.

References

1. Tikhonov, A.N. Regularization of incorrectly posed problem. *Sov. Math. Dokl.* **1963**, *4*, 1624–1627.
2. Phillips, D.L. A technique for the numerical solution of certain integral equations of the first kind. *J. ACM* **1962**, *9*, 84–97. [CrossRef]
3. Strakhov, V.N. The solution of linear ill-posed problems in Hilbert space. *Differ. Uravn.* **1970**, *6*, 1490–1495.
4. Lavrent'ev, M.M.; Romanov, V.G.; Shishatskii, S.P. *Ill—Posed Problems of Mathematical Physics and Analysis*, 1st ed.; American Mathematical Society: Providence, RI, USA, 1986; pp. 290–299.
5. Tikhonov, A.N.; Arsenin, V.Y. *Solutions of Ill—Posed Problems*; Wiley: Hoboken, NJ, USA, 1977.
6. Zhdanov, M.S. *Geophysical Inverse Theory and Regularization Problems*, 1st ed.; Elsevier: Amsterdam, The Netherlands, 2002; Volume 36, pp. 42–186.
7. Zhdanov, M.S. *Inverse Theory and Applications in Geophysics*, 2nd ed.; Elsevier: Amsterdam, The Netherlands, 2015; Volume 36, pp. 46–172.
8. Zhang, R.; Li, T.; Zhou, S.; Deng, X. Joint MT and gravity inversion using structural constraints: A case study from the Linjiang copper mining area, Jilin, China. *Minerals* **2019**, *9*, 407. [CrossRef]
9. Li, Y.; Oldenburg, D.W. 3-D inversion of gravity data. *Geophysics* **1998**, *63*, 109–119. [CrossRef]
10. Li, Y. 3-D inversion of gravity gradiometer data. In *SEG Technical Program Expanded Abstracts*; Society of Exploration Geophysicists: Tulsa, OK, USA, 1999; pp. 1470–1473.
11. Van Zon, T.; Roy-Chowdhury, K. Structural inversion of gravity data using linear programming. *Geophysics* **2006**, *71*, 41–50. [CrossRef]
12. Martins, C.M.; Lima, W.A.; Barbosa, V.C.; Silva, J.B. Total variation regularization for depth-to-basement estimate: Part 1—Mathematical details and applications. *Geophysics* **2011**, *76*, I1–I12. [CrossRef]
13. Lima, W.A.; Martins, C.M.; Silva, J.B.; Barbosa, V.C. Total variation regularization for depth-to-basement estimate: Part 2—Physicogeologic meaning and comparisons with previous inversion methods. *Geophysics* **2011**, *76*, I13–I20. [CrossRef]
14. Last, B.J.; Kubik, K. Compact gravity inversion. *Geophysics* **1983**, *48*, 713–721. [CrossRef]
15. Portniaguine, O.; Zhdanov, M.S. Focusing geophysical inversion images. *Geophysics* **1999**, *64*, 874–887. [CrossRef]
16. Farquharson, C.G.; Oldenburg, D.W. A comparison of automatic techniques for estimating the regularization parameter in non-linear inverse problems. *Geophys. J. Int.* **2004**, *156*, 411–425. [CrossRef]
17. Wahba, G. Spline models for observational data. *Technometrics* **1990**, *34*, 113–114.
18. Hansen, P.C. *Rank-Deficient and Discrete Ill—Posed Problems: Numerical Aspects of Linear Inversion*, 2nd ed.; Society for Industrial and Applied Mathematics: Philadelphia, PA, USA, 1998; pp. 69–87.
19. Liu, S.; Baniamerian, J.; Fedi, M. Imaging Methods versus inverse methods: An option or an alternative? *IEEE Trans. Geosci. Remote Sens.* **2020**, *58*, 3484–3494. [CrossRef]

20. Zhdanov, M.S.; Liu, X.; Wilson, G.A. Rapid imaging of gravity gradiometry data using 2D potential field migration. In *SEG Technical Program Expanded Abstracts*; Society of Exploration Geophysicists: Tulsa, OK, USA, 2010; pp. 1132–1136.
21. Zhdanov, M.S.; Liu, X.; Wilson, G. Potential field migration for rapid 3D imaging of entire gravity gradiometry surveys. *First Break* **2010**, *28*, 47–51. [CrossRef]
22. Zhdanov, M.S.; Liu, X.; Wan, L.; Čuma, M.; Wilson, G.A. 3D potential field migration for rapid imaging of gravity gradiometry data—A case study from Broken Hill, Australia, with comparison to 3D regularized inversion. In *SEG Technical Program Expanded Abstracts*; Society of Exploration Geophysicists: Tulsa, OK, USA, 2011; pp. 825–829.
23. Zhdanov, M.S.; Liu, X.; Wilson, G.A.; Wan, L. Potential field migration for rapid imaging of gravity gradiometry data. *Geophys. Prospect.* **2011**, *59*, 1052–1071. [CrossRef]
24. Wan, L.; Zhdanov, M.S. Iterative migration of gravity and gravity gradiometry data. In *SEG Technical Program Expanded Abstracts*; Society of Exploration Geophysicists: Tulsa, OK, USA, 2013; pp. 1211–1215.
25. Wan, L.; Han, M.; Zhdanov, M.S. Joint iterative migration of surface and borehole gravity gradiometry data. In *SEG Technical Program Expanded Abstracts*; Society of Exploration Geophysicists: Tulsa, OK, USA, 2016; pp. 1607–1611.
26. Wolfe, P. Convergence conditions for ascent methods. *SIAM Rev.* **1969**, *11*, 226–235. [CrossRef]
27. Powell, M.J.D. *Some Global Convergence Properties of a Variable Metric Algorithm for Minimization without Exact Line Searches*, 1st ed.; SIAM publications: Philadelphia, PA, USA, 1976; pp. 53–72.
28. Forsberg, R.; Tscherning, C.C. The use of height data in gravity field approximation by collocation. *J. Geophys. Res. Solid Earth* **1981**, *86*, 7843–7854. [CrossRef]
29. Zhdanov, M.S.; Tolstaya, E. Minimum support nonlinear parametrization in the solution of a 3D magnetotelluric inverse problem. *Inverse Probl.* **2004**, *20*, 937. [CrossRef]
30. Lemaréchal, C. A view of line-searches. In *Optimization and Optimal Control*, 1st ed.; Springer: Berlin/Heidelberg, Germany, 1981; pp. 59–78.
31. Portniaguine, O.; Zhdanov, M.S. 3-D magnetic inversion with data compression and image focusing. *Geophysics* **2002**, *67*, 1532–1541. [CrossRef]
32. Zhdanov, M.S.; Ellis, R.; Mukherjee, S. Three-dimensional regularized focusing inversion of gravity gradient tensor component data. *Geophysics* **2004**, *69*, 925–937. [CrossRef]
33. Cheng, X.; Wang, J.; Wang, J.; Zhai, D.; Cai, T. Geological characteristics and prospecting potential of Dazhang iron deposit in Qihe County of Shangdong Province. *Shandong Land Resour.* **2017**, *33*, 24–29.
34. Guo, Y.; Hao, X.; Zhong, W.; Ru, L.; Zhao, C. Application of high-precision magnetic measurement in exploration of hidden iron deposit: Setting iron deposits in Litun area of Yuchen City in Shandong Province as an example. *Shandong Land Resour.* **2017**, *33*, 52–56.
35. Hao, X.; Yang, Y.; Li, Y.; Gao, H.; Chen, L. Ore-controlling characteristics and prospecting criteria of iron deposits in Qihe area of western Shandong. *J. Jilin Univ.* **2019**, *49*, 982–991.
36. Kong, Q.; Zhang, T. *Deposits of Shandong*, 1st ed.; Shandong Science and Technology Press: Jinan, China, 2006; pp. 64–71.
37. Hao, X.; Liu, W.; Zang, K.; Zhang, X. Primary study on metallogenic regularity of skarn type iron deposit in Pandian area in western Shandong Province. *Shandong Land Resour.* **2018**, *34*, 27–33.
38. Syberg, F.J.R. A Fourier method for the regional—Residual problem of potential fields. *Geophys. Prospect.* **1972**, *20*, 47–75. [CrossRef]

© 2020 by the authors. Licensee MDPI, Basel, Switzerland. This article is an open access article distributed under the terms and conditions of the Creative Commons Attribution (CC BY) license (http://creativecommons.org/licenses/by/4.0/).

Article

The Efficient 3D Gravity Focusing Density Inversion Based on Preconditioned JFNK Method under Undulating Terrain: A Case Study from Huayangchuan, Shaanxi Province, China

Qingfa Meng [1,2], Guoqing Ma [1,2], Taihan Wang [1,2,*] and Shengqing Xiong [3,*]

[1] College of Geo-Exploration Sciences and Technology, Jilin University, Changchun 130026, China; mengqf19@mails.jlu.edu.cn (Q.M.); maguoqing@jlu.edu.cn (G.M.)

[2] Institute of National Development and Security Studies, Jilin University, Changchun 130026, China

[3] China Aerogeophysical Survey and Remote Sensing Center for Land and Resources, Beijing 100083, China

[*] Correspondence: wangtaihan@jlu.edu.cn (T.W.); xsq@agrs.cn (S.X.)

Received: 2 July 2020; Accepted: 21 August 2020; Published: 22 August 2020

Abstract: Since polymetallic ores show higher anomalies in gravity exploration methods, we usually obtain the position and range of ore bodies by density inversion of gravity data. The three-dimensional (3D) gravity focusing density inversion is a common interpretation method in mineral exploration, which can directly and quantitatively obtain the density distribution of subsurface targets. However, in actual cases, it is computation inefficient. We proposed the preconditioned Jacobian-free Newton-Krylov (JFNK) method to accomplish the focusing inversion. The JFNK method is an efficient algorithm in solving large sparse systems of nonlinear equations, and we further accelerate the inversion process by the preconditioned technique. In the actual area, the gravity anomalies are distributed on the naturally undulating surface. Nowadays, the gravity inversion under undulating terrain was mainly achieved by discretizing the ground into unstructured meshes, but it is complicated and time-consuming. To improve the practicality, we presented an equivalent-dimensional method that incorporates unstructured meshes with structured meshes in gravity inversion, and the horizontal size is determined by the gradient of observed gravity and terrain data. The small size meshes are adopted at the position where the terrain or gravity gradient is large. We used synthetic data with undulating-terrain to test our new method. The results indicated that the recovered model obtained by this method was similar to the inversion method of unstructured meshes, and the new method computes faster. We also applied the method to field data in Huayangchuan, Shaanxi Province. The survey area has complicated terrain conditions and contains multiple polymetallic ores. Based on the high-density characteristics of polymetallic ore bodies in the area, we calculate the field data into 3D density models of the subsurface by the preconditioned JFNK method and infer six polymetallic ores.

Keywords: preconditioned jacobian-free Newton-Krylov (JFNK) method; undulating terrain; gravity focusing density inversion; adaptive equivalent-dimension; polymetallic minerals; unstructured mesh and structured mesh

---

1. Introduction

The gravity exploration method is used to detect polymetallic ores because of its higher density feature, and they can use the density inversion method of gravity anomaly to obtain the approximate horizontal position of the ores. The most commonly gravity inversion method is three-dimensional gravity inversion, which can directly obtain the subsurface density distribution to interpret the three-dimensional area of ores [1–3]. The high-density borders of general gravity inversion are not sharp enough to match the ores. Last and Kubic presented compact gravity inversion, which is the precursor of gravity focusing inversion [4–6]. The focusing inversion uses the minimum support functional to increase the high-density value with the iteration of solution, so the volume of high-density results of inversion is minimal. In this way, the focusing inversion has a higher resolution and better convergence speed than the standard smooth inversion. However, the gravity method is suitable for the interpretation of mass data over a large area. Therefore, many geophysics make an effort to improve the computational efficiency of gravity inversion [7,8].

The JFNK method is designed for solving the large implicit nonlinear equations, and it is the combinations of Newton's method and Krylov subspace methods. This method uses the differential of vector instead of the jacobian matrix, so it does not need to form and store the elements of the true jacobian matrix. The jacobian matrix consists of the first-order partial derivatives of objective equation, which represents the optimal linear approximation of the equation to a given solution, so memory is intensive in solving large sparse nonlinear equations. Therefore, the JFNK method has a much faster convergence and smaller computation cost. This method has been successfully applied in fluid computing and other fields [9,10]. The Krylov subspace method is a kind of common iterative algorithm, and the conjugate gradient (CG) is a representative method of the Krylov subspace method which is commonly used in gravity inversion. Vandecar and Snieder used the preconditioned conjugate gradient method for solving the inversion problem of large data [11].

The study on the gravity inversion method considering the undulating terrain can be divided into two categories. The first type is to reduce the curved surface into a horizontal plane to offset the effects of undulating terrain, e.g., the equivalent source method, finite element method, boundary element method, Taylor series method and iterative method [12–16]. Yao also used the data of field and vertical gradient to reduce the curved surface into a horizontal plane to obtain more accurate and stable results [17]. To improve the direct solution method, Liang used the gradient data and boundary element method to obtain the potential field of the horizontal plane directly from the curved surface data [18,19]. Unfortunately, this method assumes that there is no field source between the undulating terrain and the converted horizontal plane. The data obtained by this way are only an approximation of the actual data, which reduces the precision of inversion results. The second type is to mesh the subsurface that corresponds to the undulating terrain. Currently, the common method is unstructured mesh technology. This method can fit the surface by dividing the subsurface into multiple triangles or tetrahedrons in two-dimensional (2D) or 3D inversions. Zhong used tetrahedrons to fit the terrain to realize terrain correction [20]. Geophysicists obtained the mesh with undulating terrain by using the triangulation technology to be applied on the inversion of the electrical prospecting with the complicated terrain [21,22]. Zhang applied triangulation technology on forward modeling and inversion of gravity and magnetic profiles [23]. In the gravity and magnetic inversion field, the triangulation was mainly used to fit geological models to obtain more accurate forward results [24]. The unstructured mesh method is complex and computationally intensive, which is not suitable for the inversion of gravity with a large amount of data. However, the common structured mesh method cannot fit the undulating terrain. Therefore, it is necessary to combine the structured and unstructured mesh methods in the gravity and magnetic inversion field.

In this paper, we presented the preconditioned JFNK gravity focusing density inversion method to obtain the density feature of the ores with accuracy and efficiency. To solve the inversion meshing under the undulating terrain, we also presented an adaptive equivalent-dimension method, which processed the subsurface by combining triangulation unstructured and rectangular structure mesh. Considering that the density variation and the terrain fitting are related to the anomaly and the change rate of the terrain, the horizontal size of each grid unit depends on the gradient of anomaly or terrain. We used the models with the undulating terrain and compared common unstructured meshing methods to gauge the performance of the algorithm and test its accuracy and efficiency. Then, we applied this approach to interpret gravity data collected in the Huayangchuan polymetallic mining area in Shaanxi Province.

2. Methodology of Preconditioned JFNK Gravity Focusing Inversion

The gravity inversion involved discretizing the subsurface into a finite number of rectangular units, and the discrete map is shown in Figure 1. There are n observation points and m rectangular units, the forward expression of the gravity anomaly could be expressed as:

$$g_{n\times1} = G_{n\times m}\rho_{m\times1},\tag{1}$$

where $g_{n\times1}$ is the gravity anomaly, $G_{n\times m}$ is the contribution to the nth datum of a unit density in the mth cell called kernel function matrix, and $\rho_{m\times1}$ is the density matrix of rectangular units.

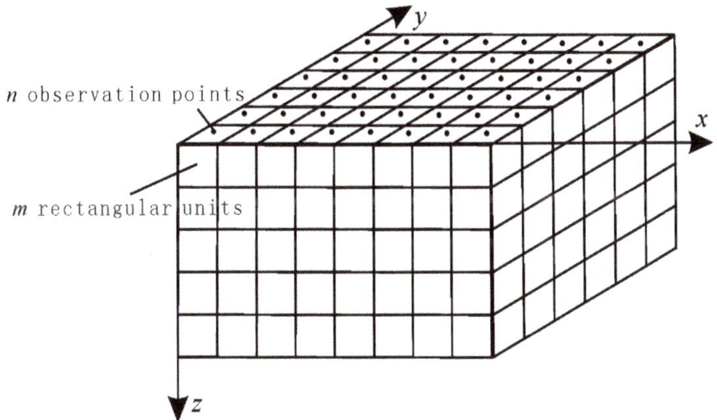

Figure 1. Discretization of underground space.

The kernel function matrix is [25]:

$$G(n,m) = -\gamma\sum_{i=1}^{2}\sum_{j=1}^{2}\sum_{k=1}^{2}\mu_{ijk}\left[x_i\ln\left(y_j + r_{ijk}\right) + y_j\ln\left(x_i + r_{ijk}\right) - z_k\arctan\left(\frac{x_iy_j}{z_kr_{ijk}}\right)\right],\tag{2}$$

$$x_i = x - \xi_i; y_j = y - \eta_j; z_k = z - \zeta_k; r_{ijk} = \sqrt{x_i^2 + y_j^2 + z_k^2}; \mu_{ijk} = (-1)^i(-1)^j(-1)^k.$$

In the expression, γ is the gravitational constant, (x, y, z) are the coordinates of the observation points, and (ξ_1, η_1, ζ_1) and (ξ_2, η_2, ζ_2) are the coordinates of the minimum and maximum corner points of the cells.

Performing gravity inversion requires solving Equation (1) for $\rho_{m\times1}$. Since the number of underground cells was greater than the data, the Equation (1) was underdetermined.

We solved this underdetermined equation in the following manner [26]:

$$\phi = \phi_g + u\phi_\rho = \left(g_{obs} - GW_z^{-1}W_z\rho\right)^T\left(g_{obs} - GW_z^{-1}W\rho\right) + u\rho^TW_e^TW_z^TW_zW_e\rho \to \min\tag{3}$$

$$W_e = \left(\rho^2 + e^2\right)^{-1/2},\tag{4}$$

where ϕ_g is the square norm of the difference between the observed anomaly(g_{obs}) and the calculated anomaly($G\rho$), and it represents the fitting functional of the data. In the expression, ϕ_ρ is the stabilizing functional that constrained the inversion results to the real conditions. u is a regularization parameter. The value of the regularization parameter is usually 10^n, and the n is determined by a "trial and error" method. W_z is the depth-weighting function which was used to counteract the inherent decay of the kernel function [27]. $W_z\rho$ is the weighted solution variable, we replaced it with ρ_w. The Equation (4) is the minimum support function [4]. In general, the e is 0.1, which is a focusing parameter to determine the sharpness of the inversion results.

The derivative formula of Equation (3) is:

$$\frac{\partial \phi}{\partial \rho} = [W_e^{-2}(GW_z^{-1})^T(GW_z^{-1}) + uI] \times \rho_w - W_e^{-2}(GW_z^{-1})^T g_{obs},$$ (5)

We can substitute the solution of Equation (3) with the minimization of Equation (5). Finally, we used the JFNK method to obtain the ρ_w of Equation (5), and the results ρ can be obtained by removing the depth-weighting function.

The JFNK method was introduced as follows. The $f(x) = 0$ can be processed as:

$$f(x) \approx f(x_k) + f'(x_k)(x - x_k) = 0,$$ (6)

$$f'(x_k)(x - x_k) = -f(x_k), k = 0, 1, \ldots,$$ (7)

The Krylov subspace method was created to solve the Equation (7), and we used the CG method which is a common solution technique with the Krylov subspace [26]. The matrix-vector product required by CG method was approximated by forward finite difference schemes:

$$f'(x)v \approx \frac{f(x + \varepsilon v) - f(x)}{\varepsilon}, \varepsilon \neq 0,$$ (8)

where v is a vector of the Krylov subspace. The process of the preconditioned CG method used to solve Equation (7) is:

$$
\begin{aligned}
&A = f'(x); \quad m = (x - x_k); \; b = -f(x), \\
&k = 0; \quad\quad\; m_0 = \text{initial solution estimate}; \; r_0 = A^T(b - Am_0), \\
&\text{while} \quad\quad r_k \neq 0 \\
&\quad\quad\quad\quad\; z_k = Sr \\
&\quad\quad\quad\quad\; k = k + 1 \\
&\quad\quad\quad\quad\; \text{if } k = 1 \\
&\quad\quad\quad\quad\; p_1 = z_0, \\
&\quad\quad\quad\quad\; \text{else} \\
&\quad\quad\quad\quad\; \beta_k = r_{k-1}^T z_{k-1} / r_{k-2}^T z_{k-2} \\
&\quad\quad\quad\quad\; p_k = z_{k-1} + \beta_k p_{k-1} \\
&\quad\quad\quad\quad\; \text{end} \\
&\quad\quad\quad\quad\; q_k = Ap_k \\
&\quad\quad\quad\quad\; \alpha_k = r_{k-1}^T z_{k-1} / q_k^T q_k \\
&\quad\quad\quad\quad\; m_k = m_{k-1} + \alpha_k p_k \\
&\quad\quad\quad\quad\; r_k = r_{k-1} - \alpha_k A^T q_k \\
&\text{end}
\end{aligned}
$$ (9)

where k is the number of iterations, and as the number of iteration increased, the residual r was minimized by moving a distance α in search direction p. The S is a preconditioner of the search direction, and we could obtain the new search direction. The optimum preconditioning is $(A^T A)^{-1}$. It produced the least-squares solution to Equation (7) in a single iteration. Since we were unable to obtain the optimum preconditioning directly, we substituted the optimum preconditioning with a depth-weighting function [10]. All the other parameters were intermediate variables with no significance.

Under the condition of the natural surface, the structured mesh method was not applicable, and the unstructured mesh method w triangulation (shown in Figure 2a). The structured method is computationally intensive. We proposed the adaptive equivalent-dimensional mesh method which combines the unstructured and structure mesh to achieve gravity inversion with the undulating terrain.

Only the surface layer (the yellow area in Figure 2b) was discretized by the triangular unstructured mesh method, and the other subsurface was discretized into structure rectangular mesh. The three blue cells in Figure 2b have the same height, and it is shown that the structure rectangular mesh was the n dimension in each horizontal position. Therefore, many judgment processes were avoided in the calculation, and the calculation efficiency was improved; the memory was not increased in the case of fitting the surface as far as possible. This method could efficiently and accurately obtain the 3D density models, so it was more applicable to the gravity inversion with a large amount of data.

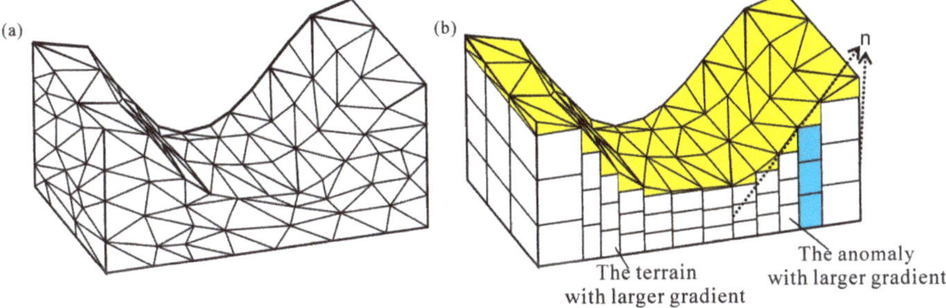

Figure 2. Mesh with triangulation and equivalent dimension in undulating terrain. (**a**) Triangulation; (**b**) equivalent dimension.

As the adaptive equivalent-dimensional mesh method, we derived the equations to determine intervals (dx, dy) of the mesh according to the terrain and anomaly:

$$dx_i = \frac{X}{2n_x} + \frac{2 - \dfrac{\sum\limits_{j=1}^{n_y} THD_g(i,j)}{\max[\sum\limits_{j=1}^{n_y} THD_g(:,j)]} - \dfrac{\sum\limits_{j=1}^{n_y} THD_h(i,j)}{\max[\sum\limits_{j=1}^{n_y} THD_h(:,j)]}}{\sum\limits_{i=1}^{n_x} [2 - \dfrac{\sum\limits_{j=1}^{n_y} THD_g(i,j)}{\max[\sum\limits_{j=1}^{n_y} THD_g(:,j)]} - \dfrac{\sum\limits_{j=1}^{n_y} THD_h(i,j)}{\max[\sum\limits_{j=1}^{n_y} THD_h(:,j)]}]} \times \frac{X}{2}, \tag{10}$$

$$dy_j = \frac{Y}{2n_y} + \frac{2 - \dfrac{\sum\limits_{i=1}^{n_x} THD_g(i,j)}{\max[\sum\limits_{i=1}^{n_x} THD_g(i,:)]} - \dfrac{\sum\limits_{i=1}^{n_x} THD_h(i,j)}{\max[\sum\limits_{i=1}^{n_x} THD_h(i,:)]}}{\sum\limits_{j}^{n_y} [2 - \dfrac{\sum\limits_{i=1}^{n_x} THD_g(i,j)}{\max[\sum\limits_{i=1}^{n_x} THD_g(i,:)]} - \dfrac{\sum\limits_{i=1}^{n_x} THD_h(i,j)}{\max[\sum\limits_{i=1}^{n_x} THD_h(i,:)]}]} \times \frac{Y}{2}, \tag{11}$$

$$THD_f = \sqrt{\left(\frac{\partial f(x,y)}{\partial x}\right)^2 + \left(\frac{\partial f(x,y)}{\partial y}\right)^2}. \tag{12}$$

In the expression, X and Y represent the total length of the measuring area along x and y directions. THD_g and THD_h represent the total horizontal derivative (THD) of gravity anomaly g and elevation of measure points h, respectively. Its equation is shown in Equation (12). In this way, the dx and dy of the cells would be 0.5~1.5 times the average length. Therefore, the intervals of mesh were close in the area where the gravity anomaly or terrain varied greatly.

3. Gravity Model Tests

We tested our method in the inversion consisting of typical models of $150 \times 150 \times 100$ m with undulating terrain, and the results are shown in Figure 3. The buried depth of the center of models was 100 m from the surface, and the central coordinates of two models were (225, 250) m and (725, 250) m. As we can see in Figure 3b, the extreme value position of the gravity effect was asymmetric because of the asymmetry of the undulating terrain. Therefore, the x-coordinates of the two extreme value positions of the gravity anomaly were 200 and 800 m, respectively. There was deviation from the x-coordinates of 225 and 725 m of the two actual models. When the terrain fitting accuracy was consistent, the terrain grid size was positively correlated with the fineness of the subsurface mesh. Figure 3c shows the run time of programs in different terrain grid sizes with twenty-five obverse points. It is shown in Figure 3c that the run time of programs with equivalent-dimensional mesh was obvious shorter than with triangulation mesh.

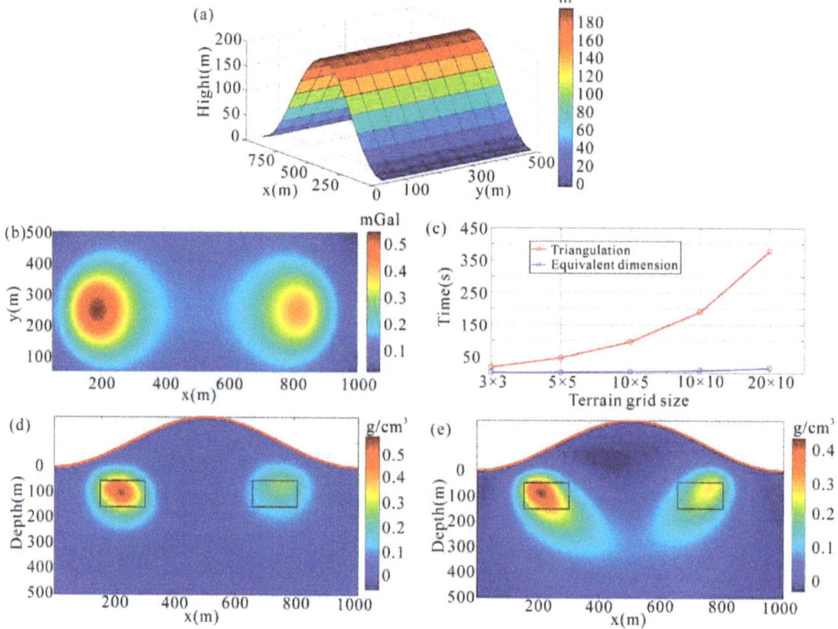

Figure 3. Gravity inversion by the preconditioned JFNK method. (**a**) Surface topography; (**b**) gravity anomalies under undulating terrain; (**c**) comparison of mesh computing time; (**d**) slice at y = 250 m of inversion result with equivalent-dimensional meshes; (**e**) slice at y = 250 m of inversion result with triangulation mesh.

The 3D gravity inversion in Figure 3d was performed based on equivalent-dimensional mesh. Figure 3d shows the slice of the inversion result at y = 250 m, and the red curve is consistent with the undulating terrain and represents the top of the subsurface, and the black rectangles represent real positions of typical models. It can be seen in Figure 3d that the high-density units in the inversion result were similar to the actual range of models. Therefore, this method could obtain the information of models directly by the inversion with equivalent-dimensional mesh. Figure 3e shows the 3D gravity inversion with triangulation mesh, and the result of the inversion with equivalent-dimensional mesh was similar to the inversion with triangulation mesh, comparing Figure 3d,e. Therefore, the new meshing method proposed by us can obtain gravity inversion results consistent with the traditional method in the regions with undulating terrain efficiently.

Figure 4a shows the gravity anomaly of Figure 3a with noise, and the anomaly showed some distortion after the addition of 30 signal noise ratio noise. Figure 4b is the slice of the inversion result, and Figure 4c is the 3D view of slices. In Figure 4d, blue lines represent the edge of the models, and the yellow and blue blocks represent densities greater than 0.28 and 0.16 g/cm^3 in the inversion results, respectively. The inversion results shown in Figure 4b–d are close to the result in Figure 3c. Therefore, the method was stabilized and could obtain an accurate range of models after adding noise.

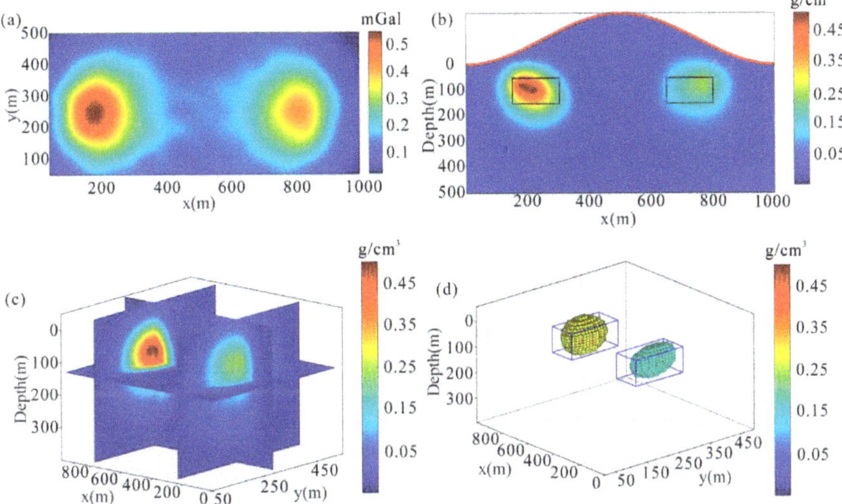

Figure 4. Gravity inversion containing noise by the preconditioned JFNK method. (**a**) Synthetic gravity anomalies under undulating terrain containing noise; (**b**) slice at y = 250 m of inversion results; (**c**) slices show of inversion results; (**d**) 3D view of inversion results.

We processed a gravity anomaly with complex terrain to verify the method of equivalent-dimensional mesh. There were two peaks and one valley—shown in Figure 5b. In the 100 m depth from the surface, there were three prisms of 100 × 100 × 100 m; the central coordinates of which were (250, 250), (250, 750), and (750, 500) m, respectively. Figure 5a is the gravity anomaly of models, and the gravity containing 30 SNR noise is shown in Figure 5c. Because of the complex terrain's effect, the coordinates of the extreme value points shown in Figure 5a were (250, 200), (250, 800), and (750, 550) m, and the coordinates were not consistent with the models' coordinates. We achieved the density inversion based on the mesh with the equivalent dimension and obtained the slice diagram of the inversion results shown in Figure 5e–h by cutting along the white dashed lines numbered 1 and 2 shown in Figure 5a–c. Figure 5d shows the high-density blocks of a density greater than 0.3 g/cm³, and the blue lines represent the edge of the models.

Although the extreme values' coordinates of the gravity anomaly deviated from the model positions at lines 1 and 2, the central position of the inversion results obtained by the gravity inversion based on equivalent-dimensional mesh were still consistent with the model position. The results shown in Figure 5d–h indicate that this method is applicable and stabilized in complex terrain.

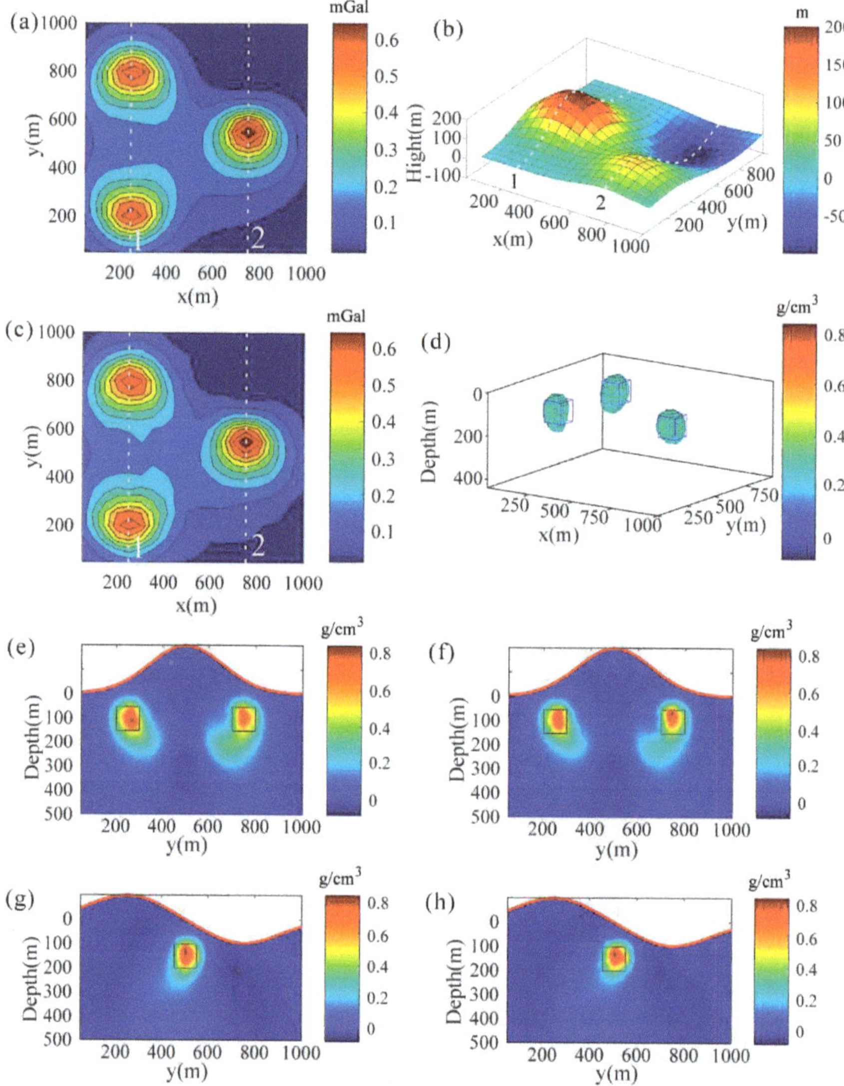

Figure 5. Gravity inversion under complex terrain. (**a**) Gravity anomalies; (**b**) surface topography; (**c**) gravity anomalies containing noise; (**d**) 3D view of inversion results; (**e**) slice at x = 250 m of inversion results; (**f**) slice at x = 250 m of inversion results computed with noise; (**g**) slice at x = 750 m of inversion results; (**h**) slice at x = 750 m of inversion results computed with noise.

4. Actual Data Processing

In order to validate the applicability of the adaptive equivalent-dimensional mesh method in the field data, we processed the gravity data of polymetallic mining areas in Shaanxi province of China. The area is located in Huayangchuan in Shaanxi province. Its tectonic location belongs to Xiaoqinling intracontinental orogenic belt in the southern margin of north China landmass. The main outcrops are the Neoarchean erathem Taihua group (deep metamorphic crystalline complex), the Mesoproterozoic erathem Xionger group (volcanic–sedimentary), Gaoshanhe formation (clasolite), and the Luonan group (carbonatite); some areas have the Sinian series and the Cambrian system clasolite. Paleogene,

Neogene, and Quaternary are deposited in Cenozoic basins, and the Ordovician, Silurian, and Paleozoic to Mesozoic stratums are missing [28,29]. The geological condition of the survey area is shown in Figure 6a.

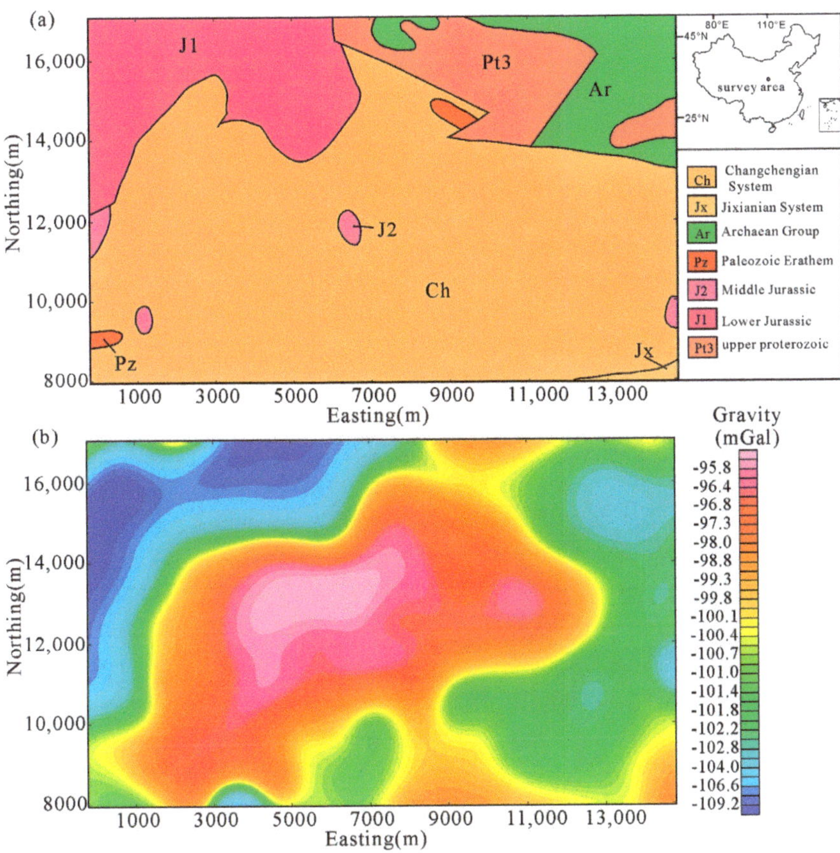

Figure 6. Geological and gravity of survey area. (**a**) The geological map; (**b**) Bouguer gravity anomaly.

Figure 6b shows the Bouguer gravity anomaly of the survey area. The center area obviously had a high-value gravity anomaly in Figure 6b, so there was a high-density body in the subsurface. There were polymetallic mines in the vicinity of this area, and the geophysical characteristics of the metal minerals were high values of density. Therefore, we inferred that the high-value gravity in this area was a polymetallic deposit. Metallic minerals were produced by magma intruding into the shallow strata of the subsurface through the faults, so the location of faults had great value for us to infer the range of minerals.

The total horizontal derivative, of which the maximum value indicates the location of the faults, is commonly used to detect faults in gravity interpretation [30]. Therefore, by comprehensively analyzing the geological map and the gravity data of this region, we finally divided the faults of this region into the total horizontal derivative and local gravity field map shown in Figure 7a.

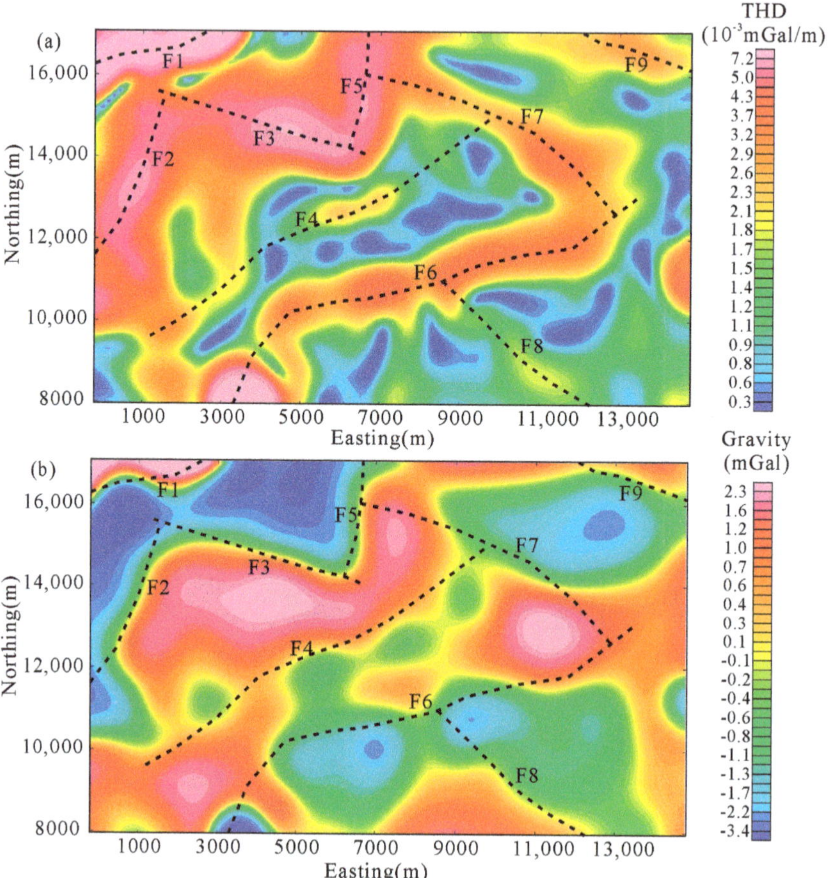

Figure 7. Fracture interpretation results on total horizontal derivative and local gravity field; the dashed lines F1–F9 represent the faults interpreted by the total horizontal derivative. (**a**) Total horizontal derivative; (**b**) field of local gravity anomaly.

The Bouguer gravity anomaly is the comprehensive response of the density at all depths. The 3D density inversion depth range was an altitude of −1500 m from the surface, so we needed to perform separation of the potential field by using the matched filter method. The expression of the matched filter is [31,32]:

$$LnE(\omega) = 2LnB - 2H\omega, \tag{13}$$

where $LnE(\omega)$ is the average logarithmic power of the potential field, H is the depth of the layer corresponding to the anomaly, ω is the wave number, and B is a constant related to the deep or shallow source anomaly.

In this way, we removed the regional field anomaly corresponding to the density below −1500 m and extracted the shallow response of the gravity anomaly, namely the local field. The result is shown in Figure 7b, indicating that the local anomalies in this area were separated by faults; this is consistent with the geological characteristics of the ores distributed around the faults, which shows the accuracy of the matched filter method. The terrain in this region was complex with an average altitude of 1480 m and a maximum height difference of 509 m, as shown in Figure 8a. Considering the regional terrain relief, our adaptive equivalent-dimensional mesh method was suitable for this real case.

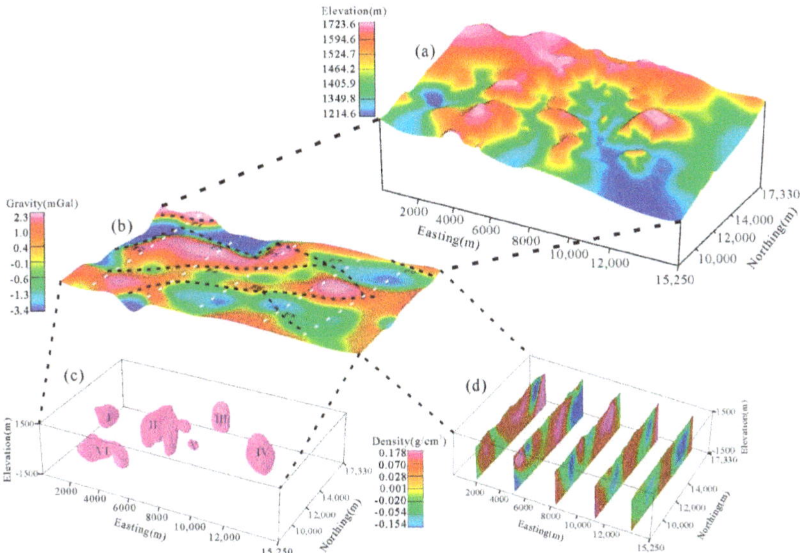

Figure 8. 3D inversion results of actual area. (**a**) Elevation of the terrain; (**b**) field of local gravity anomaly; (**c**) 3D view of inversion results greater than 0.12 g/cm^3; (**d**) slices of inversion results.

The depth of −1500 m extending to surface of this region was meshed by the proposed method in this paper, and 3D gravity inversion was performed to obtain the density in the subsurface, as shown in Figure 8c,d. Figure 8b is the local field of gravity, in which the white dotted lines represent the horizontal position of the slices, and Figure 8d shows the slices of the 3D density model obtained from the gravity inversion results shown in Figure 8c. The density model in each section show the high-density units, which represent the position of the explained minerals.

Through the recognition of high-density units in the 3D density inversion model, the range of metallic ore bodies in this region was interpreted in Figure 9 by a red range. The ore bodies distributed near the faults that conform to the principle of the metal minerals were generated in the magma which intruded along the faults. There were six ore bodies explained in this area, and the number shown in Figure 9 around the red range is the height above sea level of these ore bodies.

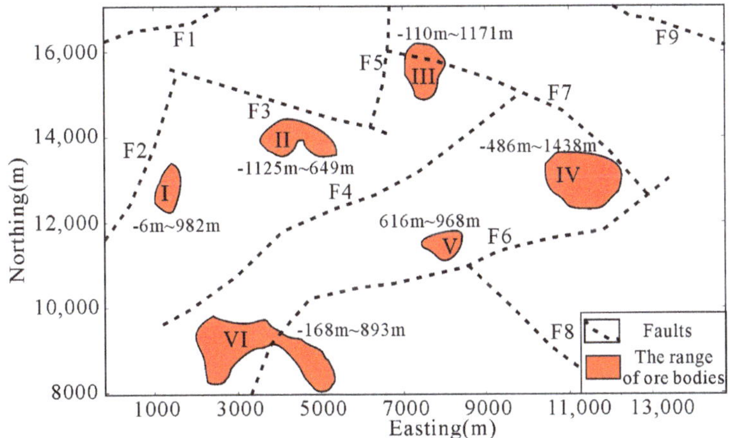

Figure 9. The range of ore bodies interpreted from 3D inversion results.

5. Conclusions

We proposed the preconditioned JFNK method to perform the intensive inversion computation; therefore, the 3D focusing inversion algorithm became more efficient. In addition, we combined unstructured and structured meshes to achieve gravity inversion with undulating terrain. Therefore, the computational complexity is less than that of the unstructured mesh method in the forward operation of the kernel function. In order to maintain the efficiency, we improved the terrain fitting and fineness of meshes by the adaptive equivalent-dimensional method. The fineness of meshes is positively correlated with the gradient of gravity and terrain. Synthetic tests showed that our method was suitable for gravity inversion under undulating terrain, and the algorithm was fast with accurate inversion results. Finally, we applied it to the field data processing in Huayangchuan, Shaanxi Province, and the distribution range and depth of the ore bodies were inferred from recovered density models, which further verified the stability and practicability of the new method.

Author Contributions: Conceptualization, Q.M. and G.M.; data curation, T.W.; formal analysis, G.M.; funding acquisition, G.M.; investigation, Q.M.; methodology, G.M.; project administration, S.X.; resources, S.X.; software, Q.M.; supervision, S.X.; validation, T.W.; visualization, Q.M.; writing—original draft, Q.M.; writing—review and editing, G.M. and T.W. All authors have read and agreed to the published version of the manuscript.

Funding: This research was funded by the national key research and development plan issue (Grant No.2017YFC0602203), the Excellent Young Talents Fund project of Jilin Province (20190103011JH), National Science and Technology Major Project task (No. 2016ZX05027-002-003), the Ministry of Education, and the China Postdoctoral Science Foundation (2019M651209).

Conflicts of Interest: The authors declare no conflict of interest.

References

1. Mosher, C.R.W.; Farquharson, C.G. Minimum-structure borehole gravity inversion for mineral exploration: A synthetic modeling study. *Geophysics* **2013**, *78*, 25–39. [CrossRef]
2. Martinez, C.; Li, Y.; Krahenbuhl, R. 3D inversion of airborne gravity gradiometry data in mineral exploration: A case study in the Quadrilátero Ferrífero, Brazil. *Geophysics* **2013**, *78*, 1–11. [CrossRef]
3. Zhdanov, M.S.; Gribenko, A.V. 3D joint inversion of geophysical data with Gramian constraints: A case study from the Carrapateena IOCG deposit, South Australia. *Lead. Edge* **2012**, *31*, 1382–1388. [CrossRef]
4. Zhdanov, M.S.; Ellis, R.; Mukherjee, S. Three-dimensional regularized focusing inversion of gravity gradient tensor component data. *Geophysics* **2004**, *69*, 925–937. [CrossRef]
5. Last, B.J.; Kubik, K. Compact gravity inversion. *Geophysics* **1983**, *48*, 713–721. [CrossRef]
6. Zhdanov, M.S. *Geophysical Inverse Theory and Regularization Problems*; Elsevier: Amsterdam, The Netherlands; New York, NY, USA; Tokyo, Japan, 2002; p. 628.
7. Chen, Z.; Meng, X.; Guo, L. GICUDA: A parallel program for 3D correlation imaging of large scale gravity and gravity gradiometry data on graphics processing units with CUDA. *Comp. Geosci.* **2012**, *46*, 119–128. [CrossRef]
8. Čuma, M.; Zhdanov, M.S. Massively parallel regularized 3D inversion of potential fields on CPUs and GPUs. *Comp. Geosci.* **2014**, *62*, 80–87. [CrossRef]
9. Bin, C.; Ren, J.; Sheng, C. A Review of JFNK methods and its applications in atmospheric non-hydrostatical model. *Chin. J. Atmos. Sci.* **2006**, *5*, 95–107.
10. Pletnyov, F.; Jeje, A.A. The Gauss-Siedel (GS) and the Jacobian-free Newton-Krylov (JFNK) methods applied to steady 2D buoyancy convection in vertical cylinders. *Contemp. Eng. Sci.* **2019**, *12*, 187–228. [CrossRef]
11. Vandecar, J.C.; Snieder, R. Obtaining smooth solutions to large, linear, inverse problems. *Geophysics* **1994**, *59*, 818–829. [CrossRef]
12. Du, V. A new method of reducing from an arbitrary surface into a plane for the three dimensional magnetic and gravity field. *Chin. J. Geophys.* **1982**, *25*, 73–83.
13. Cheng, Z. The finite element method for reduce curved surface into a horizontal plane of gravity and magnetic fields. *Geophys. Geochem. Explor.* **1981**, *3*, 153–158.
14. Ma, Q.Z. The boundary element method for 3D dc resistivity modeling in layered earth. *Geophysics* **2002**, *67*, 610–617. [CrossRef]

15. Liu, J.; Wang, W.; Yu, C. Reduction of potential field data to a horizontal plane by a successive approximation procedure. *Chin. J. Geophys.* **2007**, *50*, 1551–1557.

16. Xu, S. The integral iteration method for continuation of potential fields. *Chin. J. Geophys.* **2006**, *49*, 1176–1182. [CrossRef]

17. Yao, C.; Huang, W.; Guan, Z. Fast splines conversion of curved-surface potential field and vertical gradient data into horizontal-plane data. *Oil Geophys. Prospect.* **1997**, *32*, 229–236.

18. Liang, J. Integrate use of potential field and its gradient for the continuation from a curved surface to a horizontal plane. *Prog. Geophys.* **2017**, *32*, 994–999.

19. Guan, Z. *Geomagnetic and Magnetic Prospecting*, 1st ed.; Geological Publishing House: Beijing, China, 2005; pp. 164–177.

20. Zhong, B.; He, C.; Jiang, Y. Polyhedral analytic method for square domain vicinity area terrain correction. *Geophys. Geochem. Explor.* **1989**, *13*, 127–135.

21. Zhang, T.; Lei, W.; Ding, S. Application of unstructured mesh generation technology in DC resistivity inversion imaging under undulating terrain. *Site Investig. Sci. Technol.* **2019**, *2*, 61–64.

22. Wu, X.; Liu, Y.; Wang, W. 3D resistivity inversion incorporating topography based on unstructured meshes. *Chin. J. Geophys.* **2015**, *58*, 2706–2717.

23. Zhang, L.; Hao, T.Y. 2-D Irregular gravity modeling and computation of gravity based on delaunay triangulation. *Chin. J. Geophys.* **2006**, *49*, 768–775. [CrossRef]

24. Zhang, Q. Research on Gravity Forward and Inversion Based on Delaunay Triangulation. Master's Thesis, Chengdu University of Technology, Chengdu, China, 2016.

25. Holstein, H.; Schürholz, P.; Starr, A.J. Comparison of gravimetric formulas for uniform polyhedra. *Geophysics* **1999**, *64*, 1438–1446. [CrossRef]

26. Pilkington, M. 3-D magnetic imaging using conjugate gradients. *Geophysics* **1997**, *62*, 1132–1142. [CrossRef]

27. Li, Y.; Oldenburg, D.W. 3-D inversion of gravity data. *Geophysics* **1998**, *63*, 109–119. [CrossRef]

28. Hui, X.; Li, Z.; Feng, Z. Research on the occurrence state of U in the Huayangchuan U-polymetallic deposit, Shanxi Province. *Acta Mieralogica Sin.* **2014**, *34*, 573–580.

29. Guo, W.; Zhou, D.; Ren, J.; Zhou, X.; Sang, H. Characteristics of the Huayangchuan ductile shear zones in the Xiao-qinling mountains, Shaanxi, China, and its regional tectonic significance. *Geol. Bull. China* **2008**, *27*, 823–828.

30. Cordell, L.; Grauch, V.J.S. Mapping basement magnetization zones from aeromagnetic data in the San Juan Basin New Mexico. *Seg Tech. Program Expand. Abstr.* **1982**, *1982*, 246–247.

31. Pawlowski, R.S. Preferential continuation for potential-field anomaly enhancement. *Geophysics* **1995**, *60*, 390–398. [CrossRef]

32. Naidu, P. Spectrum of the potential field due to randomly distributed sources. *Geophysics* **1968**, *33*, 337–345. [CrossRef]

© 2020 by the authors. Licensee MDPI, Basel, Switzerland. This article is an open access article distributed under the terms and conditions of the Creative Commons Attribution (CC BY) license (http://creativecommons.org/licenses/by/4.0/).

Article

Recovering Magnetization of Rock Formations by Jointly Inverting Airborne Gravity Gradiometry and Total Magnetic Intensity Data

Michael Jorgensen [1,2] and Michael S. Zhdanov [1,2,*]

[1] TechnoImaging, LLC, Salt Lake City, UT 84107, USA
[2] Consortium for Electromagnetic Modeling and Inversion (CEMI), University of Utah,
     Salt Lake City, UT 84112, USA; mike.jorgensen@utah.edu
*   Correspondence: michael.zhdanov@utah.edu

Citation: Jorgensen, M.; Zhdanov, M.S. Recovering Magnetization of Rock Formations by Jointly Inverting Airborne Gravity Gradiometry and Total Magnetic Intensity Data. *Minerals* **2021**, *11*, 366. https://doi.org/10.3390/min11040366

Academic Editors: Stanisław Mazur and Amin Beiranvand Pour

Received: 5 March 2021
Accepted: 29 March 2021
Published: 31 March 2021

Publisher's Note: MDPI stays neutral with regard to jurisdictional claims in published maps and institutional affiliations.

Copyright: © 2021 by the authors. Licensee MDPI, Basel, Switzerland. This article is an open access article distributed under the terms and conditions of the Creative Commons Attribution (CC BY) license (https://creativecommons.org/licenses/by/4.0/).

Abstract: Conventional 3D magnetic inversion methods are based on the assumption that there is no remanent magnetization, and the inversion is run for magnetic susceptibility only. This approach is well-suited to targeting mineralization; however, it ignores the situation where the direction of magnetization of the rock formations is different from the direction of the induced magnetic field. We present a novel method of recovering a spatial distribution of magnetization vector within the rock formation based on joint inversion of airborne gravity gradiometry (AGG) and total magnetic intensity (TMI) data for a shared earth model. Increasing the number of inversion parameters (the scalar components of magnetization vector) results in a higher degree of non-uniqueness of the inverse problem. This increase of non-uniqueness rate can be remedied by joint inversion based on (1) Gramian constraints or (2) joint focusing stabilizers. The Gramian constraints enforce shared earth structure through a correlation of the model gradients. The joint focusing stabilizers also enforce the structural similarity and are implemented using minimum support or minimum gradient support approaches. Both novel approaches are applied to the interpretation of the airborne data collected over the Thunderbird V-Ti-Fe deposit in Ontario, Canada. By combining the complementary AGG and TMI data, we generate jointly inverted shared earth models that provide a congruent image of the rock formations hosting the mineral deposit.

Keywords: inversion; gravity; magnetics; magnetization vector; remanent magnetization; three-dimensional; joint inversion

1. Introduction

In mineral exploration, magnetic data have traditionally been inverted to produce magnetic susceptibility models, which represent magnetization induced by the current magnetic field. This does not take into account, however, the remanent magnetization of the rocks produced by the ancient magnetic field. More information about rock formations and geological processes can be obtained by inverting magnetic data for magnetization vector, as opposed to magnetic susceptibility only. At the same time, increasing the number of inversion parameters (i.e., inverting for the scalar components of magnetization vector) increases the non-uniqueness of the inverse problem. We overcome this problem by jointly inverting multiple geophysical data sets. The total magnetic intensity (TMI) and airborne gravity gradiometry (AGG) data are typically gathered in coincident surveys, making them a natural choice for joint interpretation.

Over the last decade, significant efforts were made by many researchers to develop the methods of magnetic data inversion in the presence of remanent magnetization. To include both induced and remanent magnetization, one needs to model the distribution of magnetization vector within the rock formation rather than the distribution of scalar susceptibility alone. This approach was used by [1,2] to invert TMI data. This enables explicit inversion of the magnetization direction and amplitude, rather than just the magnetization amplitude only (e.g., [3]). Ellis et al. [4] reported further progress in the solution of this problem; they introduced a technique for regularized inversion for the magnetization vector. From the magnetization vector, one can recover information about both the remanent and induced magnetization. A detailed overview of the published papers can be found in the comprehensive review paper by Y. Li [5], where one can find a long list of publications on the subject. We reference the interested readers to this publication for more information about the recent advances in 3D inversion of magnetic data in the presence of significant remanent magnetization

The results presented in the cited papers, however, illustrate the practical difficulties of the inversion related to the fact that, in this case, one has to determine from the observed scalar TMI data three unknown components of the magnetization vector within each cell instead of one unknown value of susceptibility. The main problem is with the increased practical non-uniqueness associated with the inverse problem.

In order to reduce the non-uniqueness of the inverse problem, we use a method of direct inversion of the magnetic data for the magnetization vector based on applying a Gramian stabilizer in the framework of the regularized inversion [6]. To this end, we also present a method of joint inversion of AGG and TMI data in the presence of remanent magnetization. We consider two different approaches to the joint inversion. The first approach is joint inversion with Gramian constraints [7], where the structural correlation of the different model gradients is enforced. The Gramian constraint is analogous to the cross-gradients method [8]; however, whereas the cross-gradients approach typically relies on an approximation, the Gramian constraint utilizes an exact analytical formula for the gradient direction of the parametric functional, which guarantees rapid convergence of the inverse solution [9].

The second approach presented is joint inversion with a joint focusing stabilizer [10,11], which enforces null-space sparsity in both models via a modified minimum support constraint. It is well known that the focusing stabilizers minimize the areas with anomalous physical properties (in the case of the minimum support stabilizer) or the areas where major changes in physical properties occurs (in the case of the minimum gradient support stabilizer). The joint focusing stabilizers force the anomalies of different physical properties to either overlap or experience a rapid change in the same areas, thus enforcing the structural correlation.

To illustrate both of these approaches, we present a case study of the joint inversion of the AGG and TMI data, which were collected over the Thunderbird V-Ti-Fe deposit in the Ring of Fire area of Ontario, Canada. The Ontario Geological Survey (OGS) and the Geological Survey of Canada (GSC) collaboratively gathered the airborne data with the Fugro Airborne Surveys gravity gradiometer and magnetic system between 2010 and 2011. Appropriate filters are applied to the data, i.e., reduction-to-pole (RTP) and deep regional trend removal. The filtered RTP data were used to invert the susceptibility model. The original TMI data with the removed regional trend were inverted for the magnetization vector. Standalone inverse models are produced using a single datum, i.e., AGG or TMI only, yielding standalone models to compare and contrast to the jointly inverted models and to produce appropriate model weights used in the joint inversions.

We compare and contrast the 3D density, magnetic susceptibility, and magnetization vector models of the Thunderbird V-Ti-Fe deposit obtained from standalone, Gramian, and joint-focused inversions. Application of both joint inversion methodologies to these data has resulted in 3D models of subsurface formations that have sharper geospatial boundaries and stronger structural correlations than the standalone inverted models. The Gramian constraint approach yielded the highest level of numerical structural correlation. The joint focusing approach provided similar results but was significantly faster and easier to implement.

2. Forward Modeling of the Gravity and Magnetic Data

2.1. Gravity Forward Modeling

The gravity field can be computed from the gradient of the gravity potential U:

$$g(r) = \nabla U(r) = \gamma \int\int\int_D \rho(r') \frac{r' - r}{|r' - r|^3} dv, \tag{1}$$

where γ is the universal gravity constant; r and r' are the observation and integration points, respectively; ρ is the density; and the gravity potential, U, has the following form:

$$U(r) = \gamma \int\int\int_D \frac{\rho(r')}{|r' - r|} dv. \tag{2}$$

The second spatial derivatives of the gravity potential U form a symmetric gravity gradient tensor,

$$\hat{g} = \begin{bmatrix} g_{xx} & g_{xy} & g_{xz} \\ g_{yx} & g_{yy} & g_{yz} \\ g_{zx} & g_{zy} & g_{zz} \end{bmatrix}, \tag{3}$$

where

$$g_{\alpha\beta}(r) = \frac{\partial^2}{\partial\alpha\partial\beta} U(r); \alpha, \beta = x, y, z. \tag{4}$$

The expressions for the gravity tensor components can be written as follows:

$$g_{\alpha\beta}(r) = \gamma \int\int\int_D \frac{\rho(r')}{|r' - r|^3} K_{\alpha\beta}(r' - r) dv, \tag{5}$$

where kernels $K_{\alpha\beta}$ are equal to:

$$K_{\alpha\beta}(r' - r) = \begin{cases} \frac{3(\alpha - \alpha')(\beta - \beta')}{|r' - r|^2}, \alpha \neq \beta \\ \frac{3(\alpha - \alpha')^2}{|r' - r|^2} - 1, \alpha = \beta \end{cases} ; \alpha, \beta = x, y, z. \tag{6}$$

One can use the point-mass approximation to calculate Formula (5) in discretized form by considering each cell as a point mass [12].

2.2. Magnetic Forward Modeling and Magnetization Vector

The earth's current magnetic field is the vector sum of contributions from two main sources: the background magnetic field due to the dynamo in the earth's liquid core, and the crustal field due to magnetic minerals. Conventional magnetic inversion methods produce a 3D magnetic susceptibility model from the magnetic vector field, **H**, or from the total magnetic intensity (TMI) data, T. This is an effective tool for targeting magnetic mineralization; however, the following assumptions must generally be made: (1) there is no remanent magnetization, (2) self-demagnetization effects are negligible, and (3) the magnetic susceptibility is isotropic (e.g., [13–16]).

Under these assumptions, the intensity of magnetization, $\mathbf{I}(r)$, is linearly proportional to the inducing magnetic field, $\mathbf{H}_0(r)$,

$$\mathbf{I}(r) = \chi(r)\mathbf{H}_0(r) = \chi(r)H_0\mathbf{l}(r), \tag{7}$$

where $\chi(r)$ is the magnetic susceptibility and $\mathbf{l}(r) = (l_x, l_y, l_z)$ is the unit vector along the direction of the inducing field. Given the inclination (I), declination (D), and azimuth (A) from the International Geomagnetic Reference Field (IGRF), the direction of the inducing magnetic field can be computed as follows (assuming that the x axis is directed eastward, the y axis has a positive direction northward, and the z axis is downward):

$$\begin{aligned} l_x &= \cos(I)\sin(D-A), \\ l_y &= \cos(I)\cos(D-A), \\ l_z &= \sin(I). \end{aligned} \tag{8}$$

Thus, the anomalous magnetic field can be presented in the following form:

$$\mathbf{H}(r) = -H_0 \int\int\int_D \frac{\chi(r')}{|r'-r|^3}\left[1 - \frac{3(1\cdot(r'-r))(r'-r)}{|r'-r|^2}\right]dv. \tag{9}$$

In airborne magnetic surveys, the total magnetic intensity (TMI) field is measured, which can be computed approximately as follows:

$$T(r) \approx 1\cdot\mathbf{H}(r) = -H_0 \int\int\int_D \frac{\chi(r')}{|r'-r|^3}\left[1 - \frac{3(1\cdot(r'-r))^2}{|r'-r|^2}\right]dv. \tag{10}$$

In the general case, however, the earth's magnetic field is nonstationary over geological time, meaning that the direction of magnetization in rock formations may differ from the direction of today's magnetic field. This is due to several factors, i.e., geomagnetic reversals and wandering poles. The ancient magnetization can be particularly useful when exploring magnetic structures such as kimberlites, dykes, iron-rich ultramafic pegmatitoids (IRUP), platinum group element (PGE) reefs, and banded iron formations (BIF). Moreover, remanence can also be used to determine the age of intrusive or alteration events.

By inverting for magnetization vector, one can take into account the effects of self-demagnetization, anisotropy, and remanent magnetization. To include both induced and remanent magnetization, we need to model on the magnetization vector rather than the scalar susceptibility. This modifies Equation (7) as follows:

$$\mathbf{I}(r) = H_0\mathbf{M}(r), \tag{11}$$

where $\mathbf{M}$ has two parts: induced, $\mathbf{M}_{ind}$, and remnant, $\mathbf{M}_{rem}$, magnetizations:

$$\mathbf{M}(r) = \mathbf{M}_{ind}(r) + \mathbf{M}_{rem}(r). \tag{12}$$

Therefore, Equation (10) should also be modified:

$$T(r) \approx 1(r)\cdot\mathbf{H}(r) =$$
$$-H_0 1(r)\cdot\int\int\int_V \frac{1}{|r'-r|^3}\left[\mathbf{M} - \frac{3(M_k\cdot(r'-r))(r'-r)}{|r'-r|^2}\right]dv, \tag{13}$$

One can compute the volume integral in Equation (13) in closed form for magnetic susceptibility. We can also evaluate the volume integral numerically with sufficient accuracy using single-point Gaussian integration with pulse basis functions provided the depth to the center of the cell exceeds twice the dimension of the cell [12].

3. Inversion for Magnetization Vector Using Gramian Constraints

In this section, we present a summary of the principles of the robust inversion for the magnetization vector using Gramian constraints.

It is well-known that the regularized solution of the geophysical inverse problem can be formulated as minimization of the Tikhonov parametric functional,

$$P^\alpha(m) = \varphi(m) + \alpha S_{MN,MS,MGS}(m) \rightarrow \min, \tag{14}$$

where $m = \{M_x, M_y, M_z\}$ is the magnetization vector and $\varphi(m)$ is a misfit functional defined as the squared L_2 norm of the difference between the predicted, $A(m)$, and observed, d, data,

$$\varphi(m) = ||A(m) - d||_{L_2}^2. \tag{15}$$

In the last formula, A is the forward modeling operator for the magnetic problem described by a discrete form of Equation (13).

Notations $S_{MN,MS,MGS}$ in Equation (14) indicate in compact form that we can use any one of three types of stabilizing functionals, S_{MN}, S_{MS}, or S_{MGS}, based on minimum norm, minimum support, and minimum gradient support constraints, respectively. These stabilizers are determined as follows [17]:

$$S_{MN}(m) = ||m - m_{apr}||_{L_2}^2 = \int_V (m - m_{apr})^2 dv, \tag{16}$$

$$S_{MS}(m) = \int_V \frac{(m - m_{apr})^2}{(m - m_{apr})^2 + e^2} dv, \tag{17}$$

and

$$S_{MGS}(m) = \int_V \frac{\nabla m \cdot \nabla m}{\nabla m \cdot \nabla m + e^2} dv, \tag{18}$$

where e is a focusing parameter, which can be selected using an L-curve method [9].

The minimum norm stabilizer, S_{MN}, minimizes the variations of the solutions, m, from the a priori model, m_{apr}. The minimum support stabilizer, S_{MS}, minimizes the volume occupied by the anomalous magnetization, while the minimum gradient support stabilizer, S_{MGS}, selects the inverse models with sharp boundaries between the formations with different magnetic properties. Thus, by selecting the proper stabilizing functionals, the user may emphasize different properties of the inverse models. The minimum norm stabilizer usually results in relatively smooth distributions of magnetization, while the focusing stabilizers, S_{MS} and S_{MGS}, generate models with sharp boundaries. We use regularized inversion with focusing stabilization, as this recovers models with sharper boundaries and higher contrasts than the regularized inversion with smooth (e.g., minimum norm) stabilization.

The minimization problem (14) can be solved using a variety of optimization methods. For improved convergence and to avoid any matrix inversions, we minimize Equation (14) using the re-weighted regularized conjugate gradient (RRCG) method. We refer the reader to [17] for further details of this method.

Inverting for the magnetization vector is a more challenging problem than inverting for scalar magnetic susceptibility, because we have three unknown scalar components of the magnetization vector for every cell. At the same time, in the case of induced magnetization, there is an inherent correlation between the different components of the magnetization vector. In order to ensure a smooth change in the direction of the magnetization vector, we impose a similar condition by requiring that the different components of the magnetization vector should be mutually correlated as well. The results of numerical model experiments and case studies show that this approach to regularization ensures a robust inversion for magnetization vector.

It was demonstrated in [7] that one can enforce the correlation between the different model parameters by using the Gramian constraints. Following the cited paper, we have included the Gramian constraint in Equation (14) as follows:

$$P^{\alpha}(m) = \varphi(m) + \lambda c_1 S_{MN,MS,MGS}(m) + \lambda c_2 \sum_{\alpha = x,y,z} S_G\left(m_{\alpha}, \chi_{eff}\right), \tag{19}$$

where m is the $3N_m$ length vector of magnetization vector components; m_α is the N_m length vector of the α component of magnetization vector, $\alpha = x, y, z$; χ_{eff} is the N_m-length vector of the effective magnetic susceptibility, defined as the magnitude of the magnetization vector,

$$\chi_{eff} = \sqrt{M_x^2 + M_y^2 + M_z^2}. \tag{20}$$

Functional S_G is the Gramian constraint,

$$S_G\left(m_\alpha, \chi_{eff}\right) = \begin{vmatrix} (m_\alpha, m_\alpha) & \left(m_\alpha, \chi_{eff}\right) \\ \left(\chi_{eff}, m_\alpha\right) & \left(\chi_{eff}, \chi_{eff}\right) \end{vmatrix}, \tag{21}$$

where $(*, *)_{L_2}$ denotes the L_2 inner product operation [9].

Using the Gramian constraint (21), we enhance a direct correlation between the scalar components of the magnetization vector with χ_{eff}, which is computed at the previous iteration of an inversion and is updated on every iteration. The advantage of using the Gramian constraint in the form of Equation (21) is that it does not require any a priori information about the magnetization vector (e.g., direction and the relationship between different components) since the amplitude, χ_{eff} is computed at the previous iteration. On the first iteration, the scalar components are determined independently

The minimization problem (19) is solved using the re-weighted regularized conjugate gradient (RRCG) method. Details of the RRCG method and conjugate gradient derivations for the parametric functional (19) can be found in [9,17].

4. Joint Inversion of Gravity and Magnetic Data

4.1. Gramian Joint Inversion

In a case of joint inversion of gravity and magnetic data, the geophysical inverse problem is given by the following set of operator Equations:

$$d^{(i)} = A^{(i)}\left(m^{(i)}\right), (i = 1, 2), \tag{22}$$

where $m^{(1)} = \rho$ is the density model, $m^{(2)} = \{M_x, M_y, M_z\}$ is the magnetization vector model, $A^{(i)}$ are the forward modeling operators, $d^{(i)}$ are the data, and the superscript $(i = 1, 2)$ indicates the gravity and magnetic problems, respectively. Operators $A^{(i)}$ are usually represented as discrete forms of the corresponding integral expressions (5) and (13) for gravity gradient and TMI fields, respectively.

We should note that the units and scales of the density and magnetization vector are very different. Therefore, in practical applications, we should operate with the normalized, dimensionless model parameters [9]:

$$\widetilde{m^{(i)}} = W_m^{(i)} m^{(i)}, i = 1, 2; \tag{23}$$

where $W_m^{(i)}, i = 1, 2$, are the corresponding linear operators of the model weighting. A trivial example of these operators could be just division of the model parameters by their upper or lower bounds, $m_{max}^{(i)}$ or $m_{min}^{(i)}$, or by the corresponding interval of the model parameters distribution, $\left(m_{max}^{(i)} - m_{min}^{(i)}\right)$. This simple normalization makes the normalized parameters vary on the same scale. In a general case, the model weights can be defined based on integrated sensitivities, which will be discussed in details below.

Similarly, different data sets, as a rule, have different physical dimensions as well. Therefore, it is convenient to consider dimensionless weighted data, $\widetilde{d^{(i)}}$, defined as follows:

$$\widetilde{d^{(i)}} = W_d^{(i)} d^{(i)}, \tag{24}$$

where $W_d^{(i)}, i = 1, 2$, are the corresponding linear operators of the data weighting.

The solutions of geophysical inverse problems (22) are typically ill-posed [9]. A corresponding well-posed problem can be constructed by applying regularization and minimizing a parametric functional using the conjugate gradient method. In this case, misfit and stabilizing terms, each corresponding to the gravity gradiometry and magnetic data, are incorporated into a joint parametric functional, which is subject to the Gramian structural constraint:

$$P\left(\widetilde{m^{(1)}}, \widetilde{m^{(2)}}\right) = \sum_{i=1}^{2} \varphi\left(\widetilde{m^{(i)}}\right) + \alpha \sum_{i=1}^{2} S_{MN,MS,MGS}\left(\widetilde{m^{(i)}}\right) + \beta G\left(\nabla \widetilde{m^{(1)}}, \nabla \widetilde{m^{(2)}}\right), \quad (25)$$

where α and β are the regularization parameters, responsible for degrees of smoothing/focusing (α) of the models and enforcing similarities (β) between the density and magnetization distributions, produced by a joint inversion. We use an adaptive algorithm of selecting these parameters, described in [9].

The misfit terms are defined as follows,

$$\varphi\left(\widetilde{m^{(i)}}\right) = \left\|\widetilde{A^{(i)}}\left(\widetilde{m^{(i)}}\right) - \widetilde{d^{(i)}}\right\|_{L_2}^{2}, i = 1, 2; \quad (26)$$

where $\widetilde{A^{(i)}}\left(\widetilde{m^{(i)}}\right)$ are the weighted predicted data:

$$\widetilde{A^{(i)}}\left(\widetilde{m^{(i)}}\right) = W_d^{(i)} A^{(i)}\left(\widetilde{m^{(i)}}\right). \quad (27)$$

The stabilizing terms, S_{MN}, S_{MS}, S_{MGS}, are based on minimum norm, minimum support, and minimum gradient support constraints, respectively, as defined by Formulas (16)–(18).

The Gramian term is defined by the following formula,

$$G\left(\nabla \widetilde{m^{(1)}}, \nabla \widetilde{m^{(2)}}\right) = G\left(\nabla \widetilde{m^{(1)}}, \nabla \widetilde{M_\gamma}\right) = \sum_{\gamma=x,y,z} \begin{vmatrix} \left(\nabla \widetilde{m^{(1)}}, \nabla \widetilde{m^{(1)}}\right) & \left(\nabla \widetilde{m^{(1)}}, \nabla \widetilde{M_\gamma}\right) \\ \left(\nabla \widetilde{M_\gamma}, \nabla \widetilde{m^{(1)}}\right) & \left(\nabla \widetilde{M_\gamma}, \nabla \widetilde{M_\gamma}\right) \end{vmatrix}, \quad (28)$$

where $\nabla m^{(1)}$ is the gradient of the density model, ∇M_γ are the gradients of the scalar components of magnetization vector, and $(*, *)_{L_2}$ denotes the L_2 inner product operation [9]. Minimization of the Gramian aligns the model gradients, which in turn enforces the structural similarity of the shared earth model.

In a case of structural Gramian constraints defined by Formula (28), one can normalize the gradient vectors by their length:

$$\widetilde{\nabla m^{(1)}} = \frac{\nabla m^{(1)}}{|\nabla m^{(1)}|}, \quad \widetilde{\nabla M_\gamma} = \frac{\nabla M_\gamma}{|\nabla M_\gamma|}, \gamma = x, y, z; \quad (29)$$

with expression for Gramian (28) taking the form:

$$G\left(\widetilde{\nabla m^{(1)}}, \widetilde{\nabla M_\gamma}\right) = \sum_{\gamma=x,y,z} \begin{vmatrix} \left(\frac{\nabla m^{(1)}}{|\nabla m^{(1)}|}, \frac{\nabla m^{(1)}}{|\nabla m^{(1)}|}\right) & \left(\frac{\nabla m^{(1)}}{|\nabla m^{(1)}|}, \frac{\nabla M_\gamma}{|\nabla M_\gamma|}\right) \\ \left(\frac{\nabla M_\gamma}{|\nabla M_\gamma|}, \frac{\nabla m^{(1)}}{|\nabla m^{(1)}|}\right) & \left(\frac{\nabla M_\gamma}{|\nabla M_\gamma|}, \frac{\nabla M_\gamma}{|\nabla M_\gamma|}\right) \end{vmatrix}. \quad (30)$$

This normalization enforces the correlations of the unit vectors in the gradient directions, thus resulting in the structural similarity between the models representing density and magnetizations of the rock formations.

4.2. Joint Focusing Inversion

Another approach to joint inversion can be based on a joint total variation functional [18], or on joint focusing stabilizers, e.g., minimum support and minimum gradient support constraints [10,11]. It is well known that the focusing stabilizers minimize the areas with anomalous physical properties (in the case of the minimum support stabilizer) or the areas where major changes in physical properties occur (in the case of the minimum gradient support stabilizer). The joint focusing stabilizers force the anomalies of different physical properties to either overlap or experience a rapid change in the same areas, thus enforcing the structural correlation. We will demonstrate these properties in the model study presented in our paper.

The AGG and TMI data misfit terms are incorporated in the joint parametric functional and subject to the joint focusing stabilizers:

$$P\left(\widetilde{m^{(1)}}, \widetilde{m^{(2)}}\right) = \sum_{i=1}^{2} \varphi\left(\widetilde{m^{(i)}}\right) + \alpha S_{JMS,JMGS}\left(\widetilde{m^{(1)}}, \widetilde{m^{(2)}}\right). \tag{31}$$

The data misfit functionals are defined as above in Equation (26). The terms S_{JMS}, and S_{JMGS} are the joint stabilizing functionals, based on minimum support and minimum gradient support constraints, respectively [10,11].

For example, a joint minimum support stabilizer can be introduced as follows:

$$S_{JMS} = \int_{V} \frac{\sum_{i=1}^{2}\left(\widetilde{m^{(i)}} - \widetilde{m_{apr}^{(i)}}\right)^2}{\sum_{i=1}^{2}\left(\widetilde{m^{(i)}} - \widetilde{m_{apr}^{(i)}}\right)^2 + e^2} dv, \tag{32}$$

where e is the focusing parameter, which can be selected using an L-curve method [19]. In a similar way, we can introduce a joint minimum gradient support functional (JMGS):

$$S_{JMGS} = \int_{V} \frac{\sum_{i=1}^{2}\left(\nabla\widetilde{m^{(i)}} \cdot \nabla\widetilde{m^{(i)}}\right)}{\sum_{i=1}^{2}\left(\nabla\widetilde{m^{(i)}} \cdot \nabla\widetilde{m^{(i)}}\right) + e^2} dv. \tag{33}$$

The joint focusing constraint is not applied until the data misfit functionals have been sufficiently minimized (e.g., $\chi^2 = 2$). This avoids the introduction of inversion artifacts arising from the focusing constraint.

4.3. Representation of a Stabilizing Functional in the Form of a Pseudo-Quadratic Functional

In this and the following sections, for simplicity, we will drop the tilde sign above the dimensionless weighted data and model parameters, assuming that they have already been normalized according to Formulas (23) and (24).

The focusing stabilizing functionals introduced above can be represented as pseudo-quadratic functionals of the model parameters [9,17]:

$$\begin{aligned} S(m) &= \left(W_e\left(m - m_{apr}\right), W_e\left(m - m_{apr}\right)\right)_{L_2} \\ &= \int_{V}\left|W_e(r)\left(m(r) - m_{apr}(r)\right)\right|^2 dv, \end{aligned} \tag{34}$$

where W_e is a linear operator of the product of the model parameters function, $m(r)$, and the weighting function, $w_e(r)$, which may depend on m. If operator W_e does not depend on $m(r)$, we obtain a quadratic functional, i.e., a minimum norm-stabilizing functional. In general cases, w_e may even be a nonlinear function of m, like the minimum support (32) or minimum gradient support (33) functionals. In these cases, the functional $S(m)$, determined by Formula (34), is not quadratic. That is why we call it a "pseudo-quadratic" functional. However, it was shown in [9,17] that presenting a stabilizing functional in a pseudo-quadratic form enables the development of a unified approach to regularization with different stabilizers and simplifies the solution of the regularization problem.

For example, the joint minimum support functional, $s_{JMS}(m)$, can be written as follows:

$$
\begin{aligned}
S_{JMS} &= \sum_{i=1}^{2} \int_V \frac{\left(m^{(i)} - m_{apr}^{(i)}\right)^2}{\sum_{j=1}^{2}\left(m^{(j)} - m_{apr}^{(j)}\right)^2 + e^2} dv \\
&= \sum_{i=1}^{2}\left(W_e\left(m^{(i)} - m_{apr}^{(i)}\right), W_e\left(m^{(i)} - m_{apr}^{(i)}\right)\right)_{L_2},
\end{aligned}
\tag{35}
$$

where W_e is a linear operator of the multiplication of the model parameter function, $m(r)$, by the following weighting function, $w_e(r)$:

$$
w_e(r) = w_e^{JMS}(r) = \frac{1}{\left[\sum_{j=1}^{2}\left(m^{(j)}(r) - m_{apr}^{(j)}(r)\right)^2 + e^2\right]^{1/2}}.
\tag{36}
$$

For the joint minimum gradient support functional, $s_{JMGS}(m)$, we assume $m_{apr} = 0$, and find

$$
\begin{aligned}
S_{JMGS} &= \sum_{i=1}^{2} \int_V \frac{\left(\nabla m^{(i)} \cdot \nabla m^{(i)}\right)}{N\left(\nabla m^{(j)} \cdot \nabla m^{(j)}\right) + e^2} dv \approx \sum_{i=1}^{2} \int_V \left| w_e^{(i)}(r) m^{(i)}(r) \right|^2 dv \\
&= \sum_{i=1}^{N}\left(W_e^{(i)} m^{(i)}, W_e^{(i)} m^{(i)}\right)_{L_2},
\end{aligned}
\tag{37}
$$

where $W_e^{(i)}$ is a linear operator of the multiplication of the model parameter function, $m(r)$, by the following weighting function, $w_e^{(i)}(r)$:

$$
w_e^{(i)}(r) = w_e^{(i)JMGS}(r) = \frac{\nabla m^{(i)}(r)}{\left(\sum_{j=1}^{2}\left(\nabla m^{(j)} \cdot \nabla m^{(j)}\right) + e^2\right)^{1/2}\left[m^{(i)2} + \varepsilon^2\right]^{1/2}},
\tag{38}
$$

and ε is a small number similar to a focusing parameter e.

Using the pseudo-quadratic form (37) of stabilizing functional, we can present the corresponding parametric functional (31) as follows:

$$
P^\alpha\left(m^{(1)}, m^{(2)}\right) = \sum_{i=1}^{2} \varphi\left(m^{(i)}\right) + \sum_{i=1}^{2} \alpha^{(i)}\left(W_e^{(i)} m^{(i)}, W_e^{(i)} m^{(i)}\right)_{L_2}.
\tag{39}
$$

Therefore, the parametric functional introduced by Equation (31) can be minimized in the same manner as a conventional Tikhonov parametric functional. The only difference is the introduction of some variable weighting operator, $W_e^{(i)}$, which depends on the model parameters. A practical technique of minimizing the parametric functional (39) is based on application of the re-weighted regularized conjugate gradient (RRCG) method. Interested readers can find a detailed description of the RRCG algorithm in [9].

4.4. Data and Model Weights

Data and model weights are critical in both joint inversion scenarios presented, because they ensure that both inverse problems (gravity and magnetic) converge towards a satisfactory solution at roughly the same rate. Appropriate data and model weights also ensure that the models do not unduly bias one another where no structural corellation exists. As such, in both joint inversion scenarios, two levels of data and model weighting are employed. Initial data weights are given by the following function of the errors:

$$
W_d^{(i)} = \frac{1}{\left(err_f^{(i)} d^{(i)} + err_{abs}^{(i)}\right)},
\tag{40}
$$

where $err_f^{(i)}$ are the relative errors, and $err_{abs}^{(i)}$ are the absolute error floors. These intial data weights are used in standalone and joint inversion; however, in the joint inversions, data weights are further scaled by the initial data misfit for each data set:

$$W_{d,j}^{(i)} = \frac{W_d^{(i)}}{\varphi_{ini}\left(m^{(i)}\right)},$$

(41)

where $\varphi_{ini}\left(m^{(i)}\right)$ are the initial misfits. The matrices of initial model weights are given by the integrated sensitivity [9,17]:

$$W_m^{(1)} = diag\left(F^{(1)\mathrm{T}}F^{(1)}\right)^{\frac{1}{4}},$$

(42)

$$W_{M_\gamma}^{(2)} = \left\{diag\left(F_\gamma^{(2)\mathrm{T}}F_\gamma^{(2)}\right)^{\frac{1}{4}}\right\}, (\gamma = x, y, z),$$

(43)

where $F^{(i)}$ is the Fréchet derivative of $A^{(i)}\left(m^{(i)}\right)$ and $F^{(i)T}$ is the transposed matrix. In the joint inversions, the model weights are further scaled by normalization with the maximum model contrast obtained from standalone inversions:

$$W_{m,j}^{(1)} = \frac{W_m^{(1)}}{\max\left(m_{sep}^{(1)} - m_b^{(1)}\right)},$$

(44)

$$W_{m,j}^{(2)} = \left\{\frac{W_{M_\gamma}^{(2)}}{\max\left(M_{sep_\gamma}^{(2)} - M_{b_\gamma}^{(2)}\right)}\right\}, (\gamma = x, y, z),$$

(45)

where $m_{sep}^{(1)}, M_{sep_\gamma}^{(i)}$ are the density and magnetization models obtained from standalone (separate) inversions, and $m_b^{(1)}, M_{b_\gamma}^{(i)}$ are the background density and magnetization values.

The constants α, β are adaptively and monotonically reduced to ensure stable convergence towards a satisfactory solution [9,12]. The inversion is halted when both misfit terms have been minimized to the noise level.

5. Model Study

To illustrate the developed methods, we present the results of their application to synthetic gravity gradiometry and TMI data generated for a two-body model shown in Figure 1. The bodies are both parallelepipeds of dimensions 200×50 m, situated at 50 m and 100 m depths. The left body has physical properties of 0.3 g/cm^3, -0.06 A/m in the X direction, and 0.06 A/m in the Z direction. The right body has physical properties of 0.4 g/cm^3, 0.06 A/m in the X direction, and 0.06 A/m in the Z direction. The magnetization vectors for both bodies have inclinations of $-45°$ and opposing declinations of $-90°$ and $90°$, or east/west. The inducing field has inclination $75.5°$ and declination $-6.2°$. The Gxx, Gyy, and Gzz gravity gradiometry components and TMI data are simulated at equidistant stations laterally spaced at 25 m with a flight height of 30 m, also shown in Figure 1. All data are contaminated with 5% Gaussian noise.

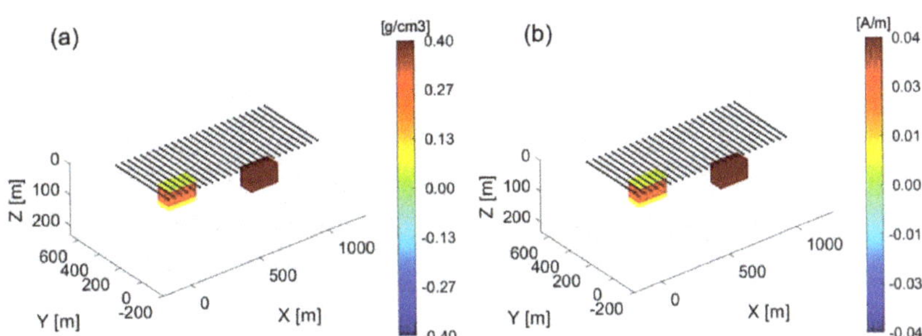

Figure 1. Panels (**a**,**b**) show 3D views of the true density model and vertical component of magnetization vector model, respectively. The black dots represent the location of the observation stations.

We have inverted the synthetic data using the standalone and joint inversions, introduced in the previous sections. Note that, in this model study and in the following case study, we have inverted magnetic data for all three scalar components of the magnetization vector, which is the principal distinguished feature of the developed method in comparison to the conventional approach based on inversion for magnetic susceptibility only. At the same time, there are several possible approaches for how to calculate the joint focusing functionals required for joint inversion. For example, we can consider three joint focusing stabilizers—each involving a scalar component of magnetization vector and density—or jointly focus the amplitude of magnetization vector and density. We have found that jointly focusing the dominant scalar component and density works the best. Therefore, we only include the vertical component of magnetization vector, M_z, in the focusing stabilizers. Based on the results of the standalone magnetization vector inversion, the vertical component of magnetization vector, M_z, is dominant in this area. As such, the joint focusing term includes only the density, $m^{(1)} = \rho$, and the vertical component of magnetization vector, $m^{(2)} = M_z$, to reduce non-uniqueness.

The inversion domain was discretized into rectangular cells with a horizontal cell size of 25×25 m and with 20 logarithmically spaced vertical layers for a total cell count of ~45,000. All inversions were terminated when both data sets reach the simulated noise level $\left(\chi^2 = 1\right)$.

Vertical sections of the true model and inverse solutions from standalone and joint inversions are shown in Figure 2. They illustrate that the joint inversions produce a better image of the model than the standalone inversions while maintaining a similar level of data misfit. Observed and predicted data for the Gzz component of gravity gradiometry and the TMI data for the various inversion scenarios are shown in Figure 3. We can see that the joint-focused inversion slightly improves data fidelity compared to the other inversion types. Between the joint inversion scenarios, the Gramian and joint-focused inversions arrive at similar models and data fits; however, inversion with joint focusing is three times faster with respect to both iterations required and computational speed.

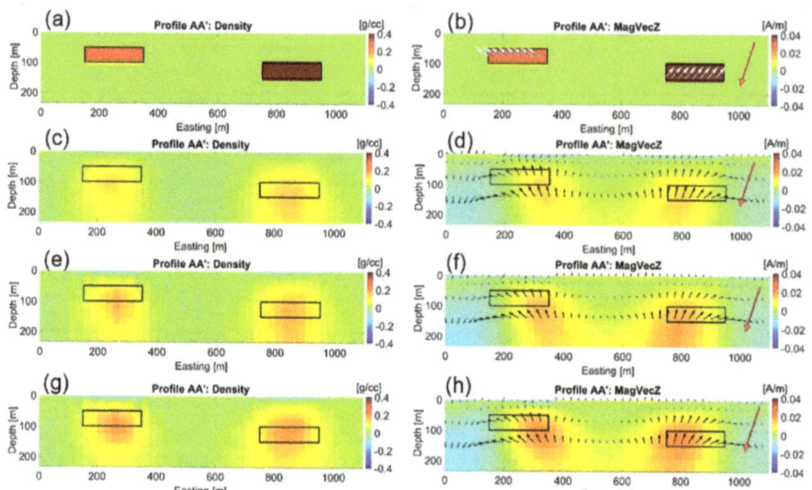

Figure 2. Panels (**a,b**) show the true density model and magnetization vector model, respectively. Panels (**c,d**) show vertical sections of the density and magnetic vector models inverted standalone, respectively. Panels (**e,f**) show vertical sections of the density and magnetic vector models inverted with Gramian constraints, respectively. Panels (**g,h**) show vertical sections of the density and magnetic vector models inverted with joint focusing constraint, respectively. Panels (**b,d,f,h**) show the vertical component of the magnetic vector in color; the full magnetic vector is shown by black and white arrows, and the direction of the inducing field is shown by red arrows.

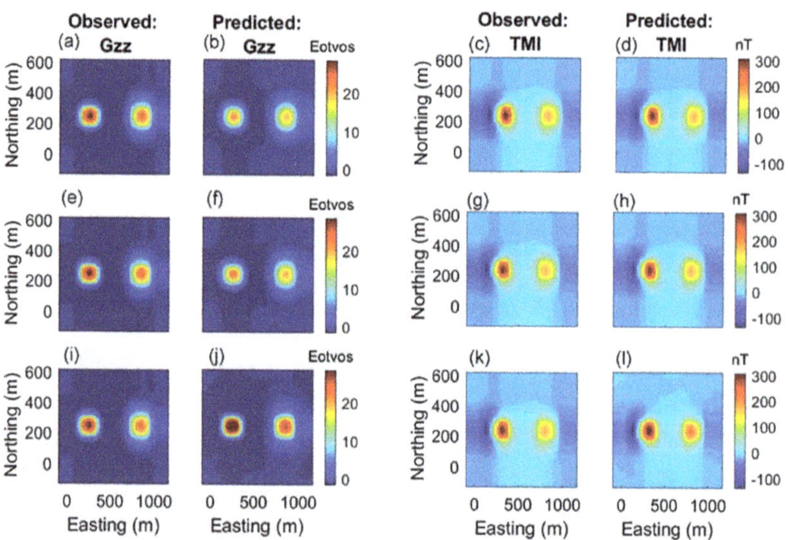

Figure 3. Model study. Panels (**a,b**) show observed and predicted Gzz gravity gradient data inverted standalone, respectively. Panels (**c,d**) show observed and predicted total magnetic intensity (TMI) data inverted standalone, respectively. Panels (**e,f**) show observed and predicted Gzz gravity gradient data inverted with Gramian constraints, respectively. Panels (**g,h**) show observed and predicted TMI data inverted with Gramian constraints, respectively. Panels (**i,j**) show observed and predicted Gzz gravity gradient data inverted with joint focusing, respectively. Panels (**k,l**) show observed and predicted TMI data inverted with joint focusing, respectively.

6. Case Study of Joint Inversions of Airborne Gravity Gradiometer (AGG) and Magnetic Data

6.1. Regional Geology in the Ring of Fire

McFaulds Lake covers the Ring of Fire intrusive complex located in the James Bay lowlands of northwestern Ontario. Ring of Fire is a roughly north–south trending Archean green belt (Figure 4) consisting of mafic and ultramafic rocks. Due to flat topography and Paleozoic carbonate rocks cover, the area was not extensively studied until the early 2000s, when, as a part of kimberlite exploration campaign, McFaulds volcanic massive sulfide (VMS) deposits were discovered. The recognition that the area hosts a greenstone belt has triggered an exploration rush which resulted in three major types of economic mineral deposit discoveries, including magmatic Ni-Cu-PGE, magmatic chromite mineralization, and volcanic massive sulfide mineralization.

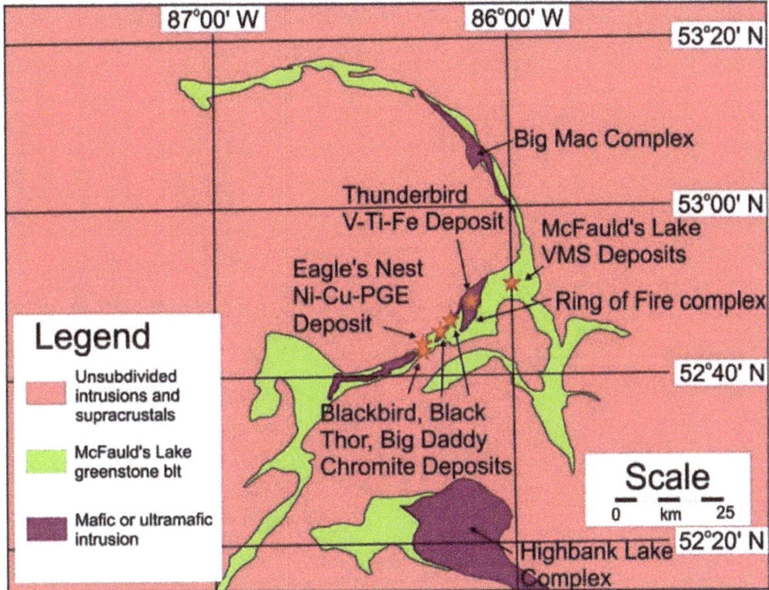

Figure 4. Geological map of the Ring of Fire area with marked known deposits (from [20]).

The Ring of Fire complex is primarily composed of mafic metavolcanic flows, felsic metavolcanic flows, and pyroclastic rocks. The Thunderbird deposit is associated with one of the numerous layered mafic-to-ultramafic intrusions, which trend subparallel with and obliquely cut the westernmost part of the belt, close to a large granitoid batholith lying west of the belt [21]. These layered intrusions are associated with high-grade Ni-Cu-PGE, Chromite, and V-Ti-Fe deposits, with Thunderbird being the latter.

6.2. Airborne Gravity and Magnetic Surveys

In the exploration rush of the mid-2000s, there were numerous smaller-scale exploration projects performed; however, the overall picture of the area was lacking. In order to map regional geology and locate further potential mineral resources, an airborne geophysical survey was carried out in the McFaulds Lake region between 2010 and 2011 using the Fugro Airborne Surveys system. Both airborne gravity gradiometer (AGG) and magnetic data were collected. This project was collaboratively operated between the Ontario Geological Survey (OGS) and the Geological Survey of Canada (GSC). Figure 5 shows the McFaulds Lake region survey block. The survey parameters included:

(1) Flight height: 100 m
(2) Line spacing: 250 m
(3) Line km: 19,733
(4) Area: 4577 m^2
(5) AGG: Falcon
(6) Magnetic: single sensor

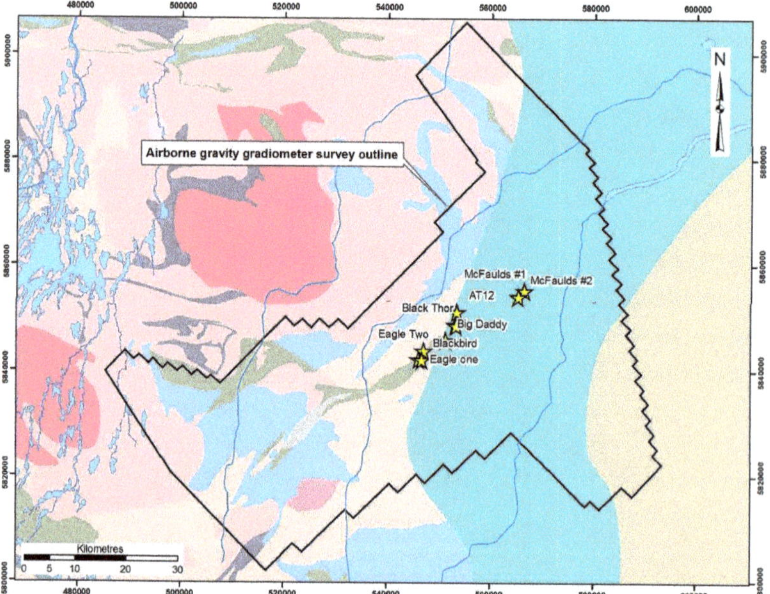

Figure 5. Airborne geophysical survey block in the McFaulds Lake region. airborne gravity gradiometry (AGG) and magnetic data were collected through the Fugro AGG and magnetic system (from [21]).

6.3. Thunderbird Deposit

The Thunderbird deposit consists of semi-massive vanadium- and titanium-enriched magnetite, which corresponds to a strong gravity and magnetic anomaly. Noront Resources, owner of the claim, estimates the ore body at 1.6 km long, 400 m wide, and 500 m deep, based on gravity and magnetic data and limited core drilling. This deposit has not been developed yet, and a more detailed analysis is lacking.

We have inverted the Gxy, Gxx, Gyy, and Gzz gravity gradient components and TMI data separately and jointly. As discussed above, the filtered RTP magnetic data were used for susceptibility inversion, while the inversion for magnetization vector was based on the TMI data with the removed regional trend. As an example, panel (a) of Figure 6 shows the Gzz component of the observed AGG data, while panel (b) presents the filtered TMI data map of the Thunderbird area. The deposit is assumed to be a semi-massive V-Ti-enriched magnetite with an approximate volume of 0.32 km, which is based on potential field data analysis and limited core drilling.

The inversion grid was discretized with a 50 × 50 m horizontal spacing and a logarithmic vertical discretization ranging from 25–150 m. The total number of cells inverted was ~250,000 cells. A 16-core Intel Xeon workstation with 128 GB memory was used to run all of the inversions. The total inversion runtime was ~10 min for the standalone AGG and TMI inversions, ~15 min for the joint inversion with joint focusing stabilization, and ~45 min for the joint inversion with Gramian constraints.

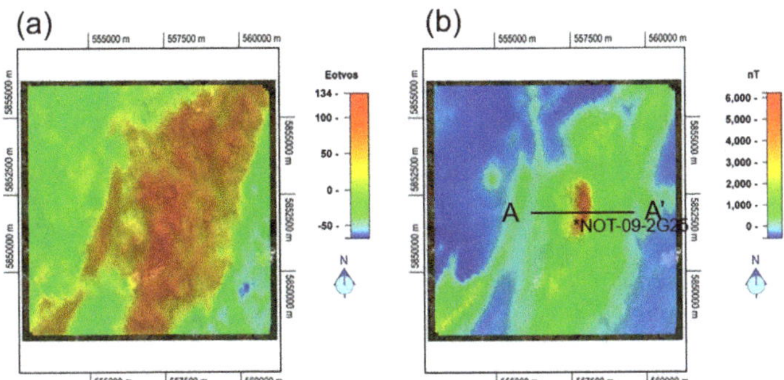

Figure 6. Panel (**a**) shows the Gzz component of the observed AGG data in UTM (16N) coordinates in panel (**a**). Panel (**b**) presents the filtered TMI data map with the removed regional trend. Profile AA′ is shown by the black line. The location of drill hole NOT-09-2G25 is shown by the black asterisk.

A comparison of vertical sections of the 3D models obtained from the various inversion methodologies is shown in Figure 7. It is clear that the models obtained from a joint inversion with either Gramian or joint focusing constraints have sharper geospatial boundaries and a higher degree of structural correlation, when compared to the models obtained from standalone inversions. It is important to note that all inverted 3D models share the same level of data misfit.

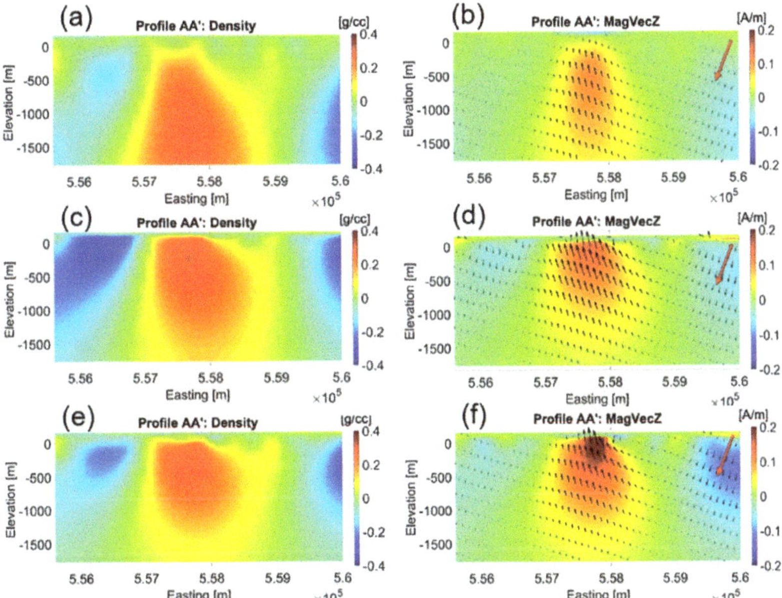

Figure 7. Panels (**a,b**) show vertical sections of the 3D density and magnetization vector models obtained from standalone inversions, respectively. Panels (**c,d**) show vertical sections of the 3D density and magnetization vector models obtained from joint inversion with Gramian constraints, respectively. Panels (**e,f**) show vertical sections of the 3D density and magnetizations vector models obtained from joint inversion with joint focusing constraint, respectively. Panels (**b,d,f**) show the vertical component of magnetization vector in color; the full magnetization vector is shown by the black arrows, and the inducing field direction is shown by the red arrows.

We have included both the full magnetization vector and the inducing field direction in the magnetic panels for clarity. Drill hole NOT-09-2G25, drilled by Noront Resources Limited in 2009 [22] and situated in the center of these profiles, found a gabbro schist unit from 21–462 m containing magnetite and titanomagnetite with concentrations up to 10% and smaller pockets containing up to 50% magnetite. This matches quite well with the jointly inverted results. Moreover, an iron formation with up to 20% magnetite concentration was found at 166–170 m which may be what the focused inversion is resolving in the near surface.

A comparison of selected observed and predicted data is shown in Figure 8. All inversion methodologies reach an excellent data fit. Figure 9 shows model parameter cross plots for the different inversion scenarios. Additionally, the correlation coefficient was calculated for the density model versus the vertical component of the magnetization vector and is summarized in Table 1.

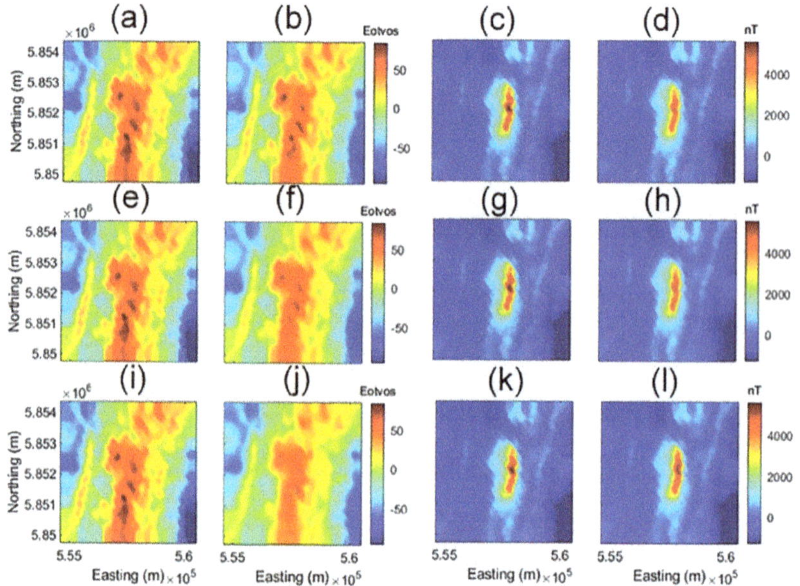

Figure 8. Case study. Panels (**a**,**b**) show observed and predicted Gzz gravity gradient data inverted standalone, respectively. Panels (**c**,**d**) show observed and predicted TMI data inverted standalone, respectively. Panels (**e**,**f**) show observed and predicted Gzz gravity gradient data inverted jointly with Gramian constraints, respectively. Panels (**g**,**h**) show observed and predicted TMI data inverted jointly with Gramian constraints, respectively. Panels (**i**,**j**) show observed and predicted Gzz gravity gradient data inverted with joint focusing, respectively. Panels (**k**,**l**) show observed and predicted TMI data inverted with joint focusing, respectively.

Figures 10 and 11 show a comparison between the results of the susceptibility-only inversion and magnetization-vector inversion. The left panel in Figure 10 shows the vertical section of the inverse model for susceptibility inversion, while the right panel shows the vertical section of the inverse model for magnetization vector inversion. The small black arrows in the right panel represent the magnetization vectors, while the color bar reflects the vertical component of magnetization.

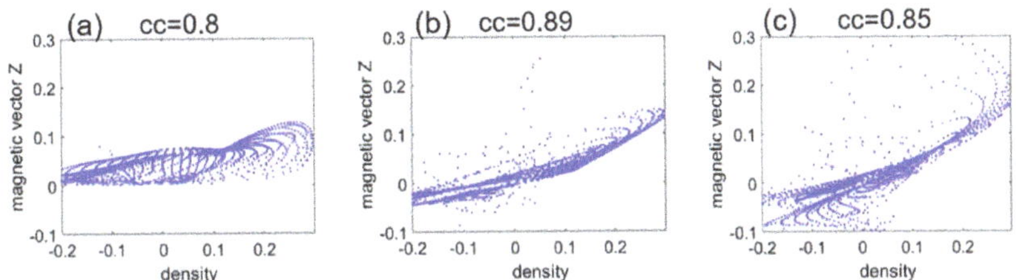

Figure 9. Model parameter cross plots shown in Panels (**a–c**) correspond to the standalone inversions, joint Gramian inversion, and joint focusing inversion, respectively.

Table 1. Correlation coefficient for different inversion scenarios.

Inversion Type	Correlation Coefficient
Standalone	8.0
Gramian	8.9
Joint focusing	8.5

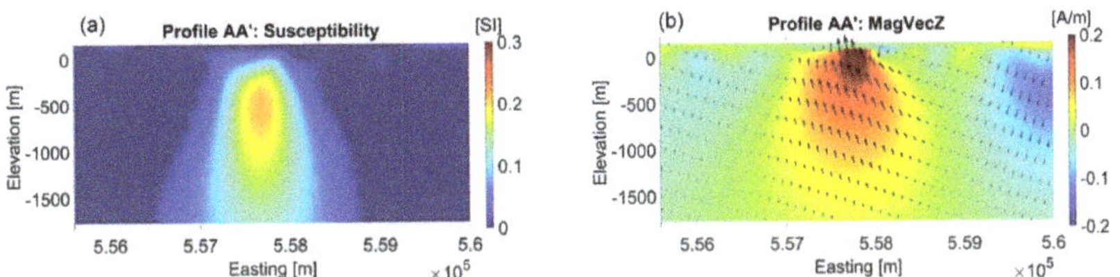

Figure 10. Left panel (**a**) shows the vertical section of the inverse model for susceptibility inversion. Right panel (**b**) shows a vertical section of the inverse model for magnetization vector inversion. The small black arrows in the right panel represent the magnetization vectors, while the color bar reflects the vertical component of magnetization.

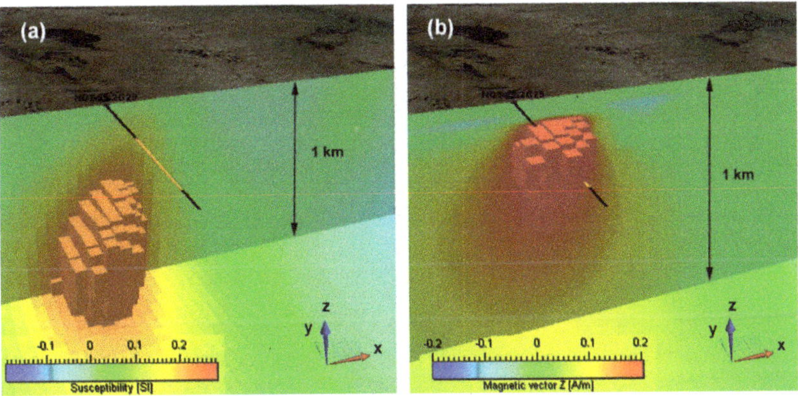

Figure 11. Comparison of 3D inverse models with drilling results. Left panel (**a**) shows the volume image of the inverse susceptibility model. Right panel (**b**) presents the volume image of the vertical component of the magnetization vector. The black–yellow–black solid line shows the location of the borehole drilled in the survey area. The yellow color indicates the mineralization zone confirmed by drilling.

Figure 11 shows a comparison of 3D inverse models with drilling results. The left panel (a) presents the volume image of the inverse susceptibility, while the right panel (b) shows a similar image of the vertical component on the magnetization vector. The black–yellow–black solid line is the borehole drilled in the survey area. The yellow color indicates the mineralization zone confirmed by drilling. One can see that the image produced by inversion for the magnetization vector correctly represents the location of the mineralization zone. In other words, the volume images also show that you can better pinpoint anomalous bounds with magnetization vector in the presence of remanent magnetization for the Thunderbird Vanadium-Titanium target in Ontario, Canada.

Finally, Figure 12 presents a comparison of the inverse density (top panel) and vertical component of magnetization models (bottom panel) recovered by joint focusing inversion. It was shown in [11] that it can be challenging to resolve a steeply dipping magnetic dike and recover a good data fit when inverting for magnetic susceptibility only. By inverting for the full magnetization vector, it is much easier to fit these structures and still retain a good data fit. Including these additional inversion parameters increases the non-uniqueness of the potential field inverse problem; however, we have remedied this with the addition of the Gramian and joint focusing constraints.

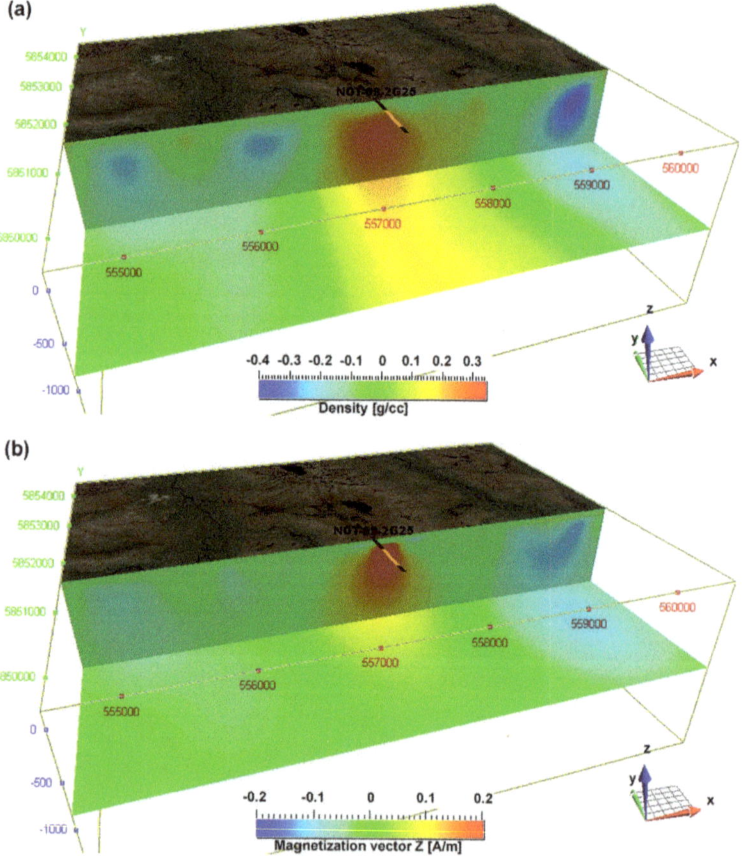

Figure 12. Top panel (**a**) shows the cut-off 3D image of the inverse density model. Bottom panel (**b**) presents the volume image of the vertical component of the magnetization vector. The black–yellow–black solid line shows the location of the borehole drilled in the survey area. The yellow color indicates the mineralization zone confirmed by drilling.

7. Conclusions

Magnetic data contain information about the remanent magnetization of rock formations that are often ignored when inverting for magnetic susceptibility only. We have introduced two novel methods of joint inversion of potential field data in the presence of remanent magnetization: (1) joint inversion subject to the Gramian constraint and (2) joint inversion subject to the joint focusing constraint. We have demonstrated the methods in a model study, and in a case study using airborne data acquired over the Thunderbird V-Ti-Fe deposit in the Ring of Fire area of Ontario, Canada. We have presented a comparison of the 3D models obtained using conventional single-datum standalone inversion and joint inversion subject to either the Gramian constraint or the joint focusing constraint. Comparison of the different inversion methodologies illustrates that 3D models obtained from joint inversion with either the Gramian or joint focusing constraint were able to resolve a shared earth model that featured sharper anomalous geospatial boundaries and a higher degree of structural correlation, as opposed to the 3D models obtained from standalone inversions. The joint inversion subject to the Gramian constraint did yield the highest degree of structural correlation; while the joint inversion subject to the joint focusing constraint provided comparable results but was much faster. By jointly inverting AGG and TMI data for density and magnetization vector, we demonstrated the efficiency of the novel methods in defining rock formations with remanent magnetization which are otherwise difficult to recover.

Author Contributions: Conceptualization, M.S.Z. and M.J.; methodology, M.S.Z.; software, M.J.; validation, M.J., and M.S.Z.; formal analysis, M.S.Z.; investigation, M.J.; resources, M.S.Z.; writing—original draft preparation, M.J.; writing—review and editing, M.S.Z.; visualization, M.J.; supervision, M.S.Z.; project administration, M.S.Z.; funding acquisition, M.S.Z. All authors have read and agreed to the published version of the manuscript.

Funding: This research was supported by Consortium for Electromagnetic Modeling and Inversion (CEMI), and TechnoImaging and received no external funding.

Data Availability Statement: The data used in this study are publicly available from the Ontario Geologic Survey.

Acknowledgments: The authors acknowledge the Consortium for Electromagnetic Modeling and Inversion (CEMI) at the University of Utah and TechnoImaging for the support of this research. The AGG and TMI data were collected by Fugro and made available by the Ontario Geological Survey, Canada.

Conflicts of Interest: The authors declare no conflict of interest.

References

1. Jorgensen, M.; Zhdanov, M.S. 3D joint inversion of potential field data in the presence of remanent magnetization. In *Expanded Abstracts, Proceedings of the 90th SEG International Exposition and Annual Meeting, Virtual Event, Houston, TX, USA, 11–16 October 2020*; Society of Exploration Geophysicists: Tulsa, OK, USA, 2020; pp. 964–968.
2. Lelièvre, P.G.; Oldenburg, D.W. A 3D total magnetization inversion applicable when significant, complicated remanence is present. *Geophysics* **2009**, *74*, L21–L30. [CrossRef]
3. Li, Y.; Shearer, S.E.; Haney, M.M.; Dannemiller, N. Comprehensive approaches to 3D inversion of magnetic data affected by remanent magnetization. *Geophysics* **2010**, *75*, L1–L11. [CrossRef]
4. Ellis, R.G.; De Wet, B.; Macleod, I.N. Inversion of magnetic data from remanent and induced sources. In *Expanded Abstracts, Proceedings of the 22nd ASEG Geophysical Conference and Exhibition, Brisbane, Australia, 26–29 February 2012*; ASEG: Crows Nest, Australia, 2012.
5. Li, Y. From susceptibility to magnetization: Advances in the 3D inversion of magnetic data in the presence of significant remanent magnetization. In Proceedings of the Exploration 17: Sixth Decennial International Conference on Mineral Exploration, Toronto, ON, Canada, 22–25 October 2017; pp. 239–260.
6. Zhu, Y.; Zhdanov, M.S.; Cuma, M. Inversion of TMI data for the magnetization vector using Gramian constraints. In *Expanded Abstracts, Proceedings of the 85th SEG International Exposition and Annual Meeting, New Orleans, LA, USA, 18–23 October 2015*; Society of Exploration Geophysicists: Tulsa, OK, USA, 2015; pp. 1602–1606.
7. Zhdanov, M.S.; Gribenko, A.V.; Wilson, G. Generalized joint inversion of multimodal geophysical data using Gramian constraints. *Geophys. Res. Lett.* **2012**, *39*. [CrossRef]

8.   Gallardo, L.A.; Meju, M.A. Characterization of heterogeneous near-surface materials by joint 2D inversion of DC resistivity and seismic data. *Geophys. Res. Lett.* **2003**, *30*, 1658. [CrossRef]

9.   Zhdanov, M.S. *Inverse Theory and Applications in Geophysics*; Elsevier: Amsterdam, The Netherlands, 2015.

10.   Molodtsov, D.; Troyan, V. Multiphysics joint inversion through joint sparsity regularization. In *Expanded Abstracts, Proceedings of the 88th SEG International Exposition and Annual Meeting, Houston, TX, USA, 29 September 2017*; Society of Exploration Geophysicists: Tulsa, OK, USA, 2017; pp. 1262–1267.

11.   Zhdanov, M.S.; Cuma, M. Joint inversion of multimodal data using focusing stabilizers and Gramian constraints. In *Expanded Abstracts, Proceedings of the 89th SEG International Exposition and Annual Meeting, Anaheim, CA, USA, 14–19 October 2018*; Society of Exploration Geophysicists: Tulsa, OK, USA, 2018; pp. 1430–1434.

12.   Zhdanov, M.S. New advances in regularized inversion of gravity and electromagnetic data. *Geophys. Prospect.* **2009**, *57*, 463–478. [CrossRef]

13.   Li, Y.; Oldenburg, D.W. 3-D inversion of magnetic data. *Geophysics* **1996**, *61*, 394–408. [CrossRef]

14.   Li, Y.; Oldenburg, D.W. Fast inversion of large-scale magnetic data using wavelet transforms and a logarithmic barrier method. *Geophys. J. Int.* **2003**, *152*, 251–265. [CrossRef]

15.   Portniaguine, O.; Zhdanov, M.S. 3-D magnetic inversion with data compression and image focusing. *Geophysics* **2002**, *67*, 1532–1541. [CrossRef]

16.   Cuma, M.; Wilson, G.; Zhdanov, M.S. Large-scale 3D inversion of potential field data. *Geophys. Prospect.* **2012**, *60*, 1186–1199. [CrossRef]

17.   Zhdanov, M.S. *Geophysical Inverse Theory and Regularization Problems*; Elsevier: Amsterdam, The Netherlands, 2002.

18.   Molodtsov, D.M.; Troyan, V.N. Generalized multiparameter joint inversion using joint total variation: Application to MT, seismic and gravity data. In *Expanded Abstracts, Proceedings of the 78th EAGE Conference & Exhibition—Workshops, Vienna, Austria, 29–30 May 2016*; EAGE: Houten, The Netherlands, 2016.

19.   Zhdanov, M.S.; Tolstaya, E. Minimum support nonlinear parameterization in the solution of 3D magnetotelluric inverse problem. *Inverse Probl.* **2004**, *20*, 937–952. [CrossRef]

20.   Mungall, J.E.; Harvey, J.D.; Balch, S.J.; Azar, B.; Atkinson, J.; Hamilton, M.A. Eagle's Nest: A magmatic Ni-Sulfide deposit in the James Bay Lowlands, Ontario, Canada. *Soc. Econ. Geol.* **2010**, *15*, 539–557.

21.   Ontario Geological Survey and Geological Survey of Canada. Ontario Geological Survey and Geological Survey of Canada. Ontario airborne geophysical surveys, gravity gradiometer and magnetic data, grid and profile data (ASCII and Geosoft formats) and vector data, McFaulds Lake area. In *Ontario Geological Survey, Geophysical Data Set 1068*; Ontario Geological Survey and Geological Survey of Canada: Sudbury, ON, Canada, 2011.

22.   Downey, M.; Atkinson, J. Report on Diamond Drill Program as Completed by Noront Resources Limited on Their Grid 2 Extension and Goodman Properties. Published Report. 2010. Available online: http://www.geologyontario.mndm.gov.on.ca (accessed on 23 May 2018).

MDPI
St. Alban-Anlage 66
4052 Basel
Switzerland
Tel. +41 61 683 77 34
Fax +41 61 302 89 18
www.mdpi.com

Minerals Editorial Office
E-mail: minerals@mdpi.com
www.mdpi.com/journal/minerals

www.ingramcontent.com/pod-product-compliance
Lightning Source LLC
LaVergne TN
LVHW071006170726
843515LV00006B/1087